VOLUME FORTY SIX

# ADVANCES IN CHEMICAL ENGINEERING

## Mesoscale Modeling in Chemical Engineering Part I

# ADVANCES IN CHEMICAL ENGINEERING

VOLUME FORTY SIX

# ADVANCES IN CHEMICAL ENGINEERING

## Mesoscale Modeling in Chemical Engineering Part I

Edited by

**GUY B. MARIN**
*Department of Chemical Engineering,*
*Ghent University,*
*Ghent, Belgium*

**JINGHAI LI**
*Institute of Process Engineering,*
*Chinese Academy of Sciences,*
*Beijing, P.R. China*

ELSEVIER

AMSTERDAM • BOSTON • HEIDELBERG • LONDON
NEW YORK • OXFORD • PARIS • SAN DIEGO
SAN FRANCISCO • SINGAPORE • SYDNEY • TOKYO
Academic Press is an imprint of Elsevier

Academic Press is an imprint of Elsevier
225 Wyman Street, Waltham, MA 02451, USA
525 B Street, Suite 1800, San Diego, CA 92101-4495, USA
The Boulevard, Langford Lane, Kidlington, Oxford OX5 1GB, UK
125 London Wall, London, EC2Y 5AS, UK

First edition 2015



ISBN: 978-0-12-801247-5
ISSN: 0065-2377

For information on all Academic Press publications
visit our website at store.elsevier.com

# CONTENTS

*Contributors* *ix*
*Preface* *xi*

**1. Unified Design Strategies for Particulate Products** **1**
Wolfgang Peukert, Doris Segets, Lukas Pflug, and Günter Leugering

1. Introduction 2
2. Hierarchical Products and Processes 6
3. Product Properties 12
4. Product Design 19
5. Processes and Properties by Design 52
6. Conclusions 70

Acknowledgments 71
References 71

**2. Multiscale Analysis of a Coating Process in a Wurster Fluidized Bed Apparatus** **83**
Stefan Heinrich, Maksym Dosta, and Sergiy Antonyuk

1. Introduction 85
2. Multiscale Simulation Framework 88
3. Experimental Estimation of the Material Parameters 103
4. Analysis of the Process Behavior 113
5. Conclusions 131

Acknowledgments 132
References 133

**3. Euler–Lagrange Modeling of the Hydrodynamics of Dense Multiphase Flows** **137**
Johan T. Padding, Niels G. Deen, E.A.J.F. (Frank) Peters, and J.A.M. (Hans) Kuipers

1. Introduction 138
2. Governing Equations for the Continuous and Discrete Phases 141
3. Dimensionless Numbers for Multiphase Flows 156
4. Flows with Particles: Discrete Particle Models (CFD–DEM) 159
5. Flows with Bubbles: DBMs 167
6. Flows with Droplets: DSMC 172

7. Summary and Outlook 180
Acknowledgments 182
Appendix A. Derivation of the Filtered Fluid Equation 182
Appendix B. Direct Numerical Simulation of Droplet Collision Outcomes 186
References 188

**4. Particle Scale Study of Heat Transfer in Packed and Fluidized Beds 193**

Qinfu Hou, Jieqing Gan, Zongyan Zhou, and Aibing Yu

1. Introduction 196
2. Model Description 198
3. Model Application 211
4. Conclusions and Future Work 235

Acknowledgment 238
References 238

**5. Mesoscale Transport Phenomena and Mechanisms in Gas–Liquid Reaction Systems 245**

Ning Yang

1. Introduction 247
2. Understanding the Mesoscale Mechanisms Through the EMMS Model 251
3. Physical Interpretation of Macroscale Behavior from Mesoscale Perspective 255
4. Consolidated Understanding of the EMMS Model for Gas–Liquid Systems 258
5. Stability-Constrained Multifluid CFD Model 268
6. Conclusions and Future Issues 276

Acknowledgments 278
References 278

**6. Mesoscale Flow Structures and Fluid–Particle Interactions 281**

Harry E.A. Van den Akker

1. Introduction 282
2. Evidence of Mesoscale Coherent Structures 284
3. The Origin of Mesoscale Structures 293
4. The Fundamentals of Fluid–Particle Interaction 303
5. The Steady-State Drag Force 308
6. The Effect of Acceleration: Added Mass and Basset Force 310

**7.** The Lift Force 313
**8.** The Impact of Turbulence 320
**9.** Multiple Particle Systems 324
**10.** Conventional CFD Simulations 328
**11.** Direct Numerical Simulations 335
**12.** Conclusions 340
References 342

*Index* *355*
*Contents of Volumes in this Serial* *367*

# CONTRIBUTORS

**Sergiy Antonyuk**
Chair of Particle Process Engineering, University of Kaiserslautern, Kaiserslautern, Germany

**Niels G. Deen**
Department of Chemical Engineering and Chemistry, Eindhoven University of Technology, Eindhoven, The Netherlands

**Maksym Dosta**
Institute of Solids Process Engineering and Particle Technology, Hamburg University of Technology, Hamburg, Germany

**Jieqing Gan**
Laboratory for Simulation and Modelling of Particulate Systems, Department of Chemical Engineering, Monash University, Clayton, Victoria, Australia

**Stefan Heinrich**
Institute of Solids Process Engineering and Particle Technology, Hamburg University of Technology, Hamburg, Germany

**Qinfu Hou**
Laboratory for Simulation and Modelling of Particulate Systems, Department of Chemical Engineering, Monash University, Clayton, Victoria, and Laboratory for Simulation and Modelling of Particulate Systems, School of Materials Science and Engineering, University of New South Wales, Sydney, New South Wales, Australia

**J.A.M. (Hans) Kuipers**
Department of Chemical Engineering and Chemistry, Eindhoven University of Technology, Eindhoven, The Netherlands

**Günter Leugering**
Institute of Applied Mathematics 2, Friedrich-Alexander-Universität Erlangen-Nürnberg (FAU), Erlangen, Germany

**Johan T. Padding**
Department of Chemical Engineering and Chemistry, Eindhoven University of Technology, Eindhoven, The Netherlands

**E.A.J.F. (Frank) Peters**
Department of Chemical Engineering and Chemistry, Eindhoven University of Technology, Eindhoven, The Netherlands

**Wolfgang Peukert**
Institute of Particle Technology (LFG), Friedrich-Alexander-Universität Erlangen-Nürnberg (FAU), Erlangen, Germany

**Lukas Pflug**
Institute of Applied Mathematics 2, Friedrich-Alexander-Universität Erlangen-Nürnberg (FAU), Erlangen, Germany

**Doris Segets**
Institute of Particle Technology (LFG), Friedrich-Alexander-Universität Erlangen-Nürnberg (FAU), Erlangen, Germany

**Harry E.A. Van den Akker**
Bernal Institute and Department of Mechanical, Aeronautical and Biomedical Engineering, University of Limerick, Limerick, Ireland, and Department of Chemical Engineering, Delft University of Technology, Delft, The Netherlands

**Ning Yang**
Center for Mesoscience and State Key Laboratory of Multiphase Complex Systems, Institute of Process Engineering, Chinese Academy of Sciences, Beijing, PR China

**Aibing Yu**
Laboratory for Simulation and Modelling of Particulate Systems, Department of Chemical Engineering, Monash University, Clayton, Victoria, Australia

**Zongyan Zhou**
Laboratory for Simulation and Modelling of Particulate Systems, Department of Chemical Engineering, Monash University, Clayton, Victoria, Australia

# PREFACE

## Opportunities and Challenges: Both at Mesoscales

Evolution of chemical engineering science features increasing generality of knowledge from "unit operations," common for different industries, to "transport phenomena," common in these operations, and is now at a transitional period with challenges in sustainable development and to its knowledge base. Designing materials, scaling-up reactors, and system optimization continue to be the main tasks for chemical engineers. However, upgraded knowledge and enabling technologies are needed to revolutionize R&D capability, which present challenges and opportunities to chemical engineering community.

Opportunities are always obscured by challenges. Among many challenges currently in chemical engineering, mesoscale phenomena at different levels of chemical engineering are recognized to be the most critical. At the material level, chemists and material scientists know molecular structures and properties of bulk materials very well, but they are still unable to manipulate the mesoscale structures in between. At the reactor level, chemical engineers have accumulated much knowledge on hydrodynamics, transport, and reaction behaviors of single particle and on global performance of reactors; however, what happen at the mesoscale between single particles and the whole reactor are still difficult to be formulated. These two mesoscales at different levels are believed to be essential not only for resolving the current global challenges but also for upgrading the knowledge base of chemical engineering science.

Currently in chemical engineering, chemists and material scientists synthesize a variety of material structures every day, most of which, however, cannot be produced massively in industry due to the lack of understanding of the formation mechanisms of these structures and the difficulty in controlling the microenvironment for their formation in reactors. In addition, scaling-up of reactors is still a big challenge for chemical engineers. Understanding of mesoscale phenomena is critical to increase our capability in resolving these issues.

In fundamental research, average approaches are widely used, such as computational fluid dynamics, which are however not sufficient enough in predictability. On the other hand, discrete approaches are blooming up, based on microscale mechanisms, which are unfortunately limited to scalability in engineering applications. As a compromising strategy between these two, coarse-graining approaches received more and more attention, aiming at bridging the scales of continuum cells and the discrete moving elements, which still suffer from the lack of understanding of intrinsic mechanisms at mesoscales. This is now a common challenge not only in chemical engineering.

Gradually, people recognized that "meso" will be a big thing. Most importantly, there are many clues that a common principle for all mesoscale problems at different levels between elemental particles and the universe may exist, leading to exploration of a transdisciplinary science, mesoscience, though great efforts are needed to further clarify its rationale and importance.

In responding to these new possibilities and challenges, *Advances in Chemical Engineering* published a thematic volume in 2011 with the title: "Charaterization of Flow, Particles and Interfaces," aiming at showing the dynamic behaviors at mesoscales. These two volumes further this effort to analyze these mesoscale phenomena and processes with the title: "Mesoscale Modeling in Chemical Engineering" so as to stimulate wide discussion and to promote transdisciplinarity between chemistry, material science, and chemical engineering, and even beyond. Volume I invited a few participants from the 4th International Conference on Multiscale Structures and Systems with the theme of "***Mesoscales: The key to multiscale problems***," and Volume II invited contributions from the major program on mesoscience of National Natural Science Foundation of China entitled "***Mechanisms and manipulation at mesoscales in multi-phase reaction systems***."

Although these contributions may be preliminary in addressing mesoscale problems, we hope these two volumes can play a role of "throwing out a minnow to catch a whale" and to stimulate thinking in the following questions: What are mesoscale phenomena? Why are they so important? How to relate mesoscale concepts to our research? And is it necessary or possible to have a common science for all mesoscales?

We would like to thank all authors for their contributions, and reviewers for their critical comments and suggestions. Help from Prof. Ning Yang and Dr. Xiaowei Wang are sincerely appreciated.

JINGHAI LI AND GUY MARIN

CHAPTER ONE

# Unified Design Strategies for Particulate Products

**Wolfgang Peukert*^,1^, Doris Segets*, Lukas Pflug†, Günter Leugering†,1**

*Institute of Particle Technology (LFG), Friedrich-Alexander-Universität Erlangen-Nürnberg (FAU), Erlangen, Germany

†Institute of Applied Mathematics 2, Friedrich-Alexander-Universität Erlangen-Nürnberg (FAU), Erlangen, Germany

[1]Corresponding authors: e-mail address: Wolfgang.Peukert@fau.de; Guenter.Leugering@fau.de

## Contents

1. Introduction 2
2. Hierarchical Products and Processes 6
3. Product Properties 12
4. Product Design 19
   4.1 Particle Formation 19
   4.2 Interactions 25
   4.3 Structure Formation 30
   4.4 Characterization Along the Process Chain 44
   4.5 Multiscale Modeling and Simulation (MSS) 45
   4.6 Integration of Particles into Devices 49
5. Processes and Properties by Design 52
   5.1 Sensitivities of Parameter-Dependent Initial Value Problems 56
   5.2 General Model-Based Optimal Control Setting 58
   5.3 Examples of Optimization 60
6. Conclusions 70
Acknowledgements 71
References 71

## Abstract

Unit operations and product design are the two most important pillars of chemical engineering. Product design is the formation, formulation, handling, manufacturing, and characterization of complex multiphase products with specific properties and is thus at the core of mesoscale science and engineering. The applications define the required product properties which cover both classical fields of process technology in the chemical industry as well as new emerging fields of electronics, energy and environmental technologies, life sciences, materials science and engineering, nanotechnology, and photonic technologies highlighting the broad relevance of mesoscale science. Unifying principles of product design are proposed which are widely applicable to many

*Advances in Chemical Engineering*, Volume 46
ISSN 0065-2377
http://dx.doi.org/10.1016/bs.ache.2015.10.004

different kinds of products including solid, liquid, and even gaseous particles. Results from the Erlangen Cluster of Excellence "Engineering of Advanced Materials" show that the joint venture of chemical engineering with materials science in concert with the basic sciences opens new prospects for all involved disciplines. In particular, chemical and biochemical engineering expands through particle technologies also in physics-related fields of technology such as electronics, photonics, or 3D printing. Rigorous mathematical optimization methods based on predictive models for products, structures, and processes catalyze new possibilities for true design of particulate products which is at the core of mesoscale science and technology.

## 1. INTRODUCTION

In recent years, chemical engineering including particle science and technology has gone through major transitions. For most of the last century, process engineering was focused on unit operations, i.e., to understand and design certain process steps by general rules that should be as independent as possible of the processed chemicals and materials. This paradigm was very successful, for instance, in the petroleum industry, where very large processes for oil refineries were developed and optimized. Teaching concepts in chemical engineering were thus almost exclusively orientated toward thermal-, mechanical-, and chemical-based unit operations plus classical disciplines including chemical thermodynamics, fluid mechanics, and transport processes for heat and mass. The unit operations approach for homogeneous systems that are mainly consisting of fluids today is mature with limited prospects for ground-breaking new developments. New momentum came by the advent of ionic liquids which opened new applications and put new emphasis on well-established lines such as characterization of fluid properties and their phase equilibria (Wasserscheid and Keim, 2000).

The understanding and rigorous modeling and design of heterogeneous systems containing interfaces and, in particular, disperse particle systems are much less developed with many open fundamental and applied challenges. In general, disperse systems consist of solid, liquid, or gaseous particles in the form of dispersions, emulsions, or foams, respectively. After many years of extensive research, we have to admit that very few predictive models exist for unit operations where disperse systems are produced and processed. Good models exist for highly diluted systems where particle interactions can be neglected. Examples are deep bed filters (Baumgartner and Löffler, 1986; Lee and Liu, 1982; Wang et al, 2007, 2008), cyclones (Mothes and Löffler, 1988), or air classier mills (Toneva et al, 2011a, 2011b) operated at low solid concentrations where the momentum transfer from the solid to the fluid phase can be neglected as first approximation. In most systems, which are operated at higher particle concentrations such as ball or stirred media mills (Kwade, 1999; Sommer et al, 2006), mixers (Portillo et al, 2008; Sommer, 1986; Vanarase et al, 2010), or granulators (Dhenge et al, 2012; Iveson et al, 2001; Rahmanian et al, 2011; Reynolds et al, 2005), the operational behavior cannot yet be predicted without free fitting parameters. Although in recent years much progress has been made in the characterization and particularly in the modeling of dense fluid particle systems such as fluidized beds (Deen et al, 2007; Dosta et al, 2012; Fries et al, 2011; Wang et al, 2010), the key challenge is related to an overall cross-cutting topic: particles in contact, i.e., the mesoscale. In Germany, a nation-wide research program was dedicated to key questions in terms of contact mechanics, adhesion, fracture, and related applications (http://www.piko.ovgu.de/).

Unit operations involving particles are best described by a fundamental unit operation function which depends on the machine and the material function (see Section 4.5). The former describes the motion of particles in the unit operation depending on the applied operational parameters such as flow rates, velocities of moving parts, concentrations, and the fluid properties. Within the unit operation, the particles react according to the applied stresses. This reaction is summarized by the material function involving all material properties such as particle density, Young's modulus, or coefficients of restitution as a measure for the energy dissipation between the particles upon contact. We will later come back to this important and too often neglected question of particle properties. The two-phase or even multiphase flow at all concentrations in the various unit operations is subject to intensive research with still many open questions to be answered. In all these examples, the key challenge is the mesoscale which is in the focus of this issue of *Advances in Chemical Engineering* and which will be addressed by the contributions of Hans Kuipers, Stefan Heinrich, and Aibing Yu.

Since the nineties of the last century, the prevailing paradigm of unit operations has been complemented and partly replaced by a product-orientated approach (Cussler and Moggride, 2011; Hill, 2004; Kind, 1999). Now product properties are in the focus; product design strategies and the engineering of advanced materials are the results of this approach.

The combination of the described developments in chemical engineering together with materials science opens unseen perspectives for both disciplines. Materials scientists are the experts in materials properties, a major challenge in modern chemical engineering. Advanced materials require sophisticated processes; thus chemical engineers come into play. Furthermore, the strong link with basic science closes the observed transfer gap between visionary ideas and real-world applications. While design of novel materials, and in particular nanomaterials, starts from a fundamental physical and chemical understanding on the atomic and molecular levels, engineering of devices and systems requires length scales that range from micro- to meso- and macroscales. Thus, multiscales need to be covered.

Advanced materials with properties tailored on the molecular and the mesoscales are expected to stimulate evolutionary advances and revolutionary breakthroughs in emerging key-technology areas, such as information and communication as well as catalysis, energy, and transportation. Hence, the ability to design novel materials and processes will have a strong impact on assuring industrial competitiveness, economic growth, and better quality of life. In Fig. 1, the five most important incentives for product development are highlighted. These are resources including energy transformation and storage and their environmental impact, modern ways of transportation such as electric cars, and information technology with a manifold of electronic and photonic devices. Other aspects are related to improving and sustaining the quality of life, for instance, in developing countries without irreversible impact on our climate. Life sciences in the century of biology in general promise and require many future applications in bio-based economies and of course in food science and medicine. Progress in all these fields requires the development of new products that can only be realized if new materials

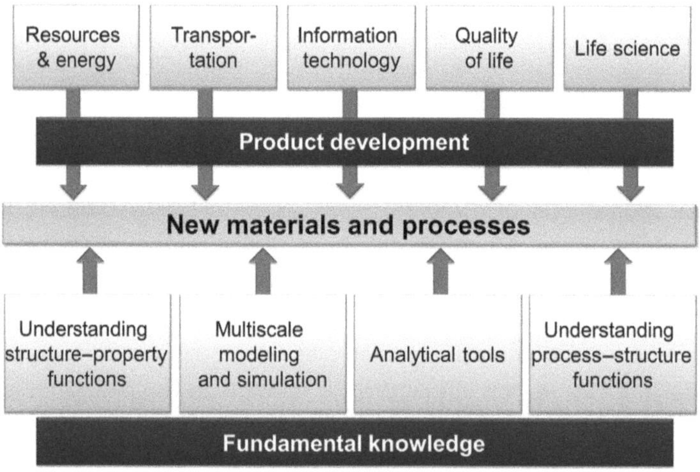


**Figure 1** Driving forces for new materials and processes.

are created together with innovative processes to produce them in an economical and sustainable way.

Despite the quite different technology fields with their manifold products, common underlying principles can be identified. They need to be implemented if new materials and processes shall be developed and put into operation. Key principles are the rigorous understanding and design of structure–property and process–structure functions. The latter strongly depend on analytical tools for materials and process characterization across all relevant length scales. In particular, multiscale modeling and simulation (MSS) are essential for the understanding and design of materials throughout their transient evolution in production processes. Of course, all these aspects rely on the application and advancement of fundamental knowledge from basic science, i.e., on biology, chemistry, mathematics, and physics. It is clear that product design is a highly inter- and transdisciplinary endeavor.

At this point, it needs to be mentioned that the relevance of chemical engineering greatly expanded along with the product- and material-oriented paradigm. Today, chemical engineers are involved in such diverse fields as the development of biomaterials for tissue engineering, the design of the next generation of catalysts for water splitting, or the deposition of thin films in printable electronics or photovoltaics. An interesting and unresolved question in this respect is how to teach chemical engineers with such a wide spectrum of possible applications. This question is particularly pressing for particulate systems which expanded from the traditional mechanical powder processing approach to an extremely broad field of particle science and technology involving gaseous, liquid, and solid particles of biological, chemical, or physical nature. Thereby sizes from below 1 nm up to meters are covered. In other words, the unifying principles beyond the classical unit operations in production and handling of disperse products, i.e., in mesoscale science and technology, need to be identified. These concepts should be reflected in any teaching curriculum to prepare young generations for challenges ahead.

*En route* to answer these pressing questions, in this review, we focus on multiscale methods for the design of particulate products. We will highlight two main routes: from molecules to particles and from single particles to particle ensembles in functional devices. Unifying principles of product design are proposed and examples will be given on how rigorous mathematical optimization can be implemented based on the predictive models for the design of particle properties and processes. Due to the very large field of applications only few selected examples can be presented. A comprehensive overview is neither intended nor possible due to space limitations. Many of the results and concepts in this review have been developed within the Erlangen Cluster of Excellence "Engineering of Advanced Materials—Hierarchical Structure Formation of Functional Devices" (www.eam.fau.de).

## 2. HIERARCHICAL PRODUCTS AND PROCESSES

The hierarchical organization of matter at different length scales, from molecular to real-world macroscopic components, offers many degrees of freedom for creating materials with multifunctional properties. The properties of an advanced material are related not just to the chemical nature of its atoms and molecules but also to their arrangement and distribution, including defects and imperfections. Similarly, the development of models to predict material properties from the atomic to the macroscopic scale and the design of related novel processes are major topics on the agenda of mesoscale science and technology. The key challenge is how to bridge the gap between material design on a molecular level and the macroscopic engineering on the level of components and systems. This requires the development of adequate approaches that are able to cover an extended range of time and length scales in modeling and simulation, processing and manufacturing, as well as structure, property, and process analyses. Figure 2 highlights these interconnections for particulate products starting from different types of molecular and

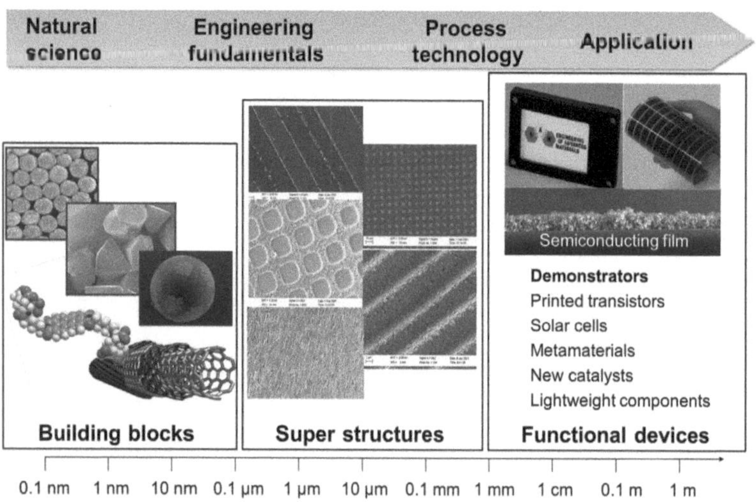


**Figure 2** From building blocks to functional devices in an interdisciplinary environment.

particulate building blocks which are integrated into superstructures, here shown as printed thin films that are finally integrated into macroscopic devices (e.g., light-emitting devices or flexible solar panels). The involved disciplines and required skills comprise basic chemistry and physics, which are mostly active at the level of building blocks, chemical engineering fundamentals, and formulation science and technology. Inks are designed, prepared and printed into superstructures under well-defined conditions in the lab and then transferred via process technologies to larger scales and to applications.

The underlying fundamental hierarchy is exemplified in Fig. 3. The electronic structure of atoms, molecules, and particles defines the molecular and particulate interactions between the objects and building blocks. Their representations are highlighted as cartoons on the right-hand side, showing on the level of building blocks the charge distribution of a nucleus that is formed during the crystallization (upper right), and the surface of a functionalized nanoparticle for printable electronics or photovoltaics. At the next level, those objects interact with each other in a continuous fluid phase until they are deposited as a thin film on a solid and maybe even flexible substrate. The film structure and the properties of the particles determine the macroscopic film properties in terms of, e.g., porosity, electrical, or thermal conductivity, their interaction with light or catalytic properties.

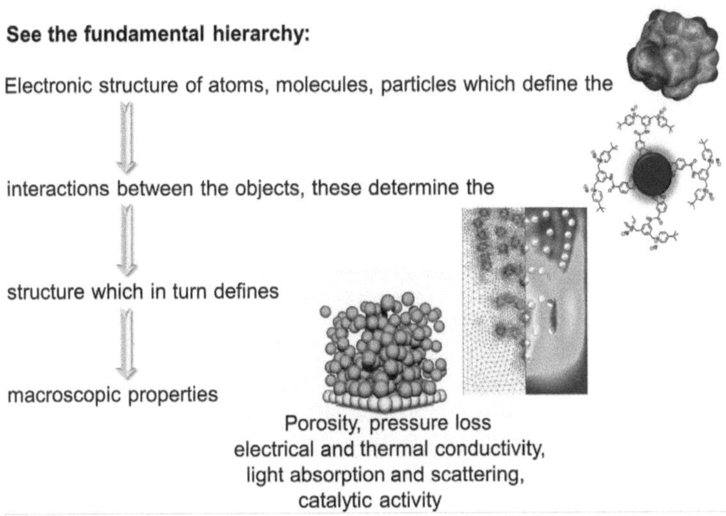


**Figure 3** Fundamental hierarchy from electronic and molecular structure to macroscopic properties.

Statistical thermodynamics already provide an excellent framework to describe and model equilibrium properties of molecular systems. Molecular interactions, summarized for instance in terms of a potential of mean force, determine correlation functions and all thermodynamic properties. The (pair) correlation function represents the material structure which can be determined by scattering experiments via the scattering function. All macroscopic properties of pure and mixed fluid systems can be derived by well-established multiphase thermodynamics. In contrast, a similar framework for particulate building blocks only partly exists and needs to be developed much further. Besides equilibrium properties, nonequilibrium effects are particularly important in most particulate systems and need to be included in a comprehensive and complete picture. We will come back to these aspects in Section 4.

Advances in multiscale processing will provide enormous benefits for more reliable, energy-, and resource-efficient fabrication of high-performance multifunctional materials and for engineering of miniaturized devices and integrated systems. Producing complex devices from designed molecular and particulate structures requires local microscopic tailoring of structure and shape to be extended to the meso- and macroscopic levels of components and systems. This approach offers chemical engineers extensive new opportunities to tackle emerging fields outside the traditional domain of industrial chemistry (Pratsinis, 2010). As highlighted in Fig. 4, functional materials design and fabrication require both in-depth

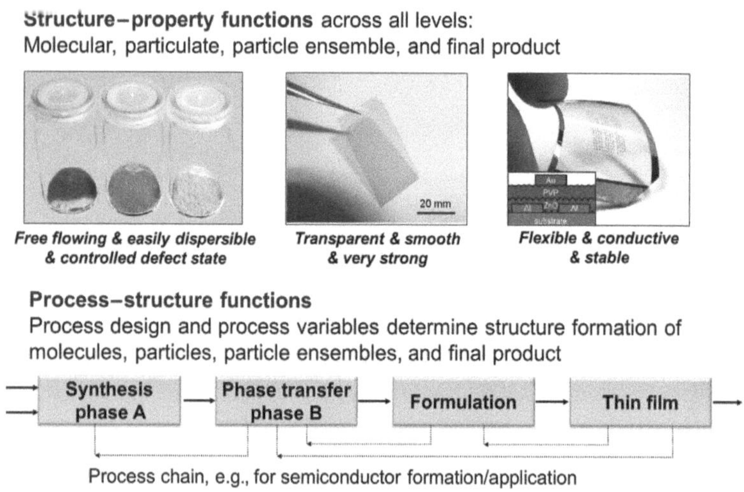


**Figure 4** Structure–property and process–structure functions.

understanding of structure–property functions and rational design strategies which are directly linked to multiscale processing based on process–structure functions.

Property functions, i.e., correlations between structure and related properties, define the targets in any product design. These must be understood across all relevant levels, ranging from molecular, single particle, and particle ensembles to the final product. Very often, multifunctionality prevails, i.e., powders should be free flowing, easily dispersible in fluids and exhibit controlled defect states. The example in the upper left of Fig. 4 shows indium tin oxide nanopowders whose electronic defect states are related to their oxygen deficiency which controls their optical and electronic properties. Thin films made of these powders are applied in opto-electronic devices such as displays. The thin film should be, e.g., transparent, smooth, and sufficiently strong and flexible to survive for instance its application on flexible substrates such as field effect transistors (FETs) as shown in the upper right of Fig. 4.

The product properties evolve from multiscale process chains as highlighted in the lower part of Fig. 4. As an example, the particles are synthesized in phase A, say the gas phase or an organic solvent, and then they are transferred to an aqueous phase B where several formulation steps such as stabilization, surface functionalization, and adjustment of the suspension rheology may be included. Finally, a functional thin film is deposited in lab scale usually by spin- or dip coating, whereas in an industrial environment the current development is targeted toward fast roll-to-roll processes. Depending on the type of product, the process chains including the involved unit operations may change of course.

The process function (process–structure functions) as defined by Krekel and Polke relates the process parameters to the product property (Krekel and Polke, 1992):

$$\text{dispersity} = f(\text{process parameters}, \text{educt concentrations}) \tag{1}$$

Process parameters are the type of unit operations, their interconnection in the process, the process conditions under which the unit operations are operated (e.g., temperature, pressure, mass flow rates, etc.), and the materials that are processed. Structure–property as well as process–structure functions must be known in order to design optimal process variables and to achieve the desired goal, i.e., to produce well-defined, often multifunctional product properties. Usually, process chains (with or without recirculation loops) have to be developed during which both handling and end-use properties have to be optimized.

While these developments of products and the related process chains have been done mostly empirical in the past, modern developments of process design make more and more use of computer-based methods including flow-sheeting software for design of the individual process steps as well as for the whole process. While such design tools are well developed for fluids, these are still in their infancy for solids processing. A larger coordinated project in Germany led to the development of SolidSim, a comparatively new flow-sheeting software tool for major unit operations including crystallization, agglomeration, comminution, gas–solid as well as solid-liquid separation and transport (Hartge et al, 2006). The outcomes of these developments are now integrated in the Aspen software package. The next step along these lines is the development of dynamic models which can handle efficiently distributed properties and time dependencies of individual unit operations and their interconnection within a whole production process. The required fundamentals are currently developed within a nation-wide priority program in Germany and will be represented by a complex flow-sheet environment (http://www.dynsim-fp.de/spp-1679.html) (Dosta et al, 2014). However, these positive developments also show that predictive models for most unit operations are rare. The reasons for this deficiency are related to (i) poorly understood turbulent multiphase flows with mass and momentum exchange between the phases, in particular at high *Re* numbers, (ii) poorly understood particle–particle interactions, and (iii) the lack of reliable single particle and powder (particulate ensemble) properties. Mesoscale science and technology evolves from molecular systems and is key to the assembly of macroscopic products. Thus, the intermediate state of disperse systems must catalyze better and more efficient solutions to pressing technological challenges of the modern world.

Molecular and nanoscale design, synthesis, and analysis of materials are becoming increasingly important for high-performance applications in emerging fields such as nanoelectronics, catalysis, and energy conversion. Molecules offer the opportunity of tailoring structure and function systematically through modern concepts of synthetic (Negishi, 2011; Suzuki, 2011) and supramolecular chemistry (Steed and Atwood, 2009; Whitesides and Boncheva, 2002). Organic molecules are particularly versatile building blocks since they can combine stability with almost limitless structural and functional variation. Inorganic nanoparticles offer a number of additional benefits, in particular high thermal stability. Individual molecules

(Lu and Lieber, 2007), hierarchically structured supramolecules (Schlundt et al, 2009), individual particles (Tenne, 2006), particle systems (Min et al, 2008; Pratsinis, 2010), and, last but not least, organic/inorganic hybrid structures (Gnichwitz et al, 2010) provide the key functionalities, for instance for devices (Lal et al, 2007) such as sensors, circuits, or photovoltaic cells. Research into fundamental processes such as charge transfer, mass transport, or interaction with light, into advanced materials, and into the formation of unprecedented structures and architectures for functional devices is of fundamental importance for many applications. The properties of multiphase components with microstructures modified on the molecular scale are governed by the interface structure and the chemical nature of the corresponding organic and inorganic components. Hence, understanding the structures of interfaces on the atomistic scale and their influence on materials processes and properties are key factors in manufacturing complex multiphase materials with optimized properties.

A prominent example for the hierarchical organization of building blocks which is decisive for the later product properties is the case of mesocrystals. Mesocrystals are a relatively new class of crystals consisting of small buildings blocks which are self-assembled to larger units (Cölfen and Antonietti, 2005; see Section 4.3). Both examples in Fig. 5 show ZnO mesocrystals. The mesocrystals in Fig. 5A consist of 5 nm ZnO quantum dots which are organized to quite uniform spherical single crystals of about 300 nm in diameter (Distaso et al, 2012). These show high UV-absorption and minimized light scattering in the region of visible light. The ZnO mesocrystals in Fig. 5B are also built from ZnO quantum dots which are then organized by multistep-oriented aggregation as rods, semi- and full ellipsoids (Klaumünzer et al, 2014). These particles show a high scattering intensity in the whole range of visible light.

Particles are seldom sole as such, and they are usually integrated in products. Paper, ceramics, and bread are just three well-known examples of everyday use. Advanced structural and functional materials consisting of particles and powders are made to be integrated into all the functional devices mentioned in this overview. The implementation of nanotechnologies is still limited due to the difficulties in preserving the beautiful nanostructures and their excellent properties produced in the lab upon transferring them into industrial practice. We have to understand how hierarchical products evolve in multiscale processes. Important questions must be answered such as where does "quality" evolve in chemical processes and how can we control electronic and structural defects of particles and particle systems across all

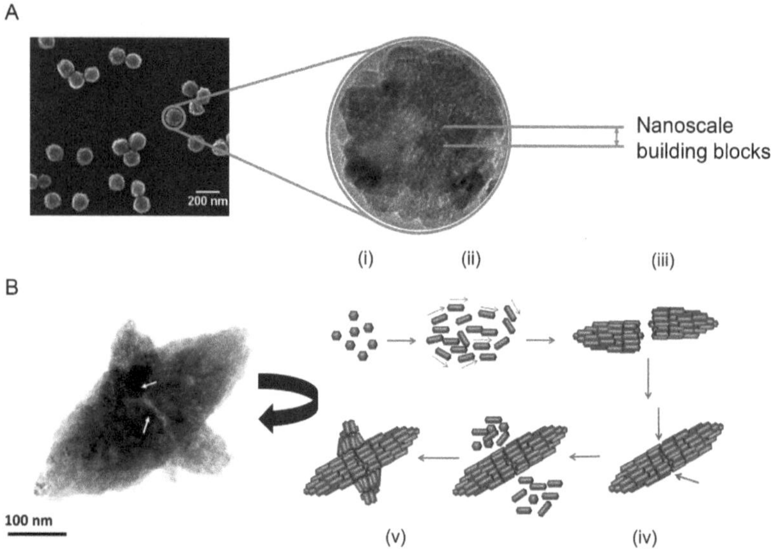


**Figure 5** Hierarchical structure of mesocrystals: (A) porous balls of ZnO quantum dots and (B) ellipsoids of ZnO nanorods and their formation mechanism. *Panel (A): Distaso et al (2012). Panel (B): Klaumünzer et al (2014). Reproduced with permission from the Royal Society of Chemistry.*

length and time scales. Key to these questions is the in-depth understanding of particle properties.

## 3. PRODUCT PROPERTIES

For particulate materials, the product properties depend on the chemical composition and on the dispersity of the material. The dispersity is characterized by the particle size distribution (PSD), the particles' shape, their morphology in terms of internal structure (amorphous, crystalline, internal pore size distributions and their defects), and their interfacial properties. This relation was called by Rumpf already in the 1960s of the last century the "property function" (Rumpf, 1967). Control of the property function is the core of product engineering or product design.

$$\text{Product property} = f(\text{dispersity, chemical composition}) \tag{2}$$

Dispersity includes:

- particle size and shape and their respective distribution,
- particle morphology including defects, and
- particle surface properties.

Thus, the property function relates the particulate structure (size, shape, morphology, and surface) to the product properties (structure–property–function). Examples of property functions are the taste of chocolate, the color of pigments, the strength of cements, or the bandgap of semiconductor nanoparticles. Particle ensembles in the form of agglomerates, thin films, or filter cakes are also included in this consideration. Modern products are often characterized by several properties which have to be achieved simultaneously: transparent *and* scratch-resistant coatings, sun blockers which are transparent for visible light *and* UV-absorbing may serve as examples. Thus, a multifunctional product space evolves which must be created through design of the related structures using the respective process technology.

Glotzer and coworkers proposed the following framework of evolving complexity for particles as shown in Fig. 6 (Glotzer and Solomon, 2007). By varying systematically geometric features such as size and shape and also surface coverage, patterns, branching, and chemical ordering, many different

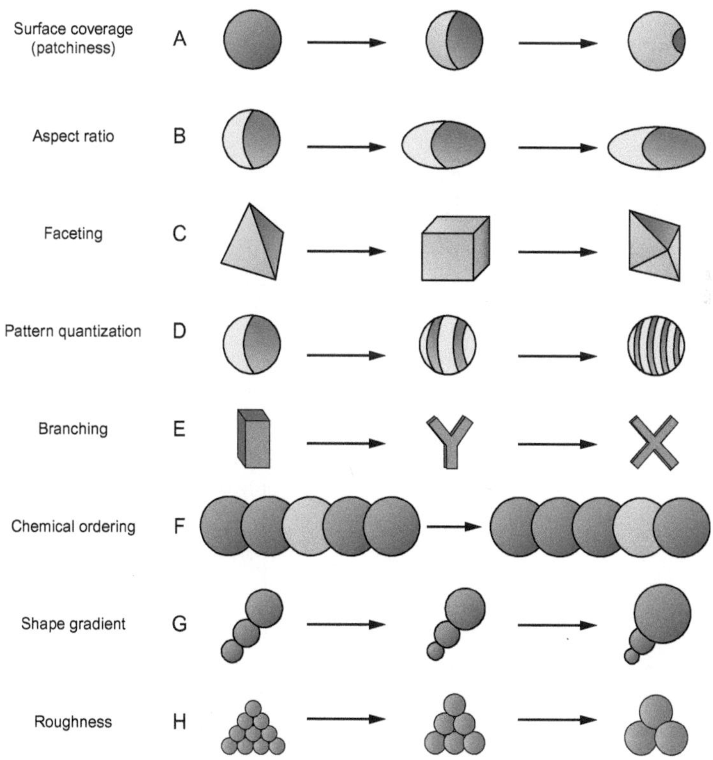


**Figure 6** Overview of complexity generation of particles. *Permission from Macmillan Publishers Ltd. Glotzer and Solomon (2007). Copyright 2007.*

types of particles can be envisioned for which the phase behavior will vary between ordered crystalline structures and disordered glassy structures. During the last years, synthesis capabilities strongly improved. Therefore, it is within reach to realize at least some of those structures within the next years. In the following, examples for electronic, optical, and mechanical properties will be presented and related to dispersity in terms of size, shape, morphology, and surface. For the former, we will discuss possibilities for property optimization (Chapter 5), and the latter are most relevant for the question of particle–particle interactions in modeling approaches of particulate dense phase systems (Chapter 4).

Many property functions in dependency of particle size are known. Today, PSDs can be measured reliably across all length scales from below 1 nm up to macroscopic sizes. In Fig. 7, we highlight a few examples of the size-dependent bandgap of semiconducting quantum dots which controls their electronic and optical properties. The latter strongly depend on the bandgap energy of the first excitonic transition: ZnO (Fig. 7A) and

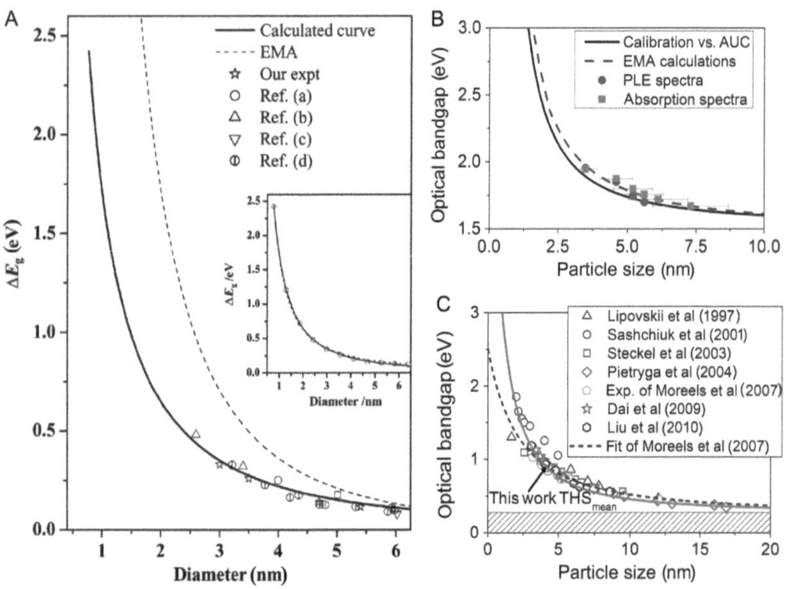


**Figure 7** Bandgap of quantum dots: (A) ZnO (Viswanatha et al, 2004), (B) CIS (Akdas et al, 2015), and (C) PbSe (Segets et al, 2012b). EMA refers to the effective mass approximation originally established by Brus (1984) which usually fails for the accurate description of the smallest particles. *Panel (A): Reproduced from Viswanatha et al (2004) with permission of The Royal Society of Chemistry. Panel (B): Reproduced from Akdas et al. (2015) with permission of The Royal Society of Chemistry. Panel (C): Reprinted with permission from Segets et al (2012b). Copyright 2012 American Chemical Society.*

$TiO_2$ are absorbing in the UV, CdS(e), CdTe, and $CuInS_2$ (CIS, Fig. 7B) in the visible, whereas lead-based materials such as PbS(e) absorb in the NIR (Fig. 7C) (Akdas et al, 2015; Segets et al, 2012b; Viswanatha et al, 2004).

In all cases, the bandgap $\Delta E$ widens due to the quantum size effect according to the following empiric equation:

$$\Delta E = \Delta E_{\mathrm{g,bulk}} + \left(\frac{x-c}{a}\right)^{1/b} \cdot \mathrm{eV} \tag{3}$$

where $x$ is the particle diameter, $\Delta E_{\mathrm{g,bulk}}$ is the bandgap of the bulk material, and $a$ (nm), $b$ (−), and $c$ (nm) are material-dependent empiric parameters. Tailoring of the opto-electronic properties of these materials is thus directly linked to the control of their size. We will come back to this issue in Chapter 5.

Only few examples of property functions as a function of particle shape are known. This lack is related to the difficulty of measuring size and shape simultaneously since at least two dimensions of the particle must be determined in the process environment. The two most prominent examples with probably the best database are the drag force of nonspherical particles (Binder et al, 2009; Clift et al, 2005; Walter et al, 2015b; Yow et al, 2005) and the optical properties of metal nanoparticles. The latter are due to surface plasmon resonances which give rise to shape-dependent absorption peaks. For a long time, particle shape determination was restricted to tedious off-line imaging. Since a few years ago, particle shapes can be measured (even inline) for particles larger than a few 10 μm in size by direct video-based shape analysis (Witt et al, 2013). For smaller particles, shape analysis is quite tedious because a statistical relevant number of particles (at least some 100) has to be measured by image analysis of deposited samples using scanning electron microscopy (SEM), transmission electron microscopy (TEM), or atomic force microscopy (AFM). New possibilities for two-dimensional characterization come from the combination of sedimentation analysis with inline absorption spectrometry (Walter et al, 2014, 2015b). In brief, analytical (ultra-)centrifugation (AUC) allows the analysis of particles ranging from 1 nm when an ultracentrifuge is used up to a few 10 μm in size when a conventional centrifuge is used (Walter et al, 2015a). The sedimentation velocity depends on the size and shape of the particles.

Figure 8A depicts as an example the frictional ratio of cylinders as a function of their aspect ratio. This relationship for shape analysis of cylinders and platelets was recently demonstrated for graphene oxide monolayers (Walter et al, 2015b). The frictional ratio plotted on the $y$-axis is the drag force of the

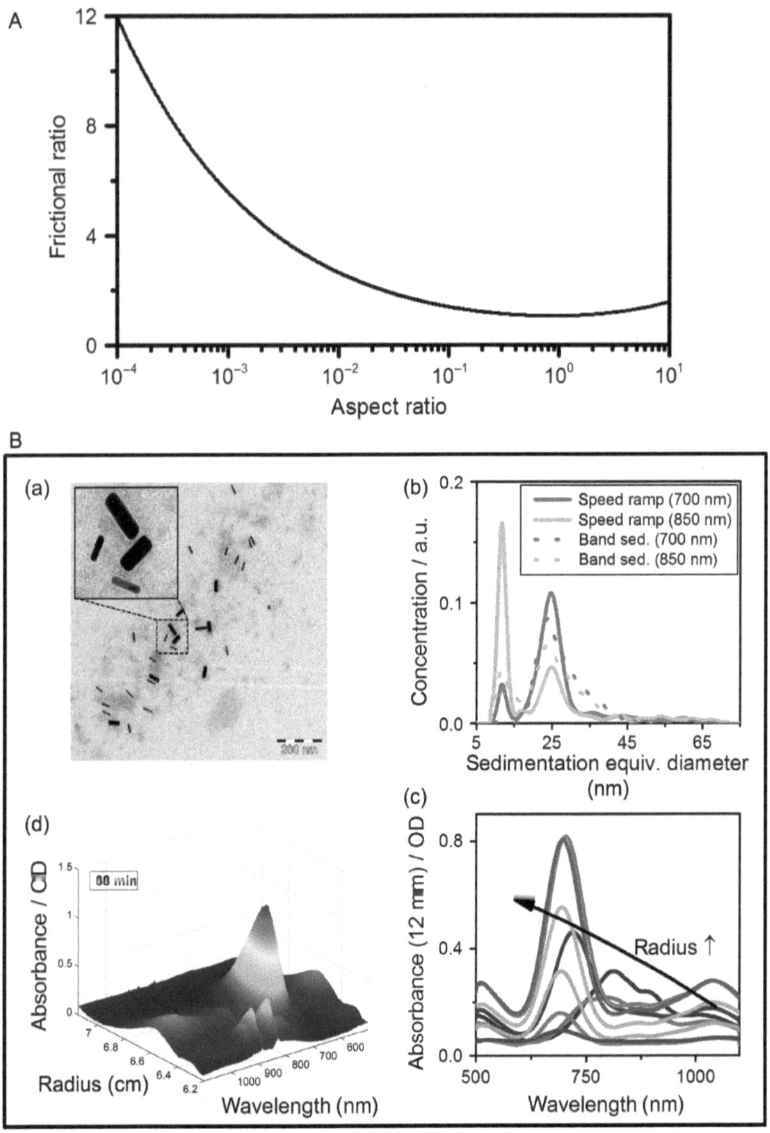


**Figure 8** (A) Frictional ratio of sphero-cylinders (ratio of drag of cylinder and volume equivalent sphere) in dependence of their aspect ratio ($Re \ll 1$) and (B) absorption sedimentation analysis of gold nanoparticles in an analytical ultracentrifuge. *Panel (A): Reprinted with permission from Walter et al (2015b). Copyright 2015 Wiley. Panel (B): Reprinted with permission from Walter et al (2014). Copyright 2014 American Chemical Society.*

cylinders related to the drag of volume equivalent spheres. During a sedimentation run, the separated particles can be analyzed with respect to their absorption properties. This is shown in Fig. 8B for gold nanorods. If the particles are small enough to show strong diffusion (typically below 20 nm), simultaneous evaluation of sedimentation and diffusion can be used to obtain shape factors.

In the next step, we discuss property functions in dependency of particle morphology. The term "morphology" includes amorphous and crystalline states, pore structures, and various defect structures of particles. Here, we give an example for mechanical particle properties. One of the grand challenges in particle technology is how to characterize the contact mechanisms and contact forces between particles. Mechanical particle properties are the key parameters in discrete element models (DEMs) which have been developed during the last decades and gradually reach "predictive power." Mechanical particle properties are urgently needed to calibrate DEMs which despite their great progress still rely on global parameters such as spring constants or coefficients of restitution. The (nano-)mechanics of particles across all length scales down to the nanoscale are highly relevant in all situations where contact mechanics of particles must be considered, i.e., in dense particle flows such as fluidization, transport, and particle storage (conveying, mixing, and silos), but also in agglomeration and granulation (adhesion), size reduction by comminution (fracture), and tribology. Particular interesting new fields are additive manufacturing and 3D printing. They "promise" a revolution in industrial manufacturing based on powder technologies where excellent flowability must be combined with dense packing structures. One key question in all these applications is how the particles react to the externally applied stress. Briefly, this reaction can be elastic or inelastic. However, in many cases neither the elastic properties (Young's modulus and Poisson ratio) are known nor the inelastic properties such as yield strength or viscoplastic properties which lead to internal defects (e.g., dislocations, twinning, and shear bands). Defect characterization inside particles requires high-end scattering or electron microscopy techniques (Vieweg et al, 2012).

In recent years, more and more systematic studies are published focusing on mechanical properties of small particles (Fuchs et al, 2014; Meier et al, 2009b; Paul et al, 2015; Paulick et al, 2014; Romeis et al, 2014, 2015; Sanimi et al, 2005; Tomas, 2004; and references within). Mechanical elastic and inelastic particle properties have been measured systematically in a customized nanoindentation device installed within a SEM to simultaneously obtain force–displacement information and SEM images of deformed and broken particles (Romeis et al, 2012). Quantitative data of intrinsic particle properties (and their respective distributions) are derived with high statistical relevance from force–displacement curves of single oxide, metal, and polymer particles. These are related to internal structural properties of the particles (Paul et al, 2014, 2015; Romeis et al, 2014). Figure 9 shows structure—property functions of "simple" amorphous Stöber–Fink–Bohn (SFB) $SiO_2$ particles (diameter 500 nm) (Stöber et al, 1968) which were densified at various temperatures. Their internal structures including internal bond–network effects have been evaluated also by nuclear magnetic resonance (NMR) spectroscopy, Fourier-transformed infrared spectroscopy (FT-IR), and Raman spectroscopy. The as-synthesized sample (in a sol–gel process) did contain many hydroxyl and dangling bonds which gradually condensed with increasing temperature and are completely removed at 1000 °C.

The last topic in this section concerns particulate surface properties. The importance of surface properties is known since a long time, e.g., in the field of colloid science. With the advent of nanotechnology, this issue is now very prominently on the agenda because of the high surface-to-volume ratio of small particles. Briefly, any technical realization of nanoparticulate products requires the careful control of interfaces (details are discussed in Section 4.2). However, since the surface roughness can influence the interaction forces by orders of magnitude, also the interactions of larger objects, for instance during adhesion and friction, are controlled by nanoscale effects (see the example on additive manufacturing in Section 4.6). Surface properties of particles are directly linked to their solubility in solvents, to their wetting properties, to their conductivity, and to their reactivity in catalysis just to mention a few of them. Conformal and nonconformal coatings on particles are used to tailor the interaction with light. The well-known and beautiful pearl luster effect pigments are based on diffraction effects such as titania-coated mica platelets.

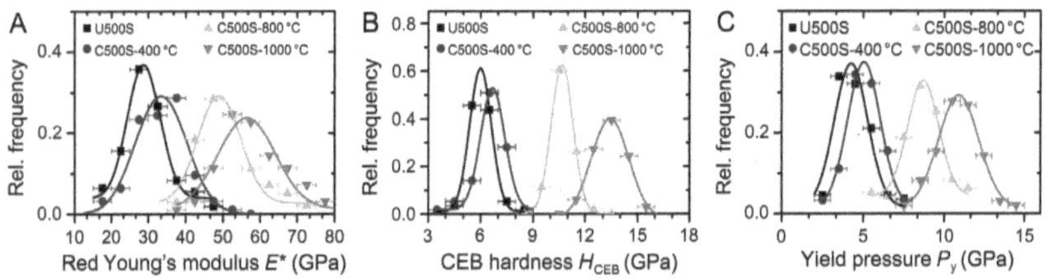


**Figure 9** Distributions of (A) Young's Modulus, (B) hardness, and (C) yield pressure for 500 nm amorphous silica beads which were tempered at temperatures of 400, 800, and 1000 °C (U500S nontreated) (Romeis, 2015). Similar data have been used to predict the packing structure of silica particles as a function of size by a DEM simulation (Parteli et al, 2014).

Nevertheless, and as discussed in more detail in Section 4.2, it is very difficult to analyze the surfaces of small particles in detail with respect to structural (e.g., roughness, steps, and kinks) and chemical properties (e.g., chemical surface moieties of the particles, adsorbed or chemisorbed layers, and surface coverage).

As an example how surfaces influence particle properties, we present the case of small nanoparticles for which the surface-to-volume ratio is quite large. The coordination number of molecules and atoms at the surface of nanoparticles is reduced relative to the bulk due to the curvature of the particles. This gives rise to the enhanced solubility and vapor pressure of small particles in solution and in the gas phase, respectively. This is expressed by the well-known Kelvin equation. In a similar way, the sintering kinetics of small nanoparticles is enhanced. This effect is important during gas phase synthesis of particles where a balance of coagulation and sintering kinetics defines the morphology of the aggregates which is often expressed by a fractal dimension. Coagulation without any sintering typically leads to a fractal dimension of around 1.8, whereas fast sintering leads to complete coalescence, i.e., to spheres with a fractal dimension of 3. The sintering kinetics (at least for amorphous silica particles) is enhanced relative to bulk values below a critical size of about 100 nm (Kirchhof et al, 2012). In the context of mechanical properties, we could recently show that the Young's modulus of small metal and salt particles (Armstrong and Peukert, 2012; Miesbauer et al, 2003) is reduced due to this size or surface effect, respectively (see Fig. 10).

# 4. PRODUCT DESIGN

Five unifying principles in product design can be identified. They will be discussed in the following under a generalized view:

1. Particle formation by top-down and bottom-up approaches.
2. Interactions between the building blocks.
3. Structure formation by self-organization and by transport.
4. Characterization along the process chain.
5. Multiscale modeling and simulation

Finally, an outlook on the integration of particles into devices will be provided.

## 4.1 Particle Formation

### *4.1.1 Top-Down Approaches*

The most common size reduction processes, comminution, emulsification, and spraying, share several common features. Eggers and Villermaux discuss

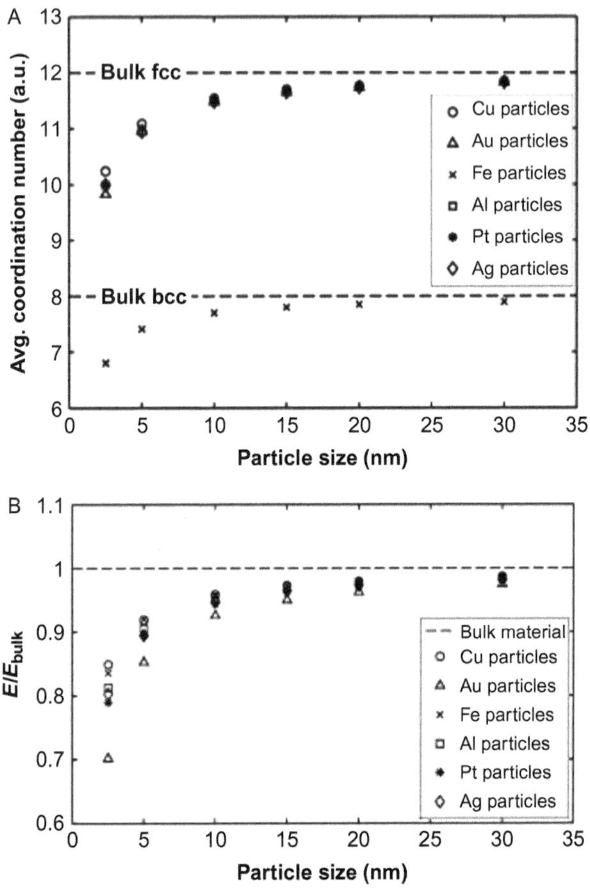


**Figure 10** (A) Average coordination number of six different metal nanoparticles and (B) their average Young's modulus $E$ with respect to the value of the bulk as obtained from molecular statics simulations. *Armstrong and Peukert (2012), with kind permission from Springer Science and Business Media.*

atomization of droplets from a unifying perspective (Eggers and Villermaux, 2008). Most importantly, size reduction occurs when sufficient external energy is supplied to the particles and made available to generate new surface. In case of solid particles, fracture is induced by stressing of the particles by impaction or compression. The transferred energy is stored in the particles as elastic energy or dissipated by viscous and/or plastic deformation leading to all sorts of structural defects. Fracture occurs when the elastically stored energy is released and sufficiently high to induce crack opening. In case of fluids, shear and more importantly elongational stresses act on the fluid particles which in turn may be disintegrated into smaller droplets if the applied stresses are sufficiently high. Surprisingly and despite all differences between these top-down methods, the mean particles size $x_{50}$ scales with the volume-specific energy input $E_V$:

$$x_{50} = a(E_V)^{-b} \tag{4}$$

where the parameters $a$ and $b$ depend on the unit operation and the type of particles (see Table 1).

The minimal particle size which can be produced depends on the balance between the local stress energy and the energy required to increase the surface. In case of solid particles, cracks are initiated at internal particle defects due to local stress enhancement. The true grinding limit is reached when the particle cannot store anymore defects (Knieke et al, 2009). It has been shown for $ZrO_2$ and $SnO_2$ that this limit is reached at around 10 nm (Knieke et al, 2011). The width of the PSD depends on the spatial distribution of $E_V$ in the "reactor" volume, i.e., on the design of the mill, on the emulsifying device, or on the nozzle and the resulting multiphase flow and thus on the machine function. In summary, the slope of Eq. (4) varies in a band of the exponent $b$ between 0.3 and 0.8 for different unit operations and materials. This indicates that unified scaling laws are valid which are based on fundamental similarities of particle formation in top-down processes.

### 4.1.2 Bottom-Up Methods

Single particles are the building blocks of mesoscale science and technology including nanotechnology and many other fields of application as discussed

**Table 1** Value of Scaling Parameter *b* for size in Top-Down Processes

| Type of Unit Operation | Value of Parameter *b* | References |
|---|---|---|
| Comminution in stirred media mills (varying sizes and operational parameters), limestone, and graphene oxide platelets | 0.8<br>0.37 | Becker et al (2001)<br>Halbig et al (2016) |
| Emulsification in colloid mill, homogenization, and ultrasound | 0.3–0.8 | Behrend et al (2000) and Schuchmann and Schubert (2001) |
| Atomization in different spray nozzles | 0.5–0.7 | Walzel (1990) |
| Ultrasound and graphene oxide platelets | 0.48 | Halbig et al (2016) |

above. It is, however, beyond the scope of this review, to present a comprehensive overview of particle synthesis and crystallization as several excellent books and review articles are available covering a broad range of materials. Therefore, we discuss here only a few guidelines for this important and very large field without claim to completeness.

Bottom-up methods for particle synthesis in gas and liquid phase can be visualized by a unified scheme as highlighted in Fig. 11. In general, mass, momentum, and heat transfer processes coupled to chemical reactions produce nucleating species. Their distribution in the reactor related to their equilibrium concentration/partial pressure (or activity/fugacity in a general sense) is defined as the supersaturation $S$. It is the thermodynamic driving force for the phase transition and thus for the formation of a new particle phase. Depending on the spatial and temporal distribution of $S$ in the reactor, nuclei form with a size distribution $q$. Noteworthy, especially in the view of classical nucleation theory $q$ strongly depends on $S$. For instance, a high supersaturation with narrow distribution in time and space would lead to small nuclei with narrow PSD since all particles "experience a similar history." After nucleation, several processes may occur sequentially or in parallel. These are growth processes for further reduction of $S$, coagulation of the particles, their stabilization against coagulation, and finally ripening effects in the liquid phase or sintering in the gas phase. This quite general

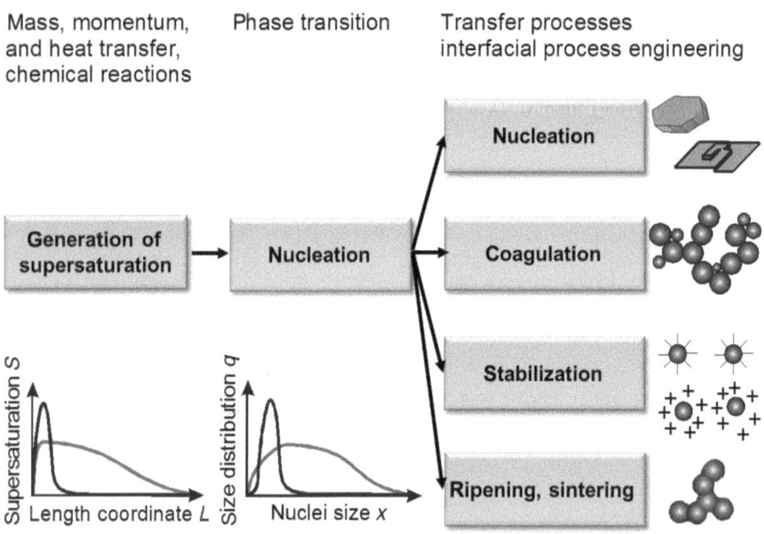


**Figure 11** Principles of particle synthesis. *Adapted from Peukert et al. (2003). Copyright 2003, with permission from Elsevier.*

framework forms the basis of any modeling approach for particle formation dynamics including mixing, global reaction kinetics, nucleation, growth, agglomeration and stabilization and even ripening and sintering. Detailed reaction mechanisms of conversion of precursor molecules into stable and growing nuclei may be included in this general scheme which are, however, too seldom known. Their full elucidation requires coupling of quantum mechanical methods such as density functional theory (DFT) with molecular dynamics (MD) simulations (see, e.g., Menz et al, 2011; Milek and Zahn, 2014; Zheng et al, 2009). A key challenge in this field is how to tailor reaction networks so that a desired target phase does form.

Particle formation processes can further be subdivided into transport- and reaction-controlled processes. Typically one assumes that the synthesis in the gas phase is not controlled by any mass transfer limitations, whereas mass transfer issues are very common in liquid phase synthesis. Prior discussing mixing and reaction-controlled systems in more detail, it needs to be mentioned that perfect stabilization is assumed for the time being. Effects of coagulation would completely change the PSD.

In transport or mixing controlled systems, the mixing intensity determines the local concentration fields and thus the supersaturation as driving force (Schwarzer and Peukert, 2002). The energy dissipation in case of a stirred tank as well as in a continuous mixer such as a T- or Y-mixer is directly related to the volume-specific energy consumption. The mixing volume will follow similar scaling laws as already discussed in Chapter 3 for the top-down processes. Higher energy dissipation will lead to successively smaller eddies in the fluid until the smallest eddy size (the Kolmogoroff length scale) is reached. Mixing is shifted from macro- to meso-mixing and finally to diffusion-controlled micromixing (Baldyga and Bourne, 1999; Bockhorn et al, 2010; Fox, 2003). The reactor design determines the residence time of fluid particles and the local distribution of supersaturation. Both effects will control the width of the obtained PSD, or in other words, the width of the PSD is a measure of the mixing energy distribution. The full PSD can be modeled (at least for well-understood precipitation reactions such as the formation of $BaSO_4$) without any free parameters by a combination of direct numerical simulation for complete resolution of the fluid flow structures and an appropriate mixing model for mass transfer at subgrid level on the one hand that is combined with a population balance model, on the other hand (Gradl and Peukert, 2009; Gradl et al, 2006).

In reaction-controlled systems, i.e., in systems where mixing is much faster than the chemical reactions leading to precursor formation, the distribution of all reactions and thus of the supersaturation is much more uniform. Therefore, the formation of narrow PSDs will mostly occur in reaction-controlled systems. First, particles form in a uniformly distributed nucleation burst which is quickly reduced due to the formation of a new phase below the threshold value where homogeneous nucleation can occur. Second, the particles grow by further reduction of the supersaturation until equilibrium is reached. This approach of separating nucleation and growth is widely applied to form quasi-monodisperse particles such as quantum dots. The term "quasi-monodisperse" is used because upon close inspection most of these systems still have a narrow PSD (Segets et al, 2015). In particular, chemical synthesis has made great progress in the last 20 years to produce beautiful and quite homogeneous semiconducting and metallic nanoparticles (Burda et al, 2005; Park et al, 2007; Reiss et al, 2009; Tao et al, 2008). These are so well controlled that they can be used to mimic crystal growth by assembling different quantum dots as particulate atoms as will be discussed in Section 4.3 in the context of Fig. 16.

The formation of crystals depends on the underlying phase diagrams and is controlled by chemical thermodyamics. This is applicable to all types of particles, i.e., to conducting, semiconducting, and nonconducting materials. However, equilibrium solubilities or equilibrium vapor pressures of even simple compounds are often unknown, not to speak of more complex particles such as three- or four-component particles like they are found in catalysis, solar absorbers, batteries, or complex molecules in pharmaceutical applications. A clear need exists to develop predictive methods for the determination of phase equilibria of particles. In other words, methods shall be developed to measure the material function of precipitation and crystallization which then can be coupled to a machine function as discussed above.

Shape control of particles is particularly important in pharmaceutical industry and in nanotechnology. The equilibrium shape of a crystal is given by Wulff's construction which is based on the minimization of the overall surface energy as the sum over all indexed surfaces of a crystal. Shape control in the liquid phase is achieved by either choosing appropriate solvents in which the crystals are grown to the desired equilibrium shape or by shape controlling surface active compounds. The latter adsorb on specific surfaces and inhibit their growth (Mann and Ozin, 1996; Mersmann, 2001).

Particle syntheses in gas and liquid phase are typically studied in quite different communities with little exchange between them. In a recent study, we have shown that the concept of separation of nucleation and growth in liquid phase can be transferred to gas phase synthesis. In particular, silicon nanoparticles were produced in a thermal hot wall reactor by decomposition of silane (Körmer et al, 2010a, 2010b). We obtained narrow PSDs down to a geometric standard deviation of 1.04 and could show by population balance modeling that nucleation and growth are indeed almost separated. This example will be described in more detail in Chapter 5 in the context of optimization.

The governing equations to model the particle formation dynamics are identical for the gas and liquid phase, whereby of course, the reaction mechanisms and the kinetic parameters (e.g., for the mass transfer) differ. The population balance equation (PBE) can be considered as the master equation for formation of particles by both top-down and bottom-up methods (Eq. 5):

$$\frac{\partial n}{\partial t} + \nabla\left(\vec{v_e} \cdot n\right) + \nabla\left(\vec{v_i} \cdot n\right) = B - D + \nabla\left(\vec{j}\right) \tag{5}$$

The number density $n$ is given as a function of the particle variables such as diameter, volume, or other properties. $v_e$ denotes the particle velocities in 3D external space; in a completely back-mixed system this term vanishes. $v_i$ is related to internal convective particle variables; in crystallization processes it defines the growth rate. The $B$- and $D$-terms denote sources (birth) and sinks (death) for nucleation, agglomeration, and breakage. The gradient of the flux $j$ denotes exchange across the boundaries of the balancing volume. In Section 5.3, we will discuss particular solutions of this equation.

In the most general way, the evolution of the whole particle property space can be included by additional variables leading to multidimensional integro-differential equations (Ramkrishna, 2000; Saad et al, 2015). Current research is directed toward efficient coupling of PBE with computational fluid dynamics (CFD) simulations (Krasnyk et al, 2012; Marchisio and Fox, 2013; Marchisio et al, 2003; Ramkrishna, 2000; Renze et al, 2014), to model increasingly complex reactions networks (Hartig et al, 2014, 2015), to take several particle coordinates into account (e.g., size and shape) (Borchert et al, 2009; Briesen, 2009; Gruy, 2011; Kruis et al, 1993), and to develop better kernels for agglomeration (e.g., complex fractal aggregates) (Khalili et al, 2010; Soos et al, 2006) or breakage (Madadi-Kandjani and Passalacqua, 2015; Patruno et al, 2009; Sommer et al, 2006).

## 4.2 Interactions

In general, interactions include aspects of particles directly in contact with each other or with external walls through "hard" mechanical forces and "soft" microscopic interactions of dispersed systems induced by molecular effects such as electrostatic, van der Waals, or hydrophobic forces. As mentioned earlier, microscopic aspects of particles in contact can induce quite large macroscopic effects. The first aspect of "hard" contact relies on contact mechanics for which the governing mechanical particle properties must be known (see Fig. 9). In this section, we will only consider "soft" interactions.

The classical picture of molecular interactions and their effect on macroscopic properties is well developed in statistical thermodynamics of phase equilibria. Briefly, molecular interactions determine the potential of mean force which is directly related to the pair correlation function from which all macroscopic (thermodynamic) properties such as virial coefficients, osmotic pressure, and so on can be derived as a function of temperature and pressure. The question arises if this well-developed framework of equilibrium thermodynamics can be expanded to particulate systems in equilibrium. Equilibrium systems can be best studied for colloids in solution on which we will focus our discussion in the following.

After these general remarks, we move now on to nanoparticle systems. All nanoparticle systems and applications have in common that the interfacial and surface properties of the particles play a central role. The ratio of van der Waals adhesion forces to particle weight scales with particle diameter $x^{-2}$ and is, for instance at 1 μm, in the order of $10^6$ (in case of smooth particles in the gas phase). To produce well-defined property functions, the particle interactions have to be carefully controlled. Macroscopic properties can only be tailored by microscopic design of the interfaces. Surface chemistry and physics determine on the one hand the particulate interactions with fluid or solid phases. The types of interactions are van der Waals, electrostatic forces and polarization effects, hydrogen bonds, and chemical bonds (Israelachvili, 2011). Particle interactions control particle and structure formation. For product engineering of (nano-)particulate systems, we start conceptually at the particle surface which "transports" the respective particle interactions, thus leading to the desired structure. Vice versa, structure formation can only be understood by considering the relevant interactions which are determined by the particle surface. This concept is illustrated in Fig. 12 for oxide particles in aqueous solution. Particle interactions can be understood in the view of well-known DLVO theory as a superposition of van der Waals and electrostatic double layer forces (Peukert et al, 2005). This general concept can also be applied to more complicated and less described types of interactions, e.g., interactions between polymer-modified particles where entropic and structural effects have to be included as well.

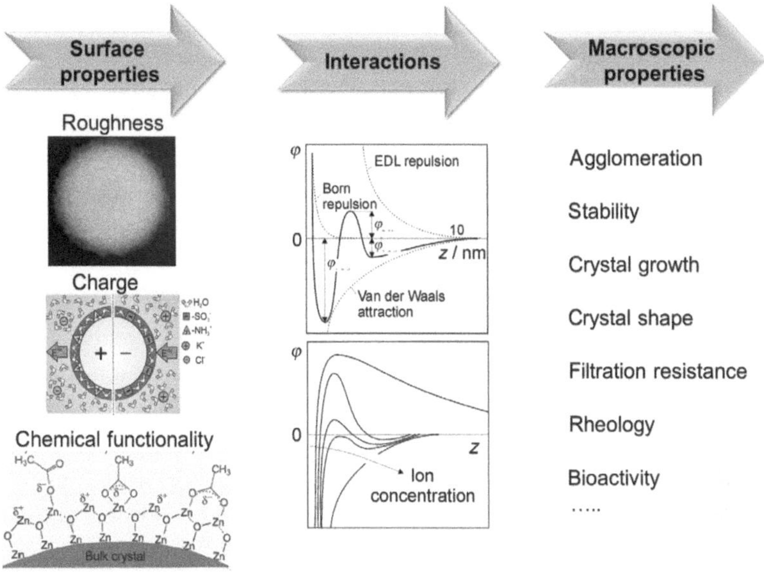


**Figure 12** From surface to macroscopic properties—the relevance of particle interactions. *Reprinted with permission from Segets et al (2011). Copyright 2011 American Chemical Society; Peukert et al (2005). Copyright 2005 Elsevier.*

This approach is based, however, on some wide-ranging preconditions. In order to bridge the gap between the microscopic molecular nature of a particle surface and the macroscopic properties, we need a multiscale approach covering several orders of magnitude in space and time. On the most basic level quantum mechanics (QM) prevails. However, it is often possible by using the Hellmann–Feynman theorem (Feynman, 1939) to transfer the intrinsic quantum mechanical nature of surfaces to the physics of molecular interactions described by classical force laws. This theorem states that once the electron density distributions have been determined, the intermolecular interactions can be calculated on the basis of classical electrostatics. The contact value theorem states that the force between two surfaces is determined by the density distribution of the molecules and particles in the space between them (Berman and Israelachvili, 1997). By using these classical interaction forces, MD and Monte Carlo simulations are nowadays able to describe and even predict mesoscopic phenomena. Thus, we use for molecular systems (without chemical reactions) the additivity of forces which is often a reasonable assumption. Macroscopic properties, therefore, evolve from the summation of the interparticle forces plus forces from external force fields such as gravity, centrifugal forces in sedimentation, and solid–liquid separation as well as electromagnetic forces. In general, there are three major possibilities to *modify the surfaces of particles*:

- chemical functionalization by means of chemisorption or physisorption of chemical components by ions, molecules (including chemical or atomic layer deposition), or macromolecules (polymers and biopolymers);
- change of surface structure, i.e., crystallinity and roughness either by coating processes or by tailoring growth processes leading to a specific surface morphology; and
- process technology, i.e., by generating specific forces acting on the particle surfaces, e.g., shear stresses in fluids, mechanochemical effects in mills, and/or time–temperature history (surface diffusion, sintering, and rounding).

Accordingly, the chemical and physical properties of the particles can be adapted to specific applications. Chemical functionalization is applied for instance (i) for the dispersion of particles in different polar or nonpolar matrices, (ii) to enhance charge transfer between particles in applications of printable electronics, (iii) to control the release of components in life sciences or for corrosion protection by design of core–shell systems and capsules, or (iv) to protect particles from oxidation and other chemical influences.

For the time being, a rigorous characterization and modeling of particle interactions is only possible for simple model systems: A nice and very recent example is the measurement of the second viral coefficient and structure factor of quantum dots in solution by three independent methods, namely small-angle X-ray scattering (SAXS), cryo-TEM, and centrifugation (Van Rijssel et al, 2014). Nevertheless, the above-described approach can be transferred to many technical applications in a qualitative and conceptual way.

One main bottleneck is the lack of characterization tools for particle surfaces and the related interactions between them. Classical tools of surface science mostly work only in high vacuum and not in liquids. For particles in liquids, several orthogonal techniques may be applied such as zeta-potential measurement, thermogravimetry (surface coverage), NMR, UV–vis, FT-IR, photoluminescence, Raman spectroscopy (chemical information), as well as AUC (interactions and core–shell structures) (Lin et al, 2015). To gain a deeper understanding of surface functionalization, highly sensitive *in situ* methods must be used. In particular, nonlinear optical techniques such as second harmonic generation (SHG) and sum frequency generation (SFG) are promising techniques for the extraction of charge and molecular structure directly from the embedded surface (Eisenthal, 1996; Roke and

Gonella, 2012). For instance, molecular brushes on colloidal particles, transformations of nanowires, and growth of metal surface patches can be measured in solution (Sauerbeck et al, 2014; Schürer et al, 2010, 2011; Wunderlich et al, 2011).

In the following, we briefly describe two examples which support our arguments for two completely different applications and materials. First, we discuss the influence of the zeta-potential of alumina suspensions which were studied in the context of nanomilling in stirred media mills. The zeta-potential is a measure of the surface charge of the particles and is here used to tailor the suspension rheology. Figure 13 shows that the magnitude of the zeta-potential controls the measured shear stress at low shear rates where the influence of interparticulate interactions can be monitored.

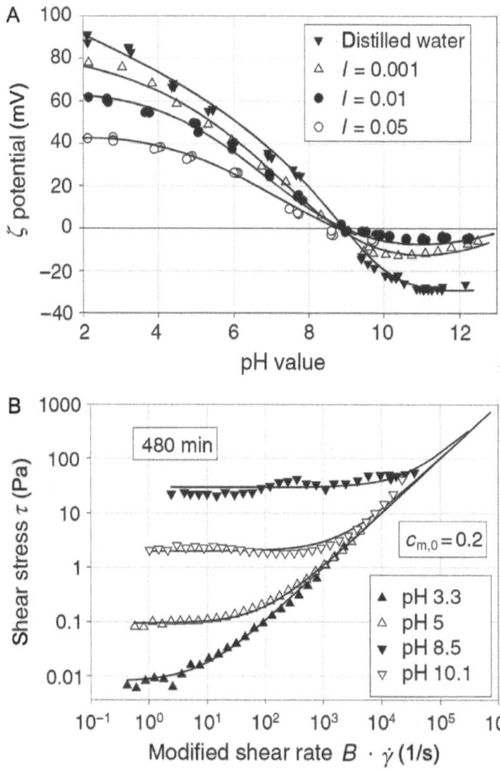


**Figure 13** (A) Zeta-potentials of aqueous alumina suspensions (at different ionic strength *I*) and (B) their rheological behavior as a function of pH at a mass concentration $c_{m,0}$ of 20%. The shear rate is modified by a shift *B* which condenses the flow data in the hydrodynamic regime into one single curve. *Reprinted with permission from Stenger and Peukert (2003). Copyright 2003 John Wiley and Sons.*

Highly repulsive interactions suppress aggregation and the suspension shows a low viscosity with just a minor yield stress, i.e., almost Newtonian behavior. At the isoelectric point electrostatic repulsion and the zeta-potential vanish, i.e., the suspension is flocculated. A considerable yield stress is obtained as an integral measure of the high attractive force between the alumina particles. Noteworthy, the change of the surface charge leads to changes of the shear stress by more than four orders of magnitude (Stenger and Peukert, 2003). Similar trends can be observed for other materials and other unit operations such as solid–liquid separation as well. For instance, the filtration resistance of a filter cake strongly depends on the surface charge of the particles.

The second example deals with fluid interfaces in protein foams. Protein foams are most relevant in food science and engineering, e.g., in the form of milk foams on a cappuccino. A foam is an aggregate of stabilized gas bubbles. These gas bubbles in turn are stabilized by adsorbed proteins and surfactants at the liquid–gas interface. In the example illustrated in Fig. 14, we present data for β-lactoglobulin (BLG) which have been obtained from SFG, ellipsometry, and surface rheology.

BLG is a main component in milk and acts as stabilizer at the gas–liquid interfaces in milk foams. We see the same hierarchy as discussed above: the zeta-potential (Fig. 14A) is a measure of the net surface charge of the proteins. At the isoelectric point (pH ~5), the tendency for protein adsorption at the interface is highest leading to a minimum of surface energy (Fig. 14B). At the interface, the proteins become unstable and aggregate since the repulsive forces vanish. This leads to a pronounced increase of layer thickness (Fig. 14C) and to a maximum of the elastic modulus of the film (Fig. 14D). At the isoelectric point, the protein foam is most stable and the size of gas bubbles directly after foaming is smallest. Thus, all macroscopic foam properties scale with the surface charge of the proteins (Engelhardt et al, 2012, 2013).

These examples show the influence of chemical surface effects on product design. The possibilities to change the surface morphology and the application of process technologies to influence surface properties will be discussed together with the example on 3D printing in the next section.

## 4.3 Structure Formation

Structure formation and self-assembly involve the organization of molecules into particles and of particles into larger particulate structures. The assembly of building blocks is common in many processes in nature and technology

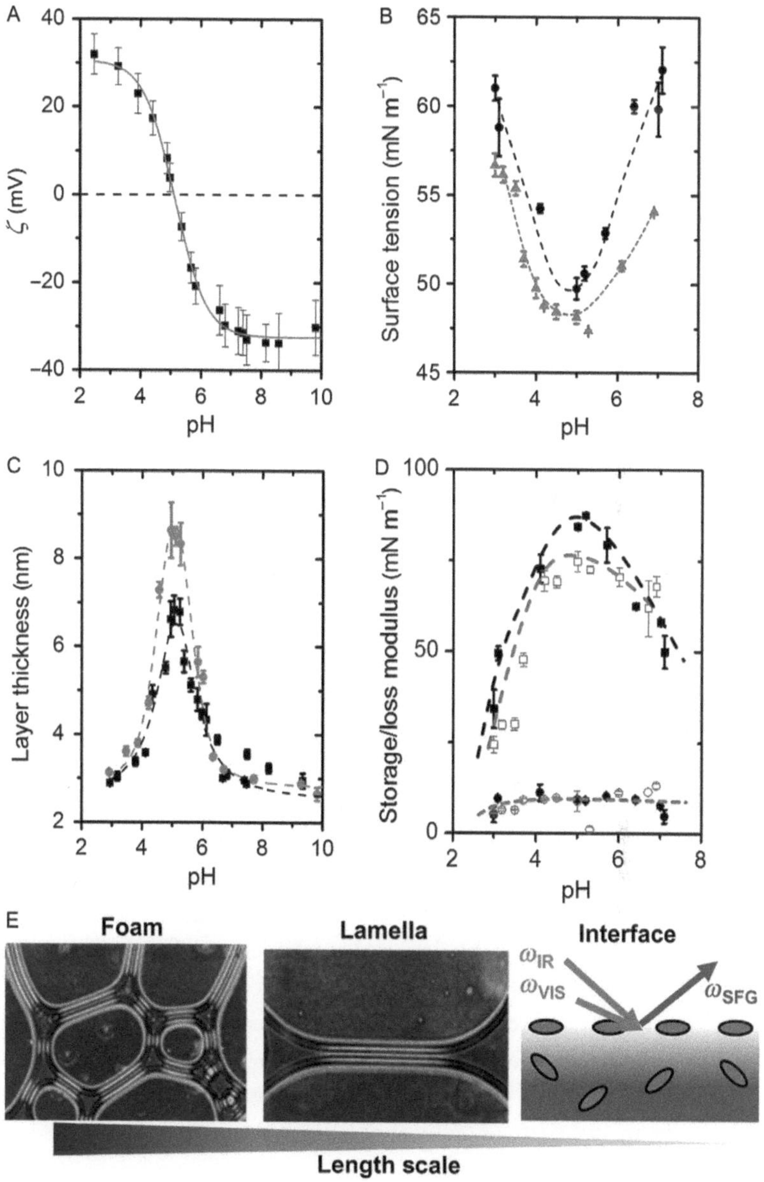


**Figure 14** (A) Zeta-potential of aqueous β-lactoglobulin (BLG) in water; (B) surface tension; (C) layer thickness; and (D) storage and loss modulus of protein foams as a function pH; black indicate 10 μM and red (gray in the print version) 50 μM (BLG); (E) foam hierarchy. *Reprinted with permission from Engelhardt et al (2013). Copyright 2013 American Chemical Society and Engelhardt et al (2012). Copyright 2012 American Chemical Society.*

including biochemical and chemical engineering and all the way up to the largest structures in the universe such as the rings of Saturn and even galaxies. Structure formation involves many different interactions, is increasingly studied in many disciplines, and is at the core of mesoscale science and technology (Li, 2015). Structure formation can be summarized by the following general "equation":

$$\text{transport} + \text{interactions} = \text{structure} \tag{6}$$

In situations of thermodynamic equilibrium, transport processes are completely neglected. The structures can be described by statistical thermodynamics as explained in the previous section based on the interactions between the building blocks. In completely transport-controlled situations nonequilibrium structures form. Transport phenomena are diffusive or convective depending on the size of the building blocks. Further, they may be influenced by field forces such as gravitational, centrifugal, or electromagnetic fields. In most cases, two- or even multiphase flows have to be controlled in order to form well-defined structures. These principles can be applied to the formation of single particles of defined size and shape such as in particle synthesis or crystallization or to the formation of particle ensembles such as filter cakes, tablets, agglomerates, ceramics (e.g., translucent nanoparticulate films made from nanoparticulate alumina as shown in Fig. 15; Vozdecky et al, 2010), or electrodes of fuel cells, batteries, etc. Structures of single particles and of particle ensembles are finally integrated into real-world products. Additive manufacturing and 3D printing are just two prominent examples with a particularly bright future and will be addressed below (Conner et al, 2014). The "trumpet" in the lower right part of Fig. 15 shows a chemical reactor made from metal powders that is designed for the release of hydrogen gas from a liquid hydrogen carrier (Peters et al, 2015).

At this point, it needs to be mentioned that a comprehensive description of these principles would be far beyond this review. Therefore, we will focus in the following only key principles that are in particular relevant for chemical engineering and particle technology.

### 4.3.1 Equilibrium Structures

Crystallization of narrowly distributed particles is known from the formation of natural and artificial opals which are in fact colloidal crystals. Their colors arise from interference of the underlying building blocks with light. Colloidal crystals in 2D and 3D became a prominent field in colloid science when quite narrow PSDs with standard deviations below 5% became accessible (Rogach et al, 2002; Sugimoto, 2001; Weidmann et al, 2014). This is a requirement of

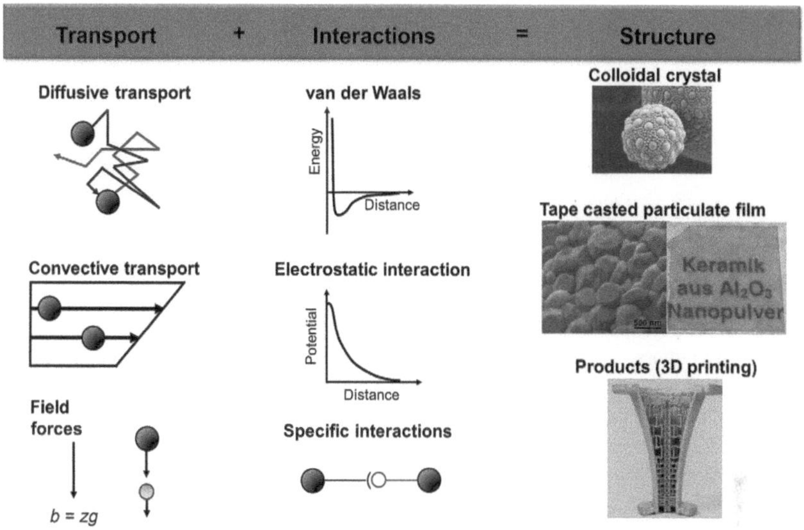


**Figure 15** Principles of structure formation. *Collidal crystal and 3D printed catalytic reactor with kind permission from the Erlangen Cluster of Excellence EAM, http://eam.fau.de/home/; tape casted slurries reprinted from Vozdecky et al (2010). Copyright 2010 The American Ceramic Society and with permission from Elsevier.*

paramount importance for highly crystalline ordering. It is the case for polymer and silica beads in the size range between 0.1 and 1 μm, but as well for quantum dots with sizes below 10 nm such as cadmium- or lead-based compounds (Rupich et al, 2010; Talapin et al, 2010).

Figure 16 shows highly regular arrangements of quantum dots on surfaces in the form of regular lattices of pure compounds but also of binary lattices and even 3D crystals composed of nanocrystals (Rupich et al, 2010; Talapin et al, 2010). These examples impressively show that nanocrystals can act as "giant molecules." The next step in these bottom-up mesoscience approaches would be to create particles with a well-defined finite-number connectivity of bonds which are able to mimic chemical coordination. Faceted structures as highlighted in Fig. 17 (Damasceno et al, 2012) show the complex phase behavior including crystalline, liquid crystalline, and various irregular structures. Janus-type and patchy particles are structures of increasing complexity *en route* toward giant molecular-type particles with spatially controlled "bonding sites." They should not only have a well-defined size and shape but must also possess a uniform number of patches at reproducible locations on the surfaces. Such Janus particles can be made in high quality by adsorbing core particles at liquid–liquid or liquid–gas interfaces with subsequent surface coating from either phase (Kaewsaneha et al, 2013; Walther and Müller, 2013). However, these

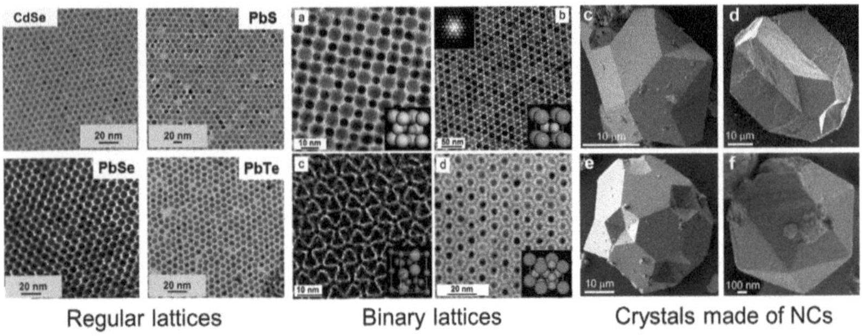


**Figure 16** Quantum dots and their self-assembly to regular lattices of single components (left), binary lattices consisting of two different types of dots (middle) and nanocrystals (NCs) made from quantum dots. *Left part panels (CdSe, PbS, PbSe, PbTe): Reprinted with permission from Talapin et al (2010). Copyright (2010) American Chemical Society. Middle part panels (a)–(c): Reprinted with permission from Shevchenko et al (2006). Copyright (2006) Nature Publishing Group. Middle part panel (d): Reprinted with permission from Chen et al (2007). Copyright (2007) American Chemical Society. Right part panels (c)–(f): Reprinted with permission from Rupich et al (2010). Copyright (2010) American Chemical Society.*

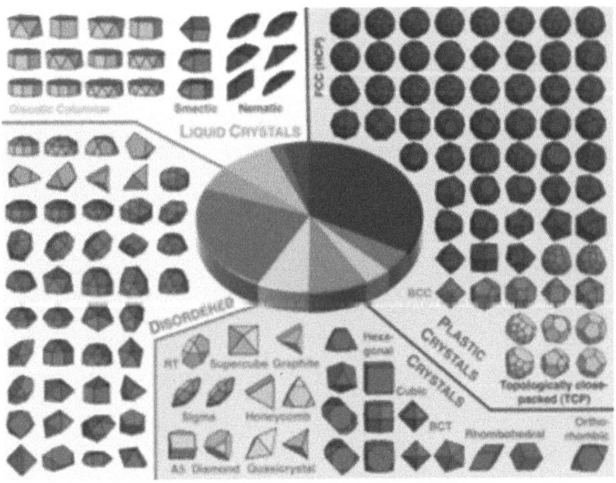


**Figure 17** Different categories of polyhedra. *From Damasceno et al (2012). Reprinted with permission from AAAS.*

methods are only applicable on a lab scale at very low quantities. Scalable approaches are rare and mostly restricted to liquid phase production (Bao et al, 2011). Recently a few promising works from the gas phase are reported as well (Bubenhofer et al, 2012; Mehringer et al, 2014; Sotiriou et al, 2011).

A recent review discusses possibilities to assemble monodisperse colloidal particles into 2D and 3D structures (Vogel et al, 2015). Topographic patterns (produced by various methods) use walls to predefine structural order. Patterns in two and three dimensions define locally distributed interaction forces which are in the simplest case just given by a hard sphere potential. Transfer of predefined structured layers is possible, e.g., via microcontact printing using polydimethylsiloxane stamps which may be understood as an intermediate step toward printing and the aforementioned roll-to-roll processes. 2D arrangements can be formed directly via the Langmuir–Blodgett technique on liquid surfaces. Thereby dense and highly regular layers are directly accessible if the wetting properties of the particles are properly adjusted. On solid substrates, the adhesion of particles on local "adhesive" spots makes the arrangement of particles in regular structures more challenging but opens additional degrees of freedom when the solid surface is modified by local surface patterns.

Negative structures are obtained when the free space of the colloidal crystal is filled with a new matrix and the original colloidal spheres are sacrificed by dissolution. Thus, reverse opal structures are formed. These can be used to generate photonic effects when the length scales of the structure are in the range of visible light. Figure 18 shows examples of highly regular zirconia-based inverse opal pigments. In brief, they are highly regular solid foams. Filling the voids with another matrix (here methanol) allows to modulate the dielectric permittivity of the voids and thus the optical properties (Aguirre et al, 2010).

In recent years, the organization of small nanoscale buildings blocks into regular superstructures came increasingly into focus under the names "oriented attachment" or "oriented aggregation." Closely related to this effect is the formation of mesocrystals (cf. Chapter 2, Fig. 5). As discussed,

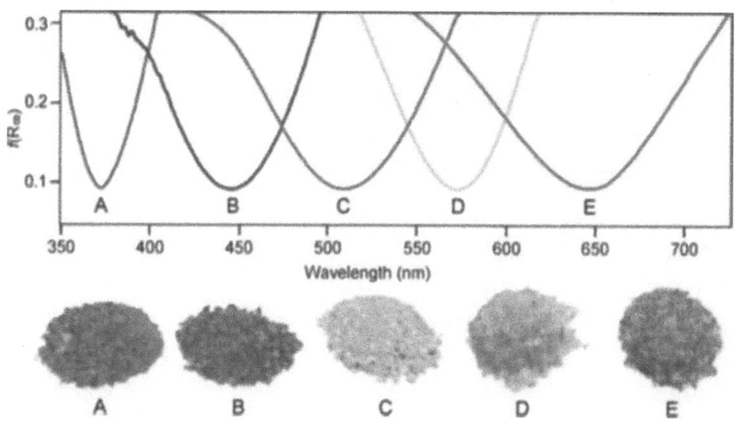


**Figure 18** Inverse opals based on $ZrO_2$ with pore sizes between (A) 200, (B) 250, (C) 285 nm, (D) sample B with methanol, and (E) sample C with methanol and the corresponding UV–vis diffuse reflection spectra. *Reprinted with permission from Aguirre et al (2010). Copyright 2010 WILEY-VCH Verlag GmbH & Co. KGaA, Weinheim.*

these are crystals made from mesoscopically structured building blocks which are formed by controlled structure formation due to a complex interplay of crystal growth and colloidal assembly (Cölfen and Antonietti, 2008; Penn and Banfield, 1998).

The following four mechanisms have been identified to explain the 3D arrangement of nanoparticles to a mesocrystal as summarized in Fig. 19 (Song and Cölfen, 2010):

**(A)** an oriented organic matrix;
**(B)** physical fields due to dipoles or polarization effects leading to mutual alignment of crystal faces;
**(C)** epitaxial growth of a nanoparticle employing a mineral bridge connecting the two nanoparticles; and
**(D)** nanoparticle alignment by spatial constraints.

This highly interesting mechanism of self-organization is observed in an increasing number of organic and inorganic systems leading to quite amazing structures with interesting properties (Bahrig et al, 2014). Biomineralization is just one prominent example with particular beautiful organic–inorganic hybrid structures where biomolecules trigger growth of inorganic phases. In Fig. 5, we have already discussed that the organization of ZnO nanoparticles can lead to spherical, highly porous structures with reduced light scattering but as well to ellipsoids consisting of nanorods with high scattering intensity. These mesocrystals comprise an intermediate stage between the formation of molecular crystals and the self-assembly of particles into larger structures. Mesocrystals are unique particles formed by one or several consecutive aggregation steps of smaller particles into equilibrium or nonequilibrium structures. The local interaction forces between the nanocrystalline building blocks which are currently far from being sufficiently understood can drive the system into quite regular structures close to equilibrium.

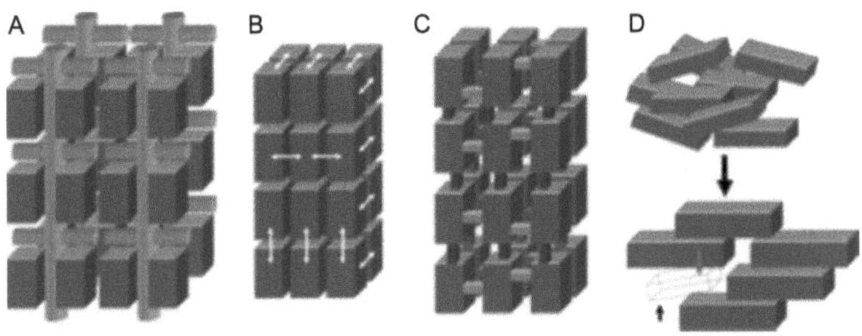


**Figure 19** Oriented aggregation of building blocks into mesocrystals by the four mechanisms. *Reprinted with permission from Song and Cölfen (2010). Copyright 2010 WILEY-VCH Verlag GmbH & Co. KGaA, Weinheim.*

As a last aspect, and although liquid crystalline structures made of macromolecules shall not be covered in this review, it needs to be mentioned that such assemblies of regular nonspherical particles are highly interesting. Typically found "nonspherical particles" are for instance uniform rods, virus particles, or platelets. Liquid crystals can act as structure-determining intermediate stage for the formation and deposition of regular, even space filling films of solid particles. Exemplarily, in Fig. 20, nanorods are illustrated which can form dense packings after drying. Smectic or nematic liquid crystalline phases form in the liquid so that the particles are already arranged

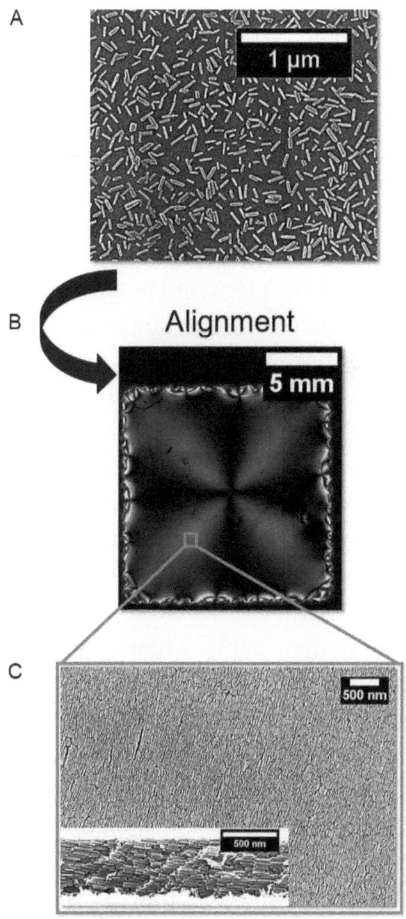


**Figure 20** (A) Single stabilized ZnO nanorods; (B) polarization microscope image of a regular nanorod layer; and (C) regular space filling thin film structures. *Adapted from Schäfer et al (2014). Copyright 2014, with permission from Elsevier.*

regularly upon drying. Once sufficient liquid is evaporated, the particles come into contact and are immobilized as thin film into regular structures (Schäfer et al, 2014).

### 4.3.2 Nonequilibirium Structures

Whereas much work was devoted to equilibrium structures, much less is known on nonequilibrium static or dynamic structures (Palma et al, 2012). They will be discussed in this section. Packings of all sorts of particles are the most widely known nonequilibrium particulate structures. They occur in packed beds, are key to transport and storage of particulate solids, and are omnipresent in nature, e.g., in soil, sand dunes, snow piles, etc. Packings form under the influence of mechanical forces such as gravity and interparticulate forces. For particles smaller than a few micrometers, the porosity increases with decreasing particle size due to van der Waals interactions. Very loose packings with porosities above 80–90% occur for packings made of nanoparticles and in filter cakes of dust filters. Today, packing structures are regularly modeled by DEM simulation (Zhu et al, 2008). Noteworthy, even quantitatively correct data can be obtained with realistic material parameters (Parteli et al, 2014). A new and potentially very huge application of functional packings is related to 3D printing and additive manufacturing as will be discussed in Section 4.6.

With nanoparticles in nonequilibrium in unconfined situations without external walls or other external fields, typically irregular fractal-like structures evolve in both, the liquid and the gas phase. Their formation can be limited by either transport or integration, i.e., by the adhesion forces between the building blocks upon contact. In a series of seminal works in the last century, Meakin distinguished accordingly between the quite universal diffusion-limited and reaction-limited colloidal aggregation (Lin et al, 1989; Meakin, 1999; Sorensen, 2011). These mechanisms are widely independent of the chemical system and can be applied to aggregation in the gas and in the liquid phase. Further, one might classify systems where growth occurs by addition of monomers bridging particles in their contact points and those where clusters do aggregate. In the most general case of reacting systems, coagulation, growth, and fusion by sintering (in the gas phase) or ripening (in the liquid phase) are the three determining kinetics.

Assuming

- characteristic coagulation times $t_c$, defined as the time to reduce the particle number concentration by half,

- characteristic growth times $t_g$, defined as the time to increase the surface area of one particle by the factor $e$, and
- characteristic fusion times $t_f$, defined as the time to reduce the surface area by a factor $e$,

we can introduce two dimensionless times, namely $\tau_f = t_c/t_f$ and $\tau_g = t_c/t_g$. In fact, they completely control the three-dimensional landscape of coagulation, growth, and fusion.

Figure 21 shows how these two parameters determine the evolution of the fractal dimensions in the gas phase as obtained from a Monte Carlo cluster–cluster aggregation model (Al Zaitone et al, 2009). Without any growth and fusion, the fractal dimension $D_f$ of 1.9 for diffusion-limited cluster–cluster aggregation was obtained (reaction-limited coagulation would lead to $D_f$ of 2.09). Growth and fusion lead to denser structures characterized by larger fractal dimensions $D_f$ until the value of three is obtained for a sphere. Thus, fractal structures can be tailored by careful control of these two dimensionless times. Together with the particle concentration which is initially determined by nucleation and the time–temperature history in the reacting system, the particle structure can be optimized.

All these data can be described reasonably well by the following process–structure "master" equation for simultaneous coagulation, growth, and fusion:

$$D_f = 1.9 + \frac{1.1}{1 + \left(\frac{0.27}{\tau_{eff}}\right)^{1.2 + 2.2\tau}} \qquad (\tau_{eff} > 0) \tag{7}$$

where $\tau_{eff}$ is defined as

$$\tau_{eff} = \tau_f + 3 \cdot \tau_g + 15 \cdot \tau_f \cdot \tau_g \tag{8}$$

Attractive forces and external energy input will lead to denser structures opening additional degrees of freedom. Densification of open fractal structures is observed in sheared systems due to restructuring of the aggregates. The properties of all these structures can be studied in considerable detail. For instance, geometric, hydrodynamic, and optical properties have been considered (Al Zaitone et al, 2009; Binder et al, 2006, 2009; Eggersdorfer et al, 2012; Schlauch et al, 2013; Sorensen, 2011).

We are now prepared to discuss the evolving field of structure formation in drying particulate (thin) films. Solid–liquid separation is a very important unit operation where the fluid flow may be directed either normal to the filter medium or, in case of cross-flow filtration, parallel to it. The filter cake

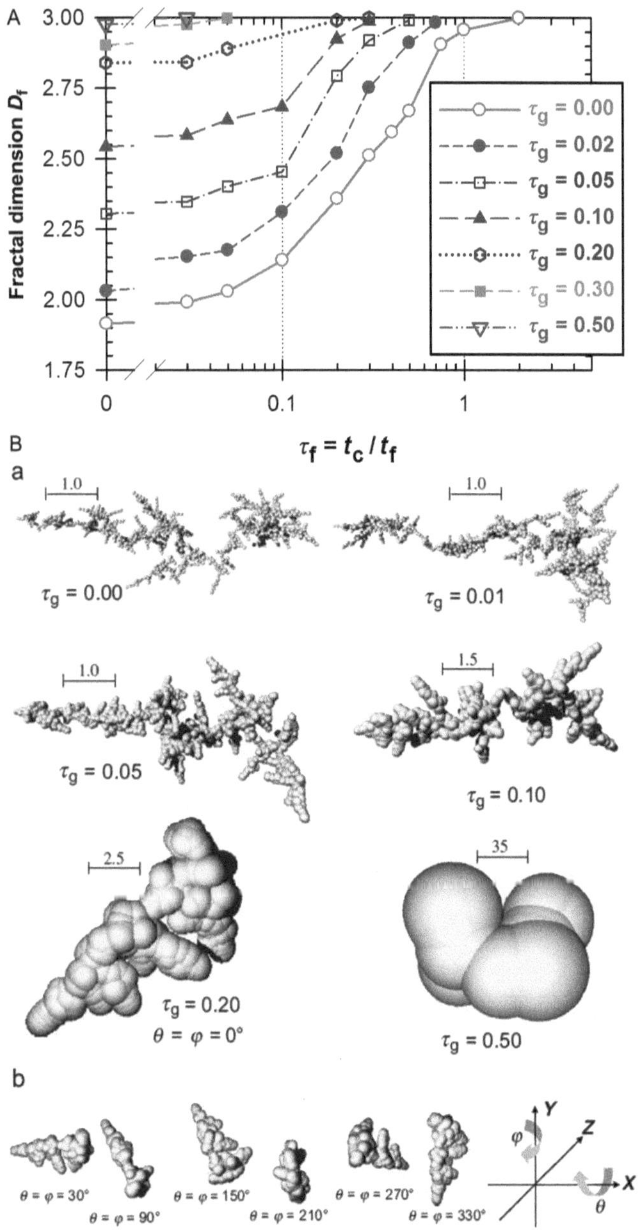


**Figure 21** (A) Evolution of the fractal dimension $D_f$ in dependence of dimensional characteristic times for fusion and growth and (B) corresponding structures at different values of $\tau_g$. *Reprinted from Al Zaitone et al (2009). Copyright 2009, with permission from Elsevier.*

structure is directly linked to the pressure loss through well-known correlations (e.g., Darcy's law or Ergun equation). However, any *a priori* prediction of cake structures is still a major challenge which is of paramount importance for the goal to efficiently separate particles from gases and liquids. Regarding the latter, in functional applications, a liquid film is deposited on a solid substrate by a number of methods, such as spin coating or dip coating, tape casting, and spraying, or by continuous roll-to-roll processes. Further, we may distinguish between reacting (e.g., via sol–gel methods) and nonreacting systems, for instance in the fabrication of ceramic tapes with preformed particles. Thin films processed from liquids find wide-ranging applications from printable electronics, next-generation solar cells, fuel cells, or batteries.

Structure formation may be induced at either the interface, i.e., at the solid–liquid, or the gas–solid interface or in the bulk of the fluid. The interfaces can be understood as templates whereas in the bulk the described coagulation processes are monitored. In the liquid phase, the interplay between drying and aggregation kinetics controls the structure evolution and the final structure (see Fig. 22). Upon drying, solvent evaporates and the concentration of the building blocks increases, while simultaneously the interactions between them change as well. Therefore, colloidal stability, aggregation, and self-organization are all a function of time, concentration, and viscosity. Noteworthy, the latter is a function of the acting shear rates. Drying kinetics

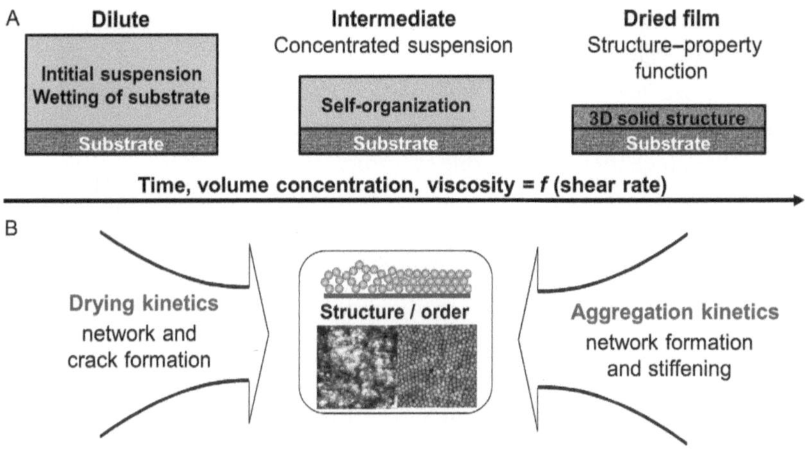


**Figure 22** Structure formation in thin films from liquids: (A) temporal evolution of concentrations, interactions, and viscosity due to drying and (B) interplay between drying and aggregation kinetics. *Reprinted from Günther and Peukert (2003). Copyright 2003, with permission from Elsevier.*

trigger aggregation kinetics and thus determine ordering and structure. Highly regular structures may form by preformation of liquid crystalline structures or templating at the interfaces. The transient interaction-dependent aggregation leads to network formation whereby stiffening mechanisms determine the strength of the network. At the end of the first drying phase, the particles are immobilized but the solvent still remains inside the pores. Upon further drying, the pores are successively emptied whereby drying stresses act on the particle network inducing compression and undesired crack formation. Thus, a major challenge is the formation of crack-free assemblies, both in 2D and in 3D. During drying, capillary forces induce drying stresses which may easily induce fracture (Günther and Peukert, 2003). As will be discussed in more detail in Section 4.6, this effect cannot yet be sufficiently controlled.

Other often used "templates" are liquid droplets in spray drying. All types of structures, irregular and regular, core–shell, and hollow, are obtained (Messing et al, 1993; Okuyama and Lenggoro, 2003). Again, structure formation is based on the careful control of drying and aggregation kinetics. Noteworthy, in case of spray drying, the latter is typically much faster than in case of thin film formation. Also here, the competition between agglomeration kinetics in the liquid phase due to the increasing particle concentrations and drying kinetics controls the evolution of the structure. Particles may adsorb at the liquid–gas interface leading to closed core–shell structures if the solvent can still evaporate. In contrast, donut-type structures are formed if the outer shell breaks. Additional chemical reactions and the temperature in the spray offer further degrees of freedom, e.g., by melting at elevated temperature.

So far, we focused on "finally static" structures which have formed through a more or less complex interplay of interactions and transport phenomena and keep their features due to immobilization. The most challenging aspect from a processing point of view is the formation of "transient dynamic" particulate structures as they form in multiphase flow situations. Examples are dense gas–solid flows in fluidization and transport reactors, i.e., in riser or downer reactors and pneumatic conveying. The dynamic formation and decay of cluster-like structures are a major challenge in the modeling of gas–solid flows. Hierarchical multiphase models must take into account the local distribution of particles which depends on drag and particle–particle impacts as well as on friction. Much progress was achieved to model more and more realistic scenarios (Lu et al, 2014; Xiong et al, 2012).

Discrete particle models are multiscale in the sense that continuum fluid mechanics is coupled with models for the motion of clusters, coarse-grained "quasi particles," and real individual particles. Coarse graining is a necessity to keep computational time and cost within reasonable limits although already world-class computational resources including graphic processors are used. As discussed throughout the previous chapters, the challenge is to include more and more realistic material functions, i.e., to understand and model particle–particle interactions. Of course, this requires hierarchical optimization with clever chosen homogenization and error estimates between the scales. Validation of these fascinating insights from modeling and simulation is so far mostly restricted to global, spatially, and temporally averaged effects. In future, highly resolved structure analysis, e.g., by advanced tomographic techniques or measurements by high speed cameras, might allow to get deeper insights and further validation (Fig. 23) (McMillan et al, 2013). Thus, for this and similar issues as they are found along many other aspects of mesoscale science and technology, innovation occurs at the interfaces between the different hierarchical levels and between the various involved disciplines.

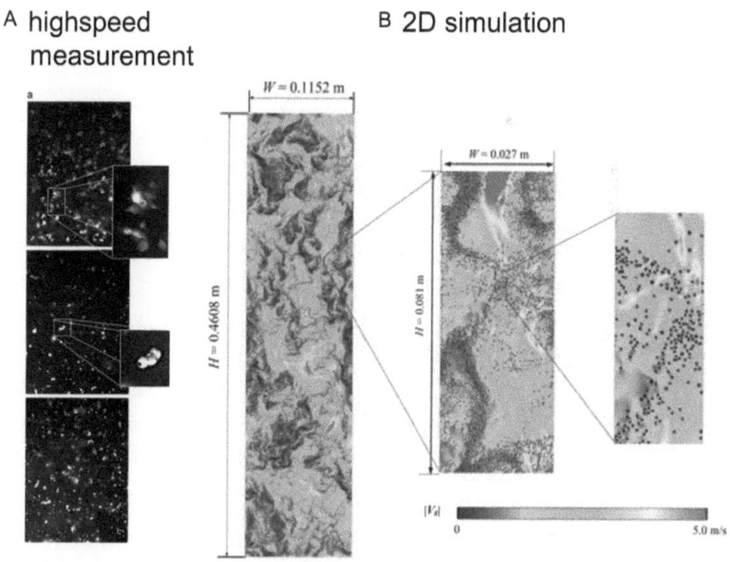


**Figure 23** (A) Particle clusters in the fluidized bed for FCC catalyst particles measured by high speed camera (20 ms exposure time) and (B) particle configuration and gas flow field of the whole region of a two-dimensional gas–solid flow. *Panel (A): Reprinted from McMillan et al (2013). Copyright 2013, with permission from Elsevier. Panel (B): Reprinted from Xiong et al (2012). Copyright 2012, with permission from Elsevier.*

## 4.4 Characterization Along the Process Chain

Keywords in the context of characterization are off-line, online, and inline measurements. Off-line measurements require proper sampling without changing the state of the particulate system to be analyzed. Representative sampling can be quite challenging and requires considerable experience. In off-line measurements, highly advanced techniques such as high-resolution TEM, for instance, can be employed to obtain detailed information on structures of particles and particle ensembles. A major challenge in these off-line measurements of particle systems is to analyze a sufficiently high number of particles which is typically very time consuming and thus expensive.

The gold standard in size characterization is AUC (cf. Chapter 3) with unprecedented accuracy and resolution (Walter et al, 2014). This technique was developed and used mostly in the life sciences to characterize protein size, aggregation, and interactions but has a huge and so far not sufficiently recognized potential for particle technology. In combination with *in situ* UV–vis absorption spectroscopy, this technique can also provide access to several dimensions such as size, shape, particle density, or spectral properties (Walter et al, 2015c). Its great advantage is due to the fact that it is a separating technique. Thus, polydisperse samples are first separated and then detected. In contrast, the purely optical standard techniques like laser diffraction or photon correlation spectroscopy (PCS) illuminate the whole ensemble. The resulting signal must be deconvoluted by solving ill-posed Fredholm integral equations in order to obtain the full PSD.

The difference between online and inline measurements is that in case of online a side stream is extracted from a larger system, e.g., a large duct or vessel, whereas inline measurements are performed directly within the system (e.g., tubular reactor). In both on- and inline cases, the particles are analyzed "on the fly" by an adequate technique. In sufficiently diluted systems, all optical techniques can be employed if adaption of the optical setup to the system under investigation is possible. Ultrasound spectroscopy and electroacoustic techniques allow the analysis of size distributions in opaque systems and give even access to zeta-potentials (see, e.g., Dukhin and Goetz, 2002). More expensive are SAXS and small-angle neutron scattering (SANS) measurements which can analyze concentration, size, shape, and local structures. An important trend goes in the direction of the simultaneous application of several techniques. A recent example is the analysis of size distributions of ZnO nanoparticles (by SAXS) and their surrounding surface layers (by SANS) at the same time (Schindler et al, 2015a,b). Having thus a strong toolbox in our hands, particles could for instance be controlled within a reactor by UV/vis spectroscopy and, if necessary small-angle scattering techniques. After completion of the reaction, the product after synthesis could be analyzed in detail by AUC, prior exposing it to repeated cycles of washing. Finally, the success of the purification procedure could be controlled by UV/vis analysis or PCS for a fast confirmation of colloidal stability and—if necessary—again by AUC to make sure that dispersity did not change (on the level of Å). If all this is fulfilled, unique insights to process–structure relationships can be derived and well-defined dispersions for further processing are obtained (e.g., layer formation). Thereby one would always start characterization using techniques that are fast, cheap, and inline or at least online applicable (e.g., UV/vis), followed by advanced techniques enabling deep knowledge gain (e.g., small-angle scattering and AUC) to selected samples whenever necessary.

Particle ensembles form structures which must be characterized. Also here unified frameworks have been developed in the past. Torquato summarizes several approaches for structure characterization such as pair distribution functions, pore size distributions, or chord length distributions (Torquato, 2002). It can be shown that the results of these methods can be converted into each other. An alternative, very general method of Minkowski tensors is applicable to many different structures and provides a unique framework of structure description as outlined in a recent review (Schröder-Turk et al, 2011). Minkowski tensors are tensorial shape indices that allow quantitative characterization of the anisotropy of complex materials. They are particularly well suited for developing structure–property relationships for tensor-valued or orientation-dependent physical properties. They are fundamental shape indices, in some sense being the simplest generalization of the concepts of volume, surface, and integral curvatures to tensor-valued quantities (Schröder-Turk et al, 2011). Applications of this very general concept include copolymer films imaged by surface force microscopy after alignment by an electric field; local cell anisotropy of spherical bead pack models for granular matter and of closed-cell liquid foam models; surface orientation in open-cell solid foams studied by X-ray tomography; and defect densities and locations in MD simulations of crystalline copper.

## 4.5 Multiscale Modeling and Simulation (MSS)

MSS aim to bridge a wide range of separated space and time scales, whereby the space and time scales are sometimes treated together and sometimes separately (Elliott, 2011). Bridging time scales is typically achieved by automatic *upscaling* from a more fundamental level to the next scale and is often complemented by adaptive *downscaling* (Anwar and Zahn, 2011; Babuska and Lipton, 2011; Weinan et al, 2007). Simulations that include several scales of description, such as *ab initio* or DFT QM (Hesselmann and Görling, 2011), atomistic force field (Özpinar et al, 2010), mesoscale and finite element, and in which information flows both upward (upscaling) and downward (downscaling), can be considered truly multiscale. Additionally, materials design and optimization involve model predictive optimization in both upscaling and downscaling steps (Haslinger et al, 2010; Hirschberger et al, 2008). Such genuine multiscale simulations are quite rare and we currently lack a generalized MSS methodology (Målqvist, 2011).

Therefore, today's methods fall into two categories: those that aim to be generally applicable and local techniques that focus on a special domain of applications (Babuska and Lipton, 2011; Elliott, 2011; Miller and Tadmor, 2009; Weinan et al, 2007). This is particularly true when materials and process optimization is the driving force. While upscaling methods (such as homogenization), heterogeneous multiscale methods, equation-free modeling, and matched asymptotics (Babuska and Lipton, 2011; Målqvist, 2011; Miller and Tadmor, 2009; Weinan et al, 2007) fall within the first category, semiempirical and *ab initio* molecular–orbital techniques, DFT, atomistic force fields, and discrete and finite element methods (e.g., Car–Parrinello molecular dynamics, hybrid QM/molecular mechanics (MM) (Cramer, 2006), lattice Boltzmann Monte Carlo (LBMC) (Landau and Binder, 2009), and quasicontinuum (Miller and Tadmor, 2002)) are more problem specific. In materials and process optimization, the two types of methods must be merged into one goal-oriented multiscale and multilevel model-based simulation and optimization framework. Current research directions rely on *a posteriori* estimates of model and state errors between the scales and are paralleled by exascale high-performance computations that serve as benchmarks to validate the multiscale methods for suitable examples.

After these general remarks, we discuss as an example the case of comminution where the mesoscale plays a dominant role. The key elements in the process chain are different unit operations as described by machine and material functions. As for any other unit operation, also in comminution it is quite helpful to separate the machine function from the material function (Fig. 24).

The machine function, here the mill function, describes the effects of external parameters such as energy input, machine geometry, rotor

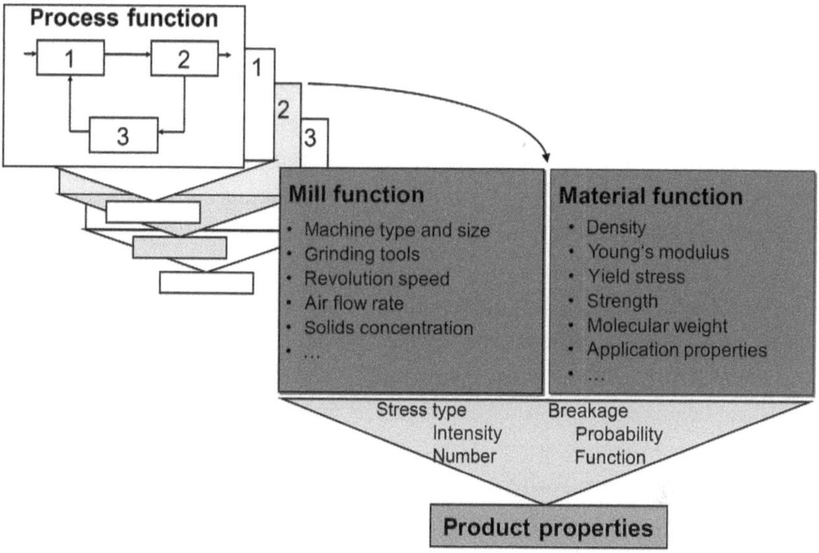


**Figure 24** Process, machine, and material function. *Reprinted from Vogel and Peukert (2003). Copyright 2003 Elsevier and Peukert (2004). Copyright 2004 Elsevier.*

revolution, two-phase flow, and stresses acting on particles in the grinding machine. As a result, the particles in the machine (but also in other unit operations such as mixers or granulators) are stressed by external forces and react accordingly depending on their material properties. The material properties are size, shape, density of the particles, and their intrinsic mechanical properties (Young's modulus, yield stress, fracture toughness, etc.). At the end, the machine function condenses all the design and operational parameters into three parameters: (i) the type of stress which is usually either by impact or by compression, (ii) the stress intensity, i.e., the energy transferred to particles per volume, and (iii) the stress number which is given by the number of stressing events per time and volume element in the mill. Of course, stress intensity and stress number are distributed with respect to space and time in the mill. The material function defines in turn the breakage probability and the breakage function. The latter is the size distribution of the fragments. Many empirical models for both the breakage probability and the breakage function have been proposed. Experimental data are best measured in situations where single particles are stressed. Few approaches exist which link these to intrinsic material properties. Prominent models were derived for impact stressing by Vogel and Peukert (Peukert, 2004; Vogel and Peukert, 2003), for attrition by Ghadiri and Zang (Ghadiri and Zhang, 2002; Zhang and Ghadiri, 2002) and independently by Gahn and

Mersmann (1999). For organic crystals, Meier was able to link fracture probability and fracture function to fracture toughness and hardness measured for single crystals (Meier et al, 2009a).

Figure 25 shows the hierarchy of milling (Peukert, 2004). The mill (in this case a sieve hammer mill) defines the distributions of stress type, intensity, and number occurring in the mill. This analysis must be done by in-depth analysis and simulation of the two-phase flow in the mill and therefore depends on the transport properties. The particles are stressed at the hammers by impact. The impact velocity and impact angle and thus the energy transferred to the particles depend on the elastic properties of particle and hammer and on the energy dissipation. The elastically stored energy can be used for crack formation which in turn depends on the defects in the particles and on particle strength. Continuum fracture mechanics is linked to the stress distribution in the particle and relates the energy release rate to the fracture energy. At the molecular level, defect formation and crack opening may be simulated by MD simulations based on empiric force fields which can be derived from DFT, opening deeper insights into the fracture mechanism.

An overall modeling strategy must include a model of the two-phase flow of the particles in the mill; currently a combined CFD–DEM is the best and most advanced choice. In order to make DEM truly predictive, intrinsic material properties such as Young's modulus and visco-plastic deformation

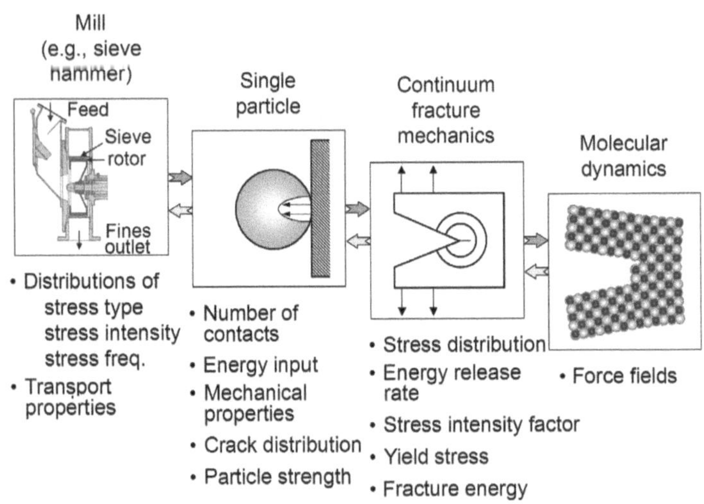


**Figure 25** Hierarchy of models in a sieve hammer mill. *Reprinted from Peukert (2004). Copyright 2004 Elsevier.*

of the particles must be known quantitatively in order to relate coefficients of restitution to intrinsic material properties. The outcome would be the distribution of impact velocities and impact angles on the hammers of the mills. For sufficiently dilute systems (some 10 g solid/m$^3$ air), Toneva et al derived distributions of impact velocities by using CFD coupled to Lagrangian particle tracking in an air classifier mill from Hosokawa Alpine (validation of particles velocities by PIV measurements) (Toneva et al, 2011a, 2011b). The most probable impact velocity was found to be 85% of the hammer speed for particles larger than roughly 100 μm. These data serve as input for modeling fracture probability and size distribution of the fragments.

Putting all these aspects together, the operational behavior could be modeled for air classifier and sieve hammer mills; i.e., the PSD could be predicted for various operating conditions such as hammer speed or screen openings. The results served as input for models which were incorporated into the SolidSim software which is now part of the Aspen software packages.

## 4.6 Integration of Particles into Devices

We now discuss the integration of particles in devices by choosing two examples, i.e., particulate thin films (case 1) and 3D printing (case 2).

### *Case 1: Particulate Thin Films*

This example is particularly helpful to highlight unifying principles. Today, thin films made of molecular or particulate buildings blocks are of high relevance in many traditional applications such as paints or catalyst layers and in several new applications related to energy conversion or energy storage: electrodes and membranes in fuel cells, electrolysis, solar absorbers, thermoelectrical devices, batteries or super-capacitors, light-emitting diode lamps, and all sorts of other printed electronics. The key question is always how to tailor the meso-structure of the film to meet a specific film property by tuning the particle interaction, the fluid dynamics of the coating processes, and the drying conditions. A typical process chain leading to optoelectronic devices is depicted in Fig. 26 exemplarily for a FET with an active layer of ZnO quantum dots (Faber et al, 2009, 2011).

The first step, particle formation by gas or liquid phase synthesis, leads to particles which are further functionalized at their surface to ensure colloidal stabilization and to allow charge transfer across particle–particle interfaces in the final layer. The surface and electronic properties of the particles are typically characterized by a whole set of techniques including time-resolved measurements of charge carrier transport. The best way of cost-efficient

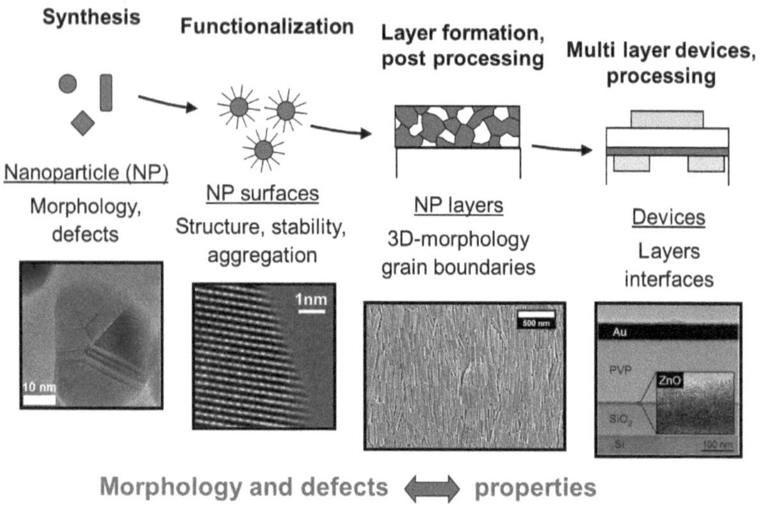


**Figure 26** Process chain for printable opto-electronics. *Adapted from Faber et al (2011) with permission from The Royal Society of Chemistry and Damm et al (2011).*

industrial production would be continuous roll-to-roll production followed by controlled drying and densification, e.g., by pulsed lasers. Due to the latter, the layer can be formed even on substrates which tolerate only temperatures below, e.g., 200 °C. This way, a layer with a 3D morphology and many grain boundaries is formed. Structure formation in the layer follows the guidelines discussed already in Section 4.3. A key question is how defect states at the surface or in the bulk influence the properties along the process chain and the quality of the final device. Thus, understanding of defect generation and control of defect states at the mesoscale is a main challenge.

### *Case 2: Additive Manufacturing (3D Printing)*

In the last few years, rapid prototyping is pushed forward toward additive manufacturing of parts and products by means of the so-called 3D printing. Indeed, 3D printing is able to trigger a revolution in production technologies since many products could be printed on-demand, allowing great flexibility in the selection of materials and in the variability of shapes. In principle, metals, ceramics, polymers, or even biomaterials can be printed opening access to many different applications for production of parts, e.g., in the aerospace or car industry, for custom-designed chemical reactors and even biomaterials for bone replacement. Here, we focus on powder-based techniques which will be described in the following. As depicted in

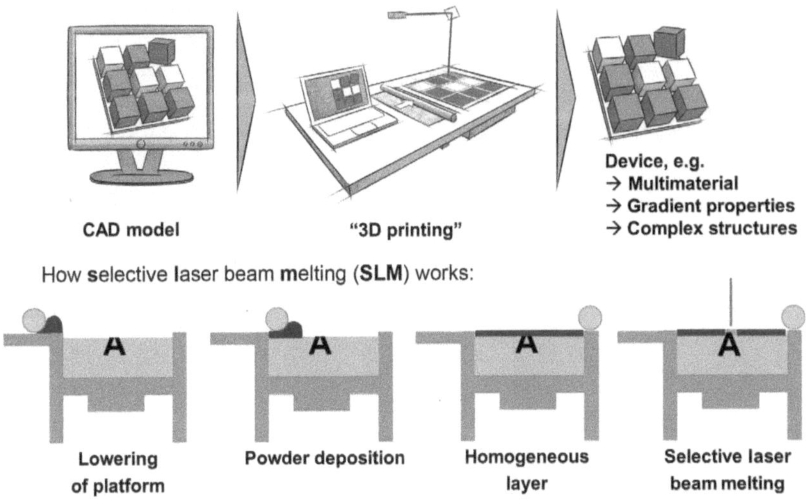


**Figure 27** Additive manufacturing by selective laser melting. *With kind permission from SFB 814,* Additive Manufacturing, *http:/www.sfb814.forschung.uni-erlangen.org/.*

Fig. 27, in a typical setup, powders are transported in a heated powder bed which is selectively molten or sintered by illumination with laser light or electron beams, the latter of course only for conductive metal powders. The loose packing is locally heated above the melting point of the powder so that the molten particles coalesce into the desired structure. These steps are repeated by lowering of the bed and adding a new layer of powder so that the designed 3D structure gradually is formed layer by layer. It is obvious that the powder properties and in particular their flowability, packing density, and ability to fuse are of major importance for the formation of dense structures.

Hence, 3D printing is situated in the core of mesoscience and technology with respect to transport and storage of functional powders. Both aspects inherently depend on the particle size, shape, and surface properties. Currently, mean powder sizes are located around several 10 μm. Good flowability and dense packings are directly related with small adhesion forces between the particles. Figure 28 shows how the tensile stress can be largely reduced for polymer powders which were produced by grinding. In particular, a simple and scalable wet cold grinding process was developed to produce polymer particles with mean diameters below 5 μm for a number of polymer materials including polystyrene, polyamide, polyetheretherketone, and polybutyltherephtalate. Wet cold grinding in a stirred media mill combines low temperature for embrittlement of the polymers with a high

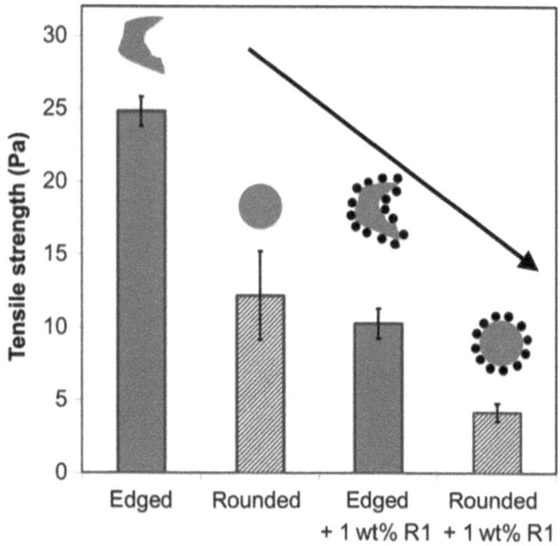


**Figure 28** Experimental optimization of the flowability and packing density of polystyrene as mimicked by the measured tensile strength of the powders for edged, rounded, and coated particles. *Reprinted from Schmidt et al (2014). Copyright 2004 Elsevier.*

number of stress events to produce unusually fine polymer powders (Schmidt et al, 2012). The particle shape after grinding is, however, irregular and therefore the tensile stress is high, the flowability is poor, and the packing density is low. In the next step, these particles are rounded in a heated downer reactor while keeping the volume of the particles constant and ensuring that the molecular polymer structure is not deteriorated. Further surface coating of these powders with nanoparticles can reduce the adhesion forces considerably, resulting in values of the tensile stress which are by a factor of 6 smaller than the tensile stress of the powder directly after grinding (Schmidt et al, 2014). Thus, these powders are very good candidates for selective laser beam melting.

## 5. PROCESSES AND PROPERTIES BY DESIGN

Product design and engineering of advanced materials are equivalent with the design of processes which lead to structures with desired properties. Too often, product design is more or less an empirical approach. Due to the extremely high complexity of particulate products and the related production processes, predictive models are still rare. However, the tools for more systematic and rigorous product and process design strategies are within reach. True product design requires rigorous mathematical optimization techniques by inverting predictive models for products and processes to derive both, the disperse properties and the related process variables. In particular, the following optimization tasks of increasing complexity need to be solved:

- Inverse problems are well known in particle technology since ensemble-based techniques such as dynamic light scattering, laser diffraction, or ultrasound spectroscopy are used for the determination of the PSD. In all these cases, the measured quantity (e.g., scattered light) comes from an ensemble of particles. The deconvolution of the measured signal which is in fact a superposition of light from all particles in the measuring volume requires the solution of ill-posed Fredholm integral equations by, e.g., regularization techniques (Colton and Kress, 2013).
- Model-based determination of difficult-to-measure properties. Material functions as discussed above are seldom known. Typical examples are, for instance, surface energies that are highly important for crystallization (Vassilev et al, 2010) or tailored particle interactions in formulation tasks, elastic, or inelastic properties such as coefficients of restitution for DEM modeling (Dosta et al, 2013; Freireich et al, 2011; Gantt et al, 2006; Schwager and Pöschel, 2007; Sutkar et al, 2015).
- Optimization of disperse properties such as PSD, shape, or structure as a function of operation parameters of a distinct unit operation. Most work is known for optimization of crystallization processes in stirred tank reactors (Kwon et al, 2014; Shi et al, 2006; Vollmer and Raisch, 2006). For many unit operations, empirical approaches prevail due to a lack of predictive forward models.
- Design of structure–property functions for single particles and particle ensembles. Both gradually become accessible if their structure is not too complex. Promising examples are the shape design of pharmaceutical crystals or optical properties of pigment particles (see Example 3 in Section 5.3 (Taylor et al, 2011)).
- Design of process–structure functions and process design including integration of several unit operations. Currently, this is seldom realized since we still lack sufficiently good models of individual unit operations. A promising example is granulation where excellent strategies have been developed by the groups of Litster (Kayrak-Talay et al, 2013) and Heinrich (Dosta et al, 2010).

At the point where we are able to model a particle production process in a mathematical sense, the question raises how to optimize these conditions in a simulation based rather than an experimental setting. Optimization in chemical engineering is in most cases based on a performance measure which implicitly depends on the solution of an ordinary differential equation (ODE), partial differential equation (PDE), or a combination thereof with integral equations altogether describing the evolution of the controlled system or its static solution. The system can, for example, be controlled by its initial state (e.g., concentrations, PSD) as well as its reaction conditions (e.g., temperature, mixing intensity, flow rates).

Figure 29 shows the flow from the input data to the output, where we have provided for the possibility that the states and the control depend on a "spatial" parameter $x$ and time $t$. Indeed, $x$ may represent the actual position in one or higher dimensional space, depending on the application, but may also include, e.g., dispersive parameters, as shape, diameter, aspect ratios, etc., i.e., the disperse properties in particle technology. Noteworthy, the output $y$ is later represented by the property density $q$ or the shape of the particles $\Omega$.

In engineering applications, the evaluation of the performance measure, also known as objective functional, is in most cases performed by some algorithms which typically consist of industrial software plus some "home-made" software interfaces. In essence, the mapping from the input (initial value, control) to the output is provided by a code. In this understanding, the problem of optimizing the performance measure with respect to the control and initial data reduces to tune these parameters within the set of admissible until some success is achieved.

Clearly, while a nonsystematic parameter search, referred to as "trial and error," is by now regarded obsolete in any serious (chemical) engineering optimization context, systematic methods for optimization with respect to the parameters $u \in U_{ad}$, where $U_{ad}$ is the set of admissible controls, or the initial process data $y_0 \in Y_{ad}$, where $Y_{ad}$ denotes the set of admissible initial data, concern a direct search without using derivative or sensitivity information. Such methods, encoded into standard algorithms, are, e.g., the Nelder and Mead (1965) method or general pattern search methods, e.g., by Torczon and Trosset (1998). One distinguishes between deterministic methods, such as the ones just mentioned, and stochastic or random based so-called meta-heuristics, such as genetic algorithms (Goldberg and

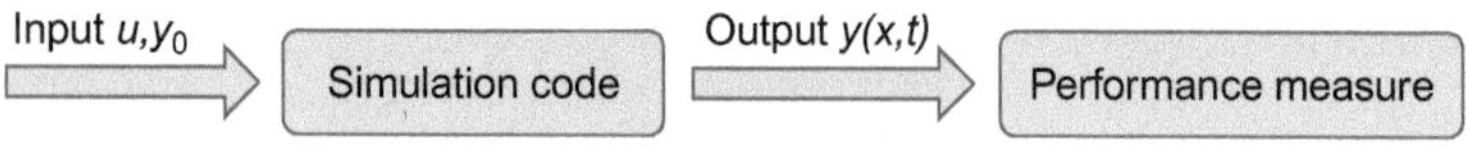


**Figure 29** Flow from input data to performance measure.

Holland, 1988) or more general evolutionary methods (Michalewicz and Schoenauer, 1996) and machine-learning methods (Freund et al, 1997). In these methods and the corresponding algorithms, state and control parameters are represented as individuals in a fictitious population that evolves according to a succession of randomly driven operations, e.g., selection, mutation, cross-over, recombination, etc. (Glover and Kochenberger, 2003). Thus, an iterative optimization of the total system involves the evolution steps of the population in a fitness-oriented "society." Fortunately enough, such building blocks of these meta-heuristic methods are often readily available which has led to the tendency that such methods are used frequently in the engineering context. The great advantage of these methods, namely that they are typically applicable to the code that is used anyway in the research (that concerns simulation data), is complemented by the fact that the meta-heuristic methods genuinely aim at global optima versus local optima.

It is obvious that the chemical engineer would rather be interested in global extrema than local ones, in particular in cases where the performance measure leads to a quite exotic optimization landscape (cf. Fig. 30). However, there are some drawbacks: typically such a code needs a very large number of iterations, where in each iteration the performance measure has to be evaluated. This causes a serious problem in those cases, where the simulation needed for the evaluation of the performance index, or the experiment that provides the data, is complex and time consuming. Whenever flow simulations and chemical reactions are involved, this may lead to inacceptable running times. But even if this is not the case, in particular if one invokes model-reduction techniques or surrogate modeling (Frangos et al, 2010), there is typically no guarantee that, once the iteration has been terminated, one is close to the global extreme value sought. No sensitivity analysis with respect to parameter changes, perturbations, and the like is possible and also no *a posteriori* estimates are available. In other

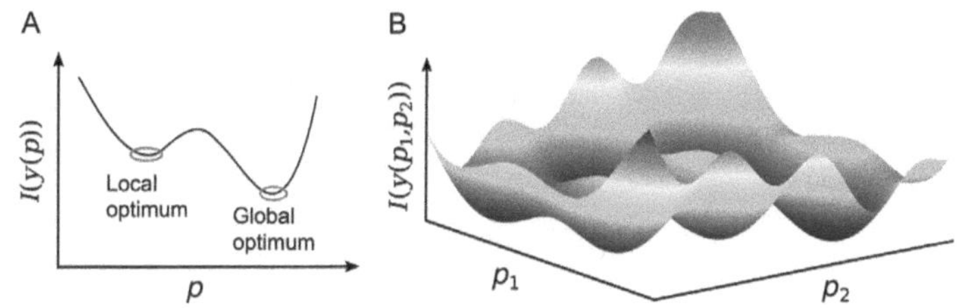


**Figure 30** Illustration of local and global optima.

words, upon termination one has to *trust* that an optimal value has been found. In a very basic understanding of optimization, one may be satisfied with such a purely forward simulation-based method, as the performance before and after is improved and possibly leads to a superior product. If, however, one is interested in a method that requires only a few iteration steps, admits *a posteriori estimates* that allow for *optimality certificates*, and is accessible to *sensitivity analysis* with respect to parameters and boundary conditions, then meta-heuristics are most often not the right choice. Even the deterministic search methods such as the Nelder–Mead and pattern search methods typically necessitate a large number of function evaluations, which may render these methods too costly in, e.g., flow-related problems as well.

A more recent and promising development aims at so-called *trust-region* methods along with *surrogates* for both the model and the performance index (Thekale, 2011). Again, also these methods tend to be rather slow. It is thus clear that sensitivities or in other words derivatives have to be taken into account. However, those derivatives do not come for free. They have to be encoded. Unfortunately, many industrial codes do not offer sensitivities, as they were not designed to do so. In order to utilize sensitivity information, one can construct *finite difference approximations* to the desired derivatives. Unfortunately, especially in the context of chemical engineering and in particular in particle technology, due to the large stiffness-related sensitivity of the system and the potential measurement noise or numerical errors, finite differencing may be very inaccurate and, hence, may lead to highly questionable results. Another way to acquire derivative information from the code has come to be known as *automatic differentiation (AD)* (Griewank and Walther, 2008). AD relies on the parse-tree representation of arithmetic expressions such that even very complicated formulae and coding steps can be retraced by elementary operations that allow for exact differentiation. Using AD, entire codes can be (and have been) automatically differentiated and thus provide the desired derivate information. While being very effective for some known benchmark examples, a typical industrial code is very hard to access by AD. In effect, many researchers in applied sciences and engineering have resorted to the direct search and heuristic methods mentioned above in the past.

## 5.1 Sensitivities of Parameter-Dependent Initial Value Problems

Aiming at derivative-based methods implies extra work both in the analysis of the model and the code. The basic idea goes back to classical methods in

ODEs depending on parameters. The question addressed there is about the sensitivity of the model with respect to parameter changes. Consider the ordinary initial value problem (IVP):

$$\begin{aligned} \dot{y}(t) &= f(t, y(t), p) \quad t \geq 0 \\ y(0) &= y_0 \end{aligned} \tag{9}$$

with the state at time $t$ now being described by $y(t)$ and a source term $f(t,y(t), p)$ with a possibly vector valued control $p$. We define the sensitivity of the state $y$ with respect to the $i$th component of the control as:

$$s_i(t) = \frac{\partial y(t)}{\partial p_i} \tag{10}$$

It satisfies the following linear IVP where $\langle \cdot | \cdot \rangle$ denotes the scalar product:

$$\begin{aligned} \dot{s}_i(t) &= \langle \nabla_y f(t, y(t), p) \,|\, s_i(t) \rangle + \partial_{p_i} f(t, y(t), p) \quad t \geq 0 \\ s_i(0) &= 0 \end{aligned} \tag{11}$$

This shows that the derivative with respect to the parameter vector can be achieved by solving an extra vector valued linear *IVP*. Thus, if one codes the original model problem together with the *linear sensitivity* model, one produces the sensitivity of the state with respect to the parameters directly. If we have a performance index, say $Q(T)$ measuring the state at the final time $T$, then the sensitivities are:

$$\frac{\mathrm{d}Q}{\mathrm{d}p_i} = \nabla_y Q \cdot \frac{\partial y}{\partial p_i} = \nabla_y Q \cdot s_i(T) \tag{12}$$

For each control variable, the sensitivity method involves an additional linear IVP to be solved. Thus for a large number of parameters, this method is not reasonable. Another method is more feasible which avoids calculating the sensitivity of the state directly, the so-called adjoint method which enables to compute the sensitivities in one shot.

Therefore, we take a multiplier $\gamma$ of the equality constraint provided by the ODE and consider the Lagrangian:

$$\begin{aligned} L(y, \gamma) &= Q(y) - \int_0^T \langle \gamma | \dot{y} - f \rangle \, \mathrm{d}t \\ &= Q(y) + \langle \nabla_y Q - \gamma | \nabla_p y \rangle |_0^T + \int_0^T \langle \dot{\gamma} | y \rangle \langle \gamma | f(t, y, p) \rangle \, \mathrm{d}t \end{aligned} \tag{13}$$

We now differentiate the Lagrangian with respect to the control $p$ and obtain:

$$\frac{\mathrm{d}L}{\mathrm{d}p}(y,\gamma)=\frac{\mathrm{d}Q}{\mathrm{d}p}$$
$$=\nabla_p Q+\langle\nabla_y Q-\gamma|\nabla_p y\rangle|_0^T+\int_0^T\langle\dot{\gamma}+\gamma\nabla_y f|\nabla_p x\rangle+\langle\gamma|\nabla_p f)\rangle\,\mathrm{d}t \tag{14}$$

Notice that the initial sensitivity of the state with respect to the control is zero. By solving the so-called *adjoint equation*, in fact, the *final value backward problem* is:

$$\begin{aligned}\dot{\gamma}(t)&=-\gamma(t)\cdot\nabla_y f(t,y(t),p)\quad t\in[0,T]\\ \gamma(T)&=\nabla_y Q\end{aligned} \tag{15}$$

We obtain the expression of the same sensitivity for Q with respect to $p$:

$$\frac{\mathrm{d}Q}{\mathrm{d}p}=\nabla_p Q+\int_0^T\langle\gamma|\nabla_p f\rangle\mathrm{d}t \tag{16}$$

By a simple change of variables, the adjoint equation mentioned above can be transferred into an IVP. This shows that the sensitivity asked for can be computed in two different ways, where the latter is more suitable for problems with large number of control parameters. As a result of this rather sketchy discussion, we realize that providing sensitivities or gradients of performance measures or cost functions with respect to parameter or control variations consists in solving an extra linear equation, the *sensitivity equation* in the first approach or the *adjoint equation* in the second.

## 5.2 General Model-Based Optimal Control Setting

In the general mathematical model-based optimal control theory, we typically distinguish between the state $y$ and the control $u$. The so-called state $y$ is a collection of parameters and functions that describe the properties of a system, e.g., concentration and PSD. The control $u$ is a collection of properties which the chemical engineer is able to control in the system and may involve parameters as in the previous subsection as well as time-dependent properties (e.g., flow rates, temperature profile, initial state). As mentioned above, the control and the state are, in most cases, coupled by an ODE, PDE, or a combination thereof with integral equations. This coupling is in a general mathematical format represented by equality and inequality constrains $h(t,y,u)=0$, $g(t,y,u)\leq_\Lambda 0$, where $\Lambda$ denotes a cone with respect to which an ordering is provided. In a "forward" or phenomenological "observation" mode, one is interested in the mapping from the input data $u$ to the output consisting of some observed quantity evaluation $I(y,u)$. This leads to the general optimization setup:

$$\begin{aligned} &\min_{u\in U_{ad}} I(y, u) \\ &\quad \text{such that} \\ &h(t, y, u) = 0 \quad t\in[0, T] \\ &g(t, y, u) \leq_{\Lambda} 0 \quad t\in[0, T] \end{aligned} \tag{17}$$

This problem formulation takes the model and the side conditions, represented by equalities and inequalities as explicit—typically vector valued—constraints, and is called *all-at-once approach.* Notice again that such an approach is typically not possible in the context of direct methods and meta-heuristics, as the side conditions are hard to implement in definition of populations. Moreover, this format is rather general also from the point of view of numerical simulations. Indeed, only once we approach the optimum in an iterative method—yet to be described—we need to satisfy the model equation and the constraints at the optimum only, whereas if we use a code to compute the state $y(t)$ for any feasible $t$, we need to implement the constraint into the code, leading to the black-box optimization problem discussed above, or proceed with an intermediate approach, where we provide a solution operator $S$ such that $S(u)$ solves the model problem for a given control $u$. Then we obtain what has come to be known as *reduced optimization problem*:

$$\begin{aligned} &\min_{u\in U_{ad}} I(S(u), u) \\ &\quad \text{such that} \\ &H(u) := h(t, S(u), u) = 0 \quad t\in[0, T] \\ &G(u) := g(t, S(u), u) \leq_{\Lambda} 0 \quad t\in[0, T] \end{aligned} \tag{18}$$

This problem is given in the standard format of nonlinear programming. For both representations of the standard optimal control problem, there are many textbooks available, exemplary the book of Sontag shall be mentioned here (Sontag, 1998). We have not made any attempt to describe and characterize the mathematical setup in terms of spaces and regularity requirements, and due to the readership we will not do so subsequently. Suffice is to note that the format of the model can be allowed to be rather general so as to include PDEs and integro-PDEs, as they are important in particle technology.

## 5.3 Examples of Optimization

It is important to notice that optimization in chemical engineering takes two steps: first, we need to validate the model with respect to a certain control and parameter range. This first phase, which as such needs optimization tools and tools from *design of experiments*, establishes the *predictive model*. This is the model we can trust during subsequent optimization. It is clear that mathematical optimization typically explores the full set of constraints and, therefore, one needs evidence that the model is valid on that admissible range in the first place. Second, the predictive model and the validated numerical simulations are utilized in the all-at-once optimization problem. With the mathematical framework described so far, we can control reactor dynamics in such a way that an optimized initial feed and optimized temperature profile yields a desired PSD. In the following, examples from gas and liquid phase as well as a shape optimization will be discussed. Models were established and validated prior being used for rigorous optimization of process conditions to derive tailored PSDs and the related functional properties such as the bandgap of quantum dots via tailoring the size distribution. The last example is an outlook in the direction of shape optimization in the context of optical properties of nanoparticles.

### *Example 1: Synthesis of Si Nanoparticles in the Gas Phase*

Our first example is on the gas phase synthesis of Si nanoparticles in a hot wall reactor via the pyrolysis of silane (temperatures between 900 and 1100 °C, low total pressure of 25 mbar, silane partial pressure of 1.0–3.2 mbar, and residence times between 80 and 420 ms). Experiments revealed narrow PSDs with geometric standard deviations below 1.1 (Körmer et al, 2010b). This at first glance astonishing finding was explained by a short nucleation burst followed by surface growth and condensation as well as negligible agglomeration at low total pressure (25 mbar) similar to approaches known for the synthesis of quasi-monodisperse colloids. Before the ultimate goal of optimized PSDs—here as narrow as possible—could be approached, a predictive population balance model needed to be established (Körmer et al, 2010a):

$$\frac{\partial}{\partial t}y(x,t) + \frac{\partial}{\partial x}\left(G(t)y(x,t)\right) = B(x,t) \tag{19}$$

where $G$ is the growth rate, nucleation is represented by the source term $B$, $y(x,t)$ describes the evolution of the number density distribution, $x$ is the particle size, and $t$ is the time. No initial seed particles or influx boundary conditions need to be assumed as all particles form within the reactor by homogeneous nucleation. Based on studies of Petersen and Crofton (2003), the highly complex reaction network of silane pyrolysis (Shekar et al, 2012) was reduced to a global equation for the reaction kinetics:

$$\frac{\mathrm{d}[\mathrm{SiH_4}]}{\mathrm{d}t} = -k_{1\mathrm{a}}[\mathrm{SiH_4}][\mathrm{Ar}] \tag{20}$$

where $k_{1\mathrm{a}}$ is the rate coefficient which contained an empiric fitting parameter $F_1$ that allowed adopting the simulation to the experimental findings:

$$k_{1\mathrm{a}} = F_1 \cdot 7.2 \times 10^9 \cdot \mathrm{e}^{\left(\frac{22,710K}{T}\right)} \frac{\mathrm{m}^3}{\mathrm{mol\,s}} \tag{21}$$

In addition to this gas phase reaction with subsequent homogeneous nucleation, heterogeneous deposition of monosilane at the surface of the Si nanoparticles as well as condensation of free monomer on the surface of existing particles was included to the population balance (PB) model (Peev et al, 1991):

$$R_\mathrm{d} = F_2 \cdot \frac{k_\mathrm{p} \cdot K_{\mathrm{SiH_4}} \cdot [\mathrm{SiH_4}]}{1 + K_{\mathrm{SiH_4}} \cdot [\mathrm{SiH_4}]} \cdot \frac{M_\mathrm{Si}}{\rho_\mathrm{Si}} \tag{22}$$

Again, an empiric parameter $F_2$ allows to adopt the simulation to the experiment. The heterogeneous rate coefficient $k_\mathrm{p}$ and the equilibrium adsorption constant of silane $K_{\mathrm{SiH_4}}$ were also adopted from Peev et al, whereas $M_\mathrm{Si}$ is the molar mass and $\rho_\mathrm{Si}$ is the density of silicon. In brief, the PB model contained two ODEs for the mass balances and the main PDE of the PBE (Gröschel et al, 2012).

Prior to the study of Gröschel et al, coupled systems of this way have rarely been investigated by advanced optimization techniques. After creation of the model, the empiric parameters were optimized for one set of experiments to get a best match between simulation and experimental data. A total of 14 iterations were required until an optimized set of unique fit parameters was obtained. The good match between simulated and measured PSDs is shown in Fig. 31.

After having established a well-validated predictive model, the inverse problem can be approached: finding optimum process parameters to adjust a PSD with certain mean size. Key process parameters are time–temperature profiles which lead to distinct PSDs. Using a trust-region algorithm and the existing population balance model, suitable time–temperature profiles were obtained after 99 iterations for the narrow PSD and 115 iterations

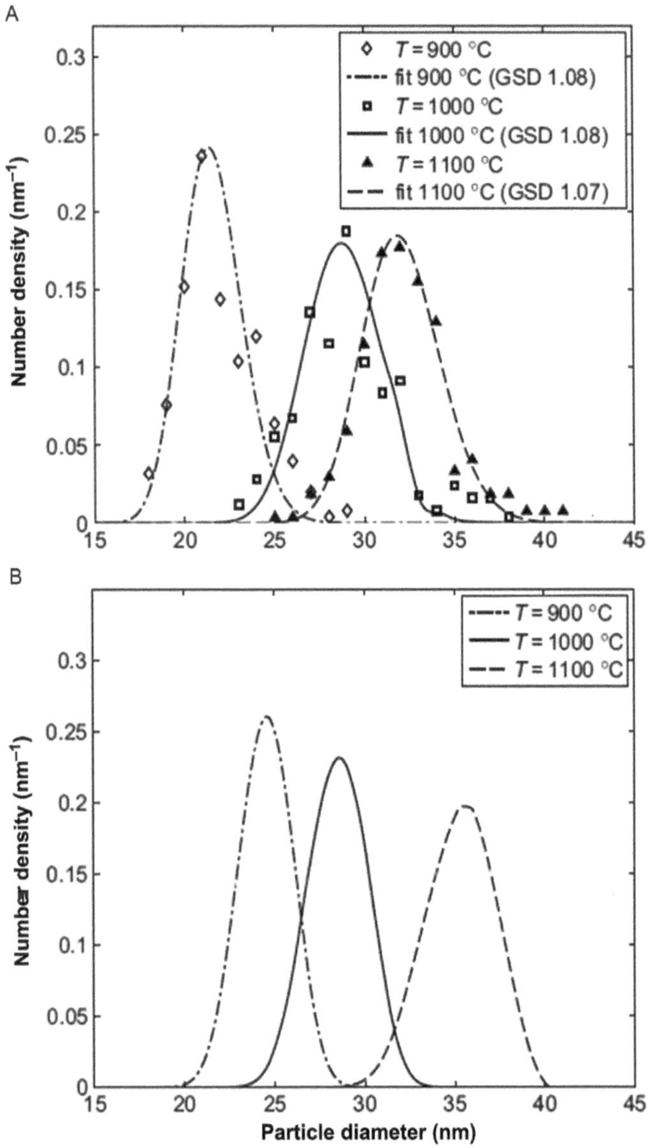


**Figure 31** (A) Experimentally determined number density distributions $q_0$ of Si nanoparticles synthesized in a hot wall reactor (Körmer et al, 2010b) in comparison to (B) number density distributions derived by PBE after optimization of empiric parameters $F_1$ and $F_2$. *Reprinted from Gröschel et al (2012). Copyright 2012, with permission from Elsevier.*

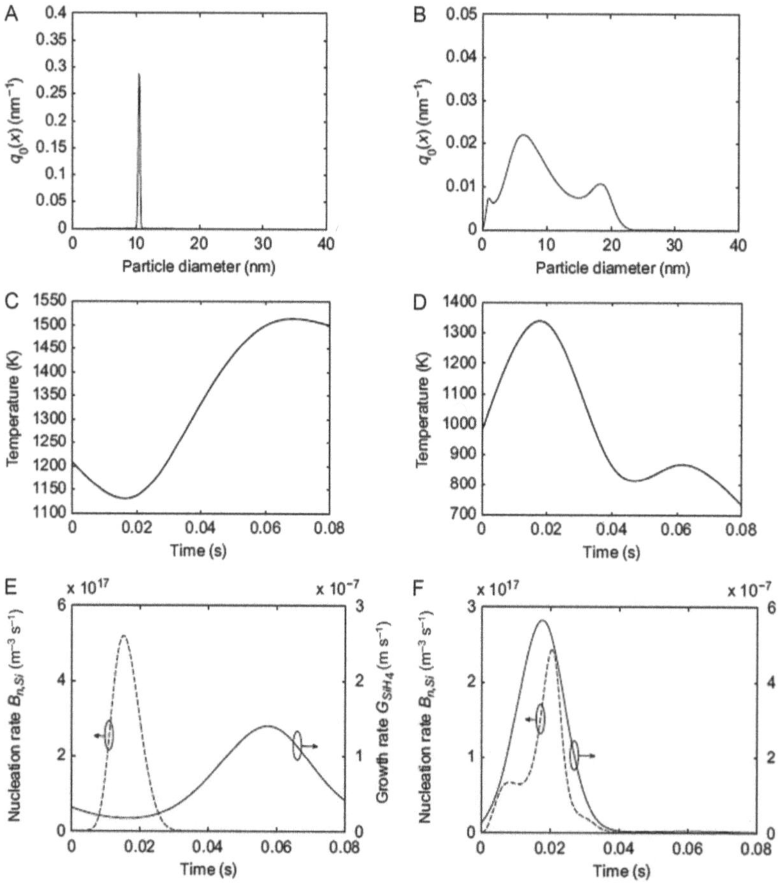


**Figure 32** Comparison between narrow (left) and broad (right) size distributions for the case of small particles showing the final PSDs, the computed temperature profile as well as the correlated nucleation and growth rate. *Reprinted from Gröschel et al (2012). Copyright 2012, with permission from Elsevier.*

for the broad PSD. From Fig. 32 in which the optimized PSDs are summarized together with the underlying temperature profiles as well as nucleation and growth rates, it becomes clear that narrow distributions with standard deviations close to 1 are obtained when nucleation and growth can be separated from each other. This can be achieved by distinct temperature–time profiles as shown in Fig. 32(C).

In case of the broad PSDs, both processes occur right at the beginning in parallel (La Mer and Dinegar, 1950). Thus, inversion of the PBE gives important insights to process design and even opens new possibilities to understand fundamentals of particle formation.

### *Example 2: Ripening of ZnO Quantum Dots in Liquid Phase*

Due to their comparatively easy synthesis by mixing two dissolved reactants (a zinc acetate precursor solution and LiOH, both dissolved in ethanol) at moderate temperatures (<50 °C) and due to high relevance of ZnO as semiconductor, a batch study on Ostwald ripening of ZnO was performed. Noteworthy, ZnO nucleation and growth are not affected by any mixing issues due to the fact that the reaction of the zinc acetate precursor to ZnO is slow (Gradl, 2010; Segets et al, 2012a). Regarding secondary particle formation, Ostwald ripening is the dominating mechanism and the PBE is the following:

$$\frac{\partial}{\partial t}q(x,t)+\frac{\partial}{\partial x}(R(x,t,c)\cdot q(x,t))=0 \tag{23}$$

It is complemented by the mass balance accounting for the concentration of ZnO in the liquid phase:

$$c_{\mathrm{L}}(t)+\frac{\pi}{6}\cdot\frac{\rho}{V}\cdot\int_{x_c}^{\infty}(x_3\cdot q(x,t)\cdot \mathrm{d}x)=c^0 \tag{24}$$

where $x$ is the particle diameter, $t$ is the time, $q$ is the number density, $c^0$ is the overall ZnO concentration, $\rho$ is the density of ZnO, $V$ is the reactor volume, and $R$ is the ripening rate. $x_c$ is the size of stable clusters, i.e., the smallest size in the system. Noteworthy, in a first study of Gradl and Segets (Gradl, 2010; Segets et al, 2012a), the exponential term in the Gibbs Thomson equation for the ripening rate needed to be approximated by the first term of a Taylor series to obtain a tractable numerical problem

$$\begin{aligned}R(x,t,c)&=\frac{4\cdot D\cdot M\cdot c_{\mathrm{L}}^{\infty}}{\rho\cdot x}\left[\frac{c(t)}{c_{\mathrm{L}}^{\infty}}-\exp\left(\frac{4\cdot\gamma_{\mathrm{SF}}\cdot V_{\mathrm{m}}}{\nu\cdot x\cdot k_{\mathrm{B}}\cdot T}\right)\right]\\&\approx\frac{4\cdot D\cdot M\cdot c_{\mathrm{L}}^{\infty}}{\rho\cdot x}\left[\frac{c(t)}{c_{\mathrm{L}}^{\infty}}-\left(1+\frac{4\cdot\gamma_{\mathrm{SF}}\cdot V_{\mathrm{m}}}{\nu\cdot x\cdot k_{\mathrm{B}}\cdot T}\right)\right]\end{aligned} \tag{25}$$

where $c$ is the molar concentration of ZnO in the solid phase, $D$ is the diffusion coefficient, $M$ is the molar mass of ZnO, $c_{\mathrm{L}}^{\infty}$ is the solubility of a flat ZnO surface, $\gamma_{\mathrm{SF}}$ is the corresponding surface energy, $V_{\mathrm{m}}$ is the molecular volume of ZnO, $\nu$ is a stoichiometric coefficient, $k_{\mathrm{B}}$ is the Boltzmann constant, and $T$ is the temperature.

For very small particles (in case of ZnO <3 nm), the full solution of the exponential term is not possible within a PBE framework when explicit solvers are used. A detailed discussion why numerics become highly unstable is found in the literature (Gröschel, 2013; Haderlein et al, 2015). Long simulation times are required even when the approximation is used. To solve this issue, Gröschel et al developed a Fully Implicit Method for Ostwald Ripening (FIMOR) (Gröschel, 2013; Gröschel et al, 2014). FIMOR overcomes the inevitable inaccuracy when using the approximation and provides efficient simulations that can be introduced to a subordinate optimization framework. As shown in Fig. 33A, by means of FIMOR, several hours of Ostwald ripening can be accurately simulated within very short times (Intel Core2 Duo CPU E8400 3.00 GHz, 4.00 GB RAM, Windows 7, Matlab 7.10.0: 10 h at 25 °C within 25 s).

Similar to the previous example, also in liquid phase synthesis parameters need to be adjusted. However, this time no empiric constants but, usually hardly accessible, material properties like solubility and surface energy are

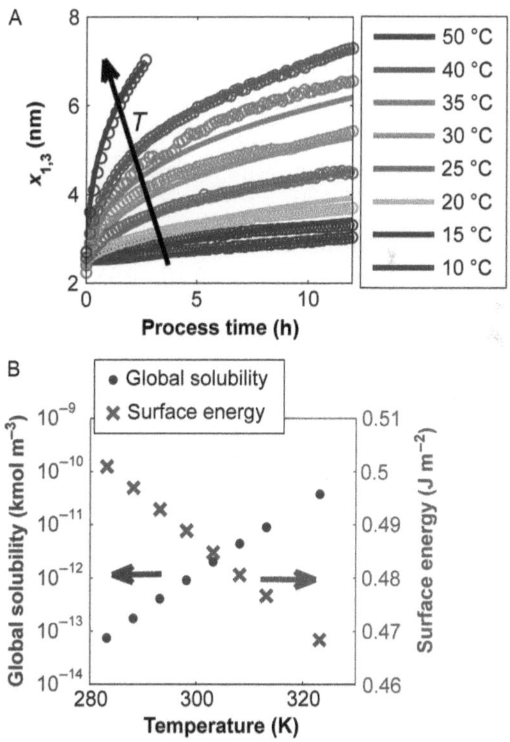


**Figure 33** (A) Experimental data and simulations using FIMOR for ZnO Ostwald ripening between 10 and 50 °C and (B) solubilities and surface energies derived from a least squares fit to the data shown in (A). *Adapted from Haderlein et al (2015). Copyright 2015, with permission from Elsevier.*

required (Fig. 33B). In brief, the experimental data set extracted from Gradl and Segets (Gradl, 2010; Segets et al, 2009) shown in Fig. 33A was matched with the simulations using a least squares approach. In general, the values obtained for different temperatures (Fig. 33B) suite well to literature data where surface energies of metal oxides are expected between 0.5 and 0.7 J m$^{-2}$. Noteworthy, once derived, the values were fixed throughout all further analyses and could be used to predict the PSD at any kind of time and synthesis temperature. Moreover, they even allowed the prediction of a PSD at the outlet of a microreactor when taking the experimentally determined residence time distribution into account (Haderlein et al, 2015).

In the next step, first optimization studies could be performed using a Nelder–Mead algorithm. As shown in Fig. 34, different target distributions with mean sizes of 3.0 (blue (black in the print version)), 4.5 (green (dark gray in the print version)), and 5.5 nm (pink (gray in the print version)) are provided together with the underlying iterations. From the initial values (indicated by a plus sign), the algorithm identifies the optimum time and temperature profile (red (gray in the print version) circles) to match the targets. At this point, it needs to be mentioned that the solution is not unique as for each temperature a time can be found where the mean particle diameter matches the desired one.

### *Example 3: Shape Optimization in Particle Technology*

In the last example, we will concentrate on the shape of particles and their optical properties. Assume that we can model, simulate, and finally control the dynamic processes in a particle production, where the shape of the particles can be adjusted and is described within the simulation. This possibility

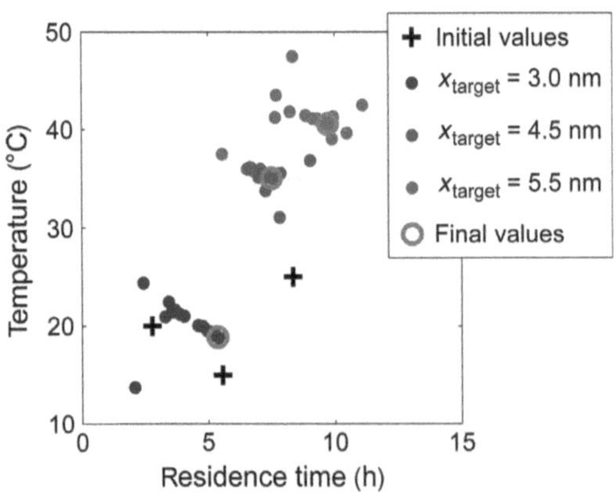


**Figure 34** Optimization of ZnO PSDs using different target distributions with mean sizes between 3.0 and 5.5 nm.

leads to the question of appropriate process conditions resulting in particles with optimal shape-dependent performance measure (Fig. 35).

For a synthesis or crystallization process where the shape of particles can be influenced but where we are up to now not able to model the mapping from the parameters to the particle morphology, the question of an optimal shape within a set of admissible shapes is also of considerable interest as the results can show promising research directions. The theory of shape optimization, known from continuum mechanical applications (Allaire, 2007), can also be applied to particle technology. In the following, the focus will be on the optical properties of particles where in most cases the scattering formulation of the time–harmonic Maxwell's equation governs the mapping from shape $\Omega$ to the desired property:

$$\nabla \times \nabla \times E_S(x) - \kappa^2 \varepsilon_\Omega(x) E_S(x) = \kappa^2 (\varepsilon_\Omega(x) - \varepsilon_0) E_I(x) \quad x \in D \tag{26}$$

With natural interface conditions, where $\varepsilon_\Omega$ changes and, for example, absorbing boundary conditions on the boundary of $D$. $E_S$ and $E_1$ denote the scattered and incident electric field. The shape dependency of the permittivity $\varepsilon$ may, in the simplest case, be given as follows:

$$\varepsilon_\Omega(x) = \begin{cases} \varepsilon_1 \text{ for } x \in \Omega \\ \varepsilon_0 \text{ for } x \notin \Omega \end{cases} \tag{27}$$

where $\varepsilon_1 = (n_1 + ik_1)^2$ denotes the squared complex refractive index (with $n$ being the refractive index and $k$ the absorption coefficient) of the particle and $\varepsilon_0 = n_0^2$ the squared refractive index of the background material. A desired property can, for example, be the extinction cross section which is denoted by:

$$W_{\mathrm{EXT}} = \int_\Omega \mathrm{Im}\left(\left(\overline{\varepsilon_\Omega(x)} - \varepsilon_0\right) E_I(x) \cdot \overline{(E_I(x) + E_S(x))}\right) \mathrm{d}x \tag{28}$$

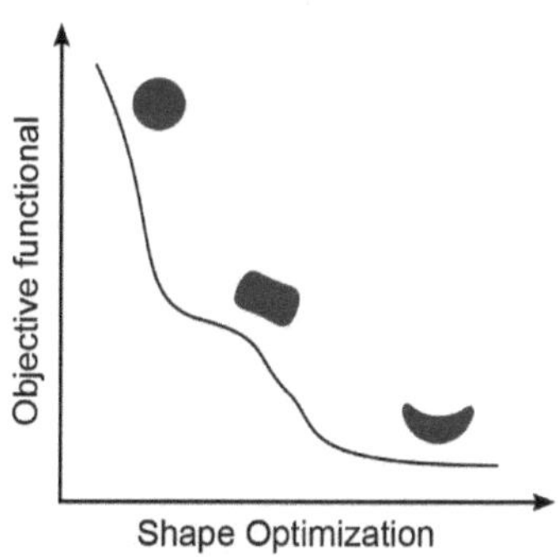


**Figure 35** Sketch of shape optimization.

This specific problem and a huge class of similar PDE-constraint shape optimization problems can be described in the following format:

$$\min_{\Omega \in \Omega_{\mathrm{ad}}} I(\Omega, y) \\ e(\Omega; y) = 0 \tag{29}$$

where the equality constraint $e$ denotes the PDE and $\Omega_{\mathrm{ad}}$ the set of admissible shapes. To be able to optimize the performance measure within a gradient-based method, we have to differentiate $I$ with respect to domain deformations. The main issue thereby is the implicit dependency of the PDE solution $y$ on the domain (Walker, 2015).

Under some assumptions on the PDE and the shape, the Hadamard–Zolesio structure theorem (Delfour and Zolésio, 1992) shows the existence of a so-called shape gradient $g$ with the following property:

$$\mathrm{d}I(\Omega, y; V) = \int_{\partial\Omega} g V \cdot n \mathrm{d}s \tag{30}$$

where $\mathrm{d}I$ denotes the derivative of $I$ in the direction of the velocity field $V$ describing the deformation of the shape on the boundary. As in the subsection above, the derivation and computation of the shape gradient $g$ involve the Lagrangian equation and the solution of an adjoint equation which in the current setting differs only from the state equation by the right-hand side of the PDE.

A small deformation of the boundary by $g$ times the normal vector $n$ leads to an improvement of the objective functional. Nevertheless, in most cases it is from numerical point of view preferable to use the so-called volume representation of the shape gradient to compute a descent direction (Hiptmair and Paganini, 2015).

To demonstrate the possibilities in nanoparticle shape optimization, two examples will be briefly discussed. First, the squared distance to a desired extinction spectrum for a defined wavelength range is minimized (see Fig. 36). The particle is homogeneous and has isotropic material properties. The random orientation of particles in many applications is taken into account as well by averaging over all orientations in 3D space. The product of extinction cross section $W_{\mathrm{ext}}$ of the particles and particle concentration in a dilute system defines via Lambert–Beer's law the resulting spectrum and the light intensity reduction. Both shape dependency on $W_{\mathrm{ext}}$ and the effect of particle concentration are optimized. Of course, anisotropic and for example layered media can be optimized in the same framework. In

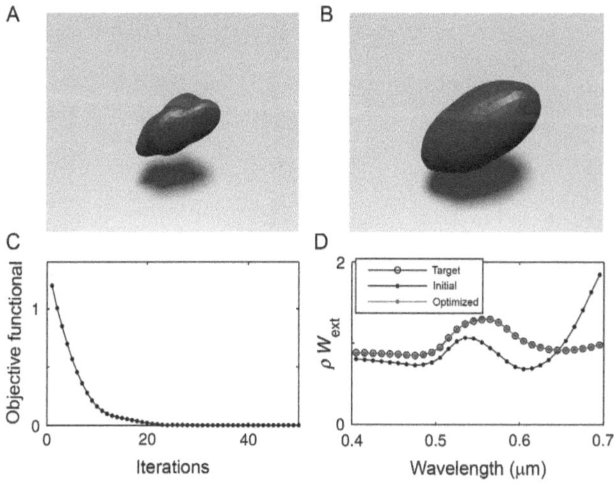


**Figure 36** Shape optimization of a nanoparticle with respect to its extinction spectrum in the visible wavelength range: (A) initial particle size; (B) optimized shape, (C) objective functional, and (D) extinction $W_{ext}$ multiplied by the particle concentration $\rho$ for the initial (red (gray in the print version)), the target (blue (black in the print version)), and the optimized (green (dark gray in the print version)) spectra (Peukert and Segets, 2015; Semmler et al, 2015).

Fig. 36A and B, one can see the initial and the optimized shape, and Fig. 36C shows the evolution of the objective functional during optimization. The initial spectrum (red (gray in the print version)), the target spectrum (blue (black in the print version)), as well as the desired spectrum (green (dark gray in the print version)) are shown in Fig. 36D.

In the second exemplary optimization setting (see Fig. 37), the shape of an absorbing spherical particle is fixed and one can see how the shape of a coating, consisting of nonabsorbing material, influences the total electric field in a plane incident wave. In Fig. 37B, the influence of the particle on the field is significantly reduced by its coating and thus for this polarized unidirectional monochromatic light, less visible for the human eye. This example points to future applications of cloaking technologies with the target to make particles invisible, at least in a specific part of the spectrum.

By using multiple wavelengths and multiple incident directions and polarizations, we are able to derive the sensitivity of the extinction spectrum of randomly oriented particles with respect to their shape (Semmler et al, 2015). Coupling the control theory of the particle production process

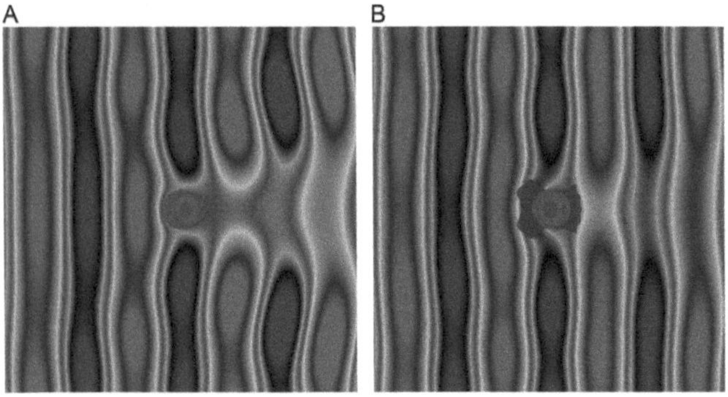


**Figure 37** Initial and optimized shape in a cloaking problem.

discussed above with the theory of shape-differential calculus, we cover the whole path from optical properties of the particles back to the process parameters of their production. Under some regularity assumptions, shape gradients can be derived for any shape-dependent objective functional where the shape influences either the objective functional direct (e.g., the volume) or implicit as in the setting above by a PDE which is either solved on the domain which is subject to optimize or where its parameters depend on the shape.

## 6. CONCLUSIONS

Mesoscale science and technology are the key for product design in chemical engineering and materials science. Structure–property and process–structure functions must be determined and optimized for (i) the design of all types of hierarchically structured products and (ii) of the processes to produce them. Within the huge universe of structured particulate systems many degrees of freedom do exist on how to tailor products with respect to their compositions, structures, and properties. This huge parameter space can, however, be condensed into the following five principles which govern structure formation: (i) particle formation by top-down and bottom-up approaches, (ii) structure formation by interactions between the building blocks, (iii) structure formation by self-organization and by transport, (iv) characterization along the process chain, and (v) multiscale modeling and simulation. These principles may be understood as the new "unit operations" of product design and may serve as guidelines for scientists in research as well as engineers in industry. All principles are based on fundamentals in basic science and engineering. Their application and the realization of real-world products require multidisciplinary teams which might be typically organized along process chains.

With the huge spectrum of (meso)structured products, the process know-how of chemical engineers is most relevant for biochemical, chemical, and physical applications. A key question remains: How to teach students this wide spectrum of methodologies? The answer can either be to focus on some parts of the spectrum, for instance in life science engineering, or, even better, to develop unifying principles which are generally applicable. Identifying the latter is the intension of this contribution.

## ACKNOWLEDGEMENTS

The authors acknowledge the funding of the Deutsche Forschungsgemeinschaft (DFG) through the Cluster of Excellence "Engineering of Advanced Materials" (EAM) as well as through CRC 814 and CRC 953.

## REFERENCES

Aguirre CI, Reguera E, Stein A: Tunable colors in opals and inverse opal photonic crystals, *Adv Funct Mater* 20:2565–2578, 2010.

Akdas T, Walter J, Segets D, et al: Investigation of the size–property relationship in $CuInS_2$ quantum dots, *Nanoscale* 7:18105–18118, 2015.

Allaire G: *Conception optimale de structures,* Berlin, 2007, Springer.

Al Zaitone B, Schmid H-J, Peukert W: Simulation of structure and mobility of aggregates formed by simultaneous coagulation, sintering and surface growth, *J Aerosol Sci* 40:950–964, 2009.

Anwar J, Zahn D: Uncovering molecular processes in crystal nucleation and growth by using molecular simulation, *Angew Chem Int Ed Engl* 50:1996–2013, 2011.

Armstrong P, Peukert W: Size effects in the elastic deformation behavior of metallic nanoparticles, *J Nanopart Res* 14:1288, 2012.

Babuska I, Lipton R: Optimal local approximation spaces for generalized finite element methods with application to multiscale problems, *Multiscale Model Simul* 9:373–406, 2011.

Bahrig L, Hickey SG, Eychmüller A: Mesocrystalline materials and the involvement of oriented attachment—a review, *CrystEngComm* 16:9408–9424, 2014.

Baldyga J, Bourne JR: *Turbulent mixing and chemical reactions,* Chichester, 1999, Wiley.

Bao H, Peukert W, Taylor RK: One-pot colloidal synthesis of plasmonic patchy particles, *Adv Mater* 23:2644–2649, 2011.

Baumgartner H-P, Löffler F: The collection performance of electret filters in the particle size range 10 nm–10 μm, *J Aerosol Sci* 17:438–445, 1986.

Becker M, Kwade A, Schwedes J: Stress intensity in stirred media mills and its effect on specific energy requirement, *Int J Miner Process* 61:189–208, 2001.

Behrend O, Ax K, Schubert H: Influence of continuous phase viscosity on emulsification by ultrasound, *Ultrason Sonochem* 7:77–85, 2000. http://www.sciencedirect.com/science/article/pii/S1350417799000292.

Berman AD, Israelachvili JN, editors: *Micro/Nanotribology and its applications, NATO Advanced Science Institute series* (Bushan B, editor), Dordrecht, 1997, Kluwer Academic, pp 317–329.

Binder C, Feichtinger C, Schmid H-J, Thürey N, Peukert W, Rüde U: Simulation of the hydrodynamic drag of aggregated particles, *J Colloid Interface Sci* 301:155–167, 2006.

Binder C, Hartig MAJ, Peukert W: Structural dependent drag force and orientation prediction for small fractal aggregates, *J Colloid Interface Sci* 331:243–250, 2009.

Bockhorn H, Mewes D, Peukert W, Warnecke H-J: *Micro and macro mixing: analysis, simulation and numerical calculation,* Heidelberg, 2010, Springer.

Borchert C, Nere N, Ramkrishna D, Voigt A, Sundmacher K: On the prediction of crystal shape distributions in a steady-state continuous crystallizer, *Chem Eng Sci* 64:686–696, 2009.

Briesen H: Two-dimensional population balance modeling for shape dependent crystal attrition, *Chem Eng Sci* 64:661–672, 2009.

Brus LE: Electron–electron and electron–hole interactions in small semiconductor crystallites: the size dependence of the lowest excited electronic state, *J Chem Phys* 80:4403–4409, 1984.

Bubenhofer SB, Schumacher CM, Koehler FM, Luechinger NA, Grass RN, Stark WJ: Large-scale synthesis of PbS–$TiO_2$ heterojunction nanoparticles in a single step for solar cell application, *J Phys Chem C* 116:16264–16270, 2012.

Burda C, Chen X, Narayanan R, El-Sayed MA: Chemistry and properties of nanocrystals of different shapes, *Chem Rev* 105:1025–1102, 2005.

Chen Z, Moore J, Radtke G, Sirringhaus H, O'Brien S: Binary nanoparticle superlattices in the semiconductor-semiconductor system: CdTe and CdSe, *J Am Chem Soc* 129:15702–15709, 2007.

Clift R, Grace JR, Weber ME: *Bubbles, drops, and particles,* New York, USA, 2005, Dover Publications.

Cölfen H, Antonietti M: Mesocrystals: inorganic superstructures made by highly parallel crystallization and controlled alignment, *Angew Chem Int Ed Engl* 44:5576–5591, 2005.

Cölfen H, Antonietti M: *Mesocrystals and nonclassical crystallization,* West Sussex, England, 2008, John Wiley & Sons Ltd.

Colton D, Kress R: *Inverse acoustic and electromagnetic scattering theory,* ed 2, Berlin, 1998, Springer.

Conner BP, Manogharan GP, Martof AN, et al: Making sense of 3-D printing: creating a map of additive manufacturing products and services, *Addit Manuf* 1–4:64–76, 2014.

Cramer CJ: *Essentials of computational chemistry,* West Sussex, England, 2006, John Wiley & Sons Ltd.

Cussler EL, Moggride GD: *Chemical product design,* ed 2, New York, 2011, Cambridge University Press.

Dai Q, Wang Y, Li X, et al: Reply to "Comment on 'size-dependent composition and molar extinction coefficient of PbSe semiconductor nanocrystals'", *ACS Nano* 3:2054, 2009.

Damasceno PF, Engel M, Glotzer SC: Predictive self-assembly of polyhedra into complex structures, *Science* 337:453–457, 2012.

Damm C, Segets D, Yang G, et al: Shape transformation mechanism of silver nanorods in aqueous solution, *Small* 7:147–156, 2011.

Deen NG, Van Sint M, Van der Hoef MA, Kuipers JAM: Review of discrete particle modeling of fluidized beds, *Chem Eng Sci* 62:28–44, 2007.

Delfour MC, Zolésio JP: Structure of shape derivatives for nonsmooth domains, *J Funct Anal* 104:1–33, 1992.

Dhenge RM, Cartwright JJ, Hounslow MJ, Salman AD: Twin screw granulation: steps in granule growth, *Int J Pharm* 438:20–32, 2012.

Distaso M, Segets D, Wernet R, Klupp Taylor R, Peukert W: Tuning the size and the optical properties of ZnO mesocrystals synthesized under solvothermal conditions, *Nanoscale* 4:864–873, 2012.

Dosta M, Heinrich S, Werther J: Fluidized bed spray granulation: analysis of the system behaviour by means of dynamic flowsheet simulation, *Powder Technol* 204:71–82, 2010.

Dosta M, Antonyuk S, Heinrich S: Multiscale simulation of the fluidized bed granulation process, *Chem Eng Technol* 35:1373–1380, 2012.

Dosta M, Antonyuk S, Heinrich S: Multiscale simulation of agglomerate breakage in fluidized beds, *Ind Eng Chem Res* 52:11275–11281, 2013.

Dosta M, Antonyuk S, Hartge EU, Heinrich S: Parameter estimation for the flowsheet simulation of solids processes, *Chem Ing Technol* 86:1073–7079, 2014.

Dukhin AS, Goetz PJ: *Characterization of liquids, nano- and microparticulates, and porous bodies using ultrasound,* ed 2, *Studies in interface science,* Amsterdam, The Netherlands, 2002, Elsevier.

Eggers J, Villermaux E: Physics of liquid jets, *Rep Prog Phys* 71:036601, 2008.

Eggersdorfer ML, Kadau D, Herrmann HJ, Pratsinis SE: Aggregate morphology evolution by sintering: number and diameter of primary particles, *J Aerosol Sci* 46:7–19, 2012.

Eisenthal KB: Liquid interfaces probed by second-harmonic and sum-frequency spectroscopy, *Chem Rev* 96:1343–1360, 1996.

Elliott JA: Novel approaches to multiscale modelling in materials science, *Int Mater Rev* 56:207–225, 2011.

Engelhardt K, Rumpel A, Walter J, et al: Protein adsorption at the electrified air-water interface: implications on foam stability, *Langmuir* 28:7780–7787, 2012.

Engelhardt K, Lexis M, Gochev G, et al: pH effects on the molecular structure of β-lactoglobulin modified air-water interfaces and its impact on foam rheology, *Langmuir* 29:11646–11655, 2013.

Faber H, Burkhardt M, Jedaa A, Kälblein D, Klauk H, Halik M: Low-temperature solution-processed memory transistors based on zinc oxide nanoparticles, *Adv Mater* 21:1–6, 2009.

Faber H, Klaumünzer M, Voigt M, et al: Morphological impact of zinc oxide layers on the device performance in thin-film transistors, *Nanoscale* 3:897–899, 2011.

Feynman RP: Forces in molecules, *Phys Rev* 56:340–343, 1939.

Fox RO: *Computational models for turbulent reacting flows,* 2003, Cambridge University Press. http://dx.doi.org/10.1017/CBO9780511610103.

Frangos M, Marzouk Y, Willcox K, van Bloemen Waanders B: Surrogate and reduced-order modeling: a comparison of approaches for large-scale statistical inverse problems. In Biegler L, Biros G, Ghattas O, Heinkenschloss M, Keyes D, Mallick B, Marzouk Y, Tenorio L, van Bloemen Waanders B, Willcox K, editors: *Large-Scale inverse problems and quantification of uncertainty*, Chichester, UK, 2010, John Wiley & Sons, Ltd. http://dx.doi.org/10.1002/9780470685853.ch7.

Freireich B, Li J, Litster J, Wassgren C: Incorporating particle flow information from discrete element simulations in population balance models of mixer-coaters, *Chem Eng Sci* 66:3592–3604, 2011.

Freund Y, Seung HS, Shamir E, Tishby N: Selective sampling using the query by committee algorithm, *Mach Learn* 28:28–133, 1997.

Fries L, Antonyuk S, Heinrich S, Palzer S: DEM-CFD modeling of a fluidized bed spray granulator, *Chem Eng Sci* 66:2340–2355, 2011.

Fuchs R, Weinhart T, Meyer J, et al: Rolling, sliding and torsion of micron-sized silica particles: experimental, numerical and theoretical analysis, *Granul Matter* 16:281–297, 2014.

Gahn C, Mersmann A: Brittle fracture in crystallization processes part A. Attrition and abrasion of brittle solids, *Chem Eng Sci* 54:1273–1282, 1999.

Gantt JA, Cameron IT, Litster J, Gatzke EP: Determination of coalescence kernels for high-shear granulation using DEM simulations, *Powder Technol* 170:53–63, 2006.

Ghadiri M, Zhang Z: Impact attrition of particulate solids. Part 1: a theoretical model of chipping, *Chem Eng Sci* 57:3659–3669, 2002.

Glotzer SC, Solomon MJ: Anisotropy of building blocks and their assembly into complex structures, *Nat Mater* 6:557–562, 2007.

Glover F, Kochenberger GA: *Handbook of metaheuristics,* Dordrecht, 2003, Kluwer Academic Publishers.

Gnichwitz J-F, Marczak R, Werner F, et al: Efficient synthetic access to cationic dendrons and their application for ZnO nanoparticles surface functionalization: new building blocks for dye-sensitized solar cells, *J Am Chem Soc* 132:17910–17920, 2010.

Goldberg DE, Holland JH: Genetic algorithms and machine learning, *Mach Learn* 3:95–99, 1988.

Gradl J: *Experimentelle und theoretische Untersuchungen der Bildungskinetik diffusions- sowie reaktionslimitierter Systeme am Beispiel der Nanopartikelfällung von Bariumsulfat und Zinkoxid,* Doktorarbeit, 2010, Friedrich-Alexander-Universität Erlangen-Nürnberg (FAU).

Gradl J, Peukert W: Simultaneous 3D observation of different kinetic subprocesses for precipitation in a T-mixer, *Chem Eng Sci* 64:709–720, 2009.

Gradl J, Schwarzer H-C, Schwertfirm F, Manhart M, Peukert W: Precipitation of nanoparticles in a T-mixer: coupling the particle population dynamics with hydrodynamics through direct numerical simulation, *Chem Eng Process Process Intensif* 45:908–916, 2006.

Griewank A, Walther A: *Evaluating derivatives: principles and techniques of algorithmic differentiation,* ed 2, Philadelphia, 2008, SIAM.

Gröschel M: *Optimization of particle synthesis—new mathematical concepts for a controlled production of functional nanoparticles,* Doktorarbeit, 2013, Friedrich-Alexander-Universität Erlangen-Nürnberg (FAU).

Gröschel M, Körmer R, Walther M, Leugering G, Peukert W: Process control strategies for the gas phase synthesis of silicon nanoparticles, *Chem Eng Sci* 73:181–194, 2012.

Gröschel M, Peukert W, Leugering G: Modeling, analysis and optimization of particle growth, nucleation and ripening by the way of nonlinear hyperbolic integro-partial differential equations. In Leugering G, Brenner P, Engell S, Griewank A, Harbrecht H, Hinze M, Rannacher R, Ulbrich S, editors: *Trends in PDE constrained optimization,* Switzerland, 2014, Springer International Publishing, pp 468–471.

Gruy F: Population balance for aggregation coupled with morphology changes, *Colloids Surf A Physicochem Eng Asp* 374:69–76, 2011.

Günther L, Peukert W: Control of coating properties by tailored particle interactions: relation between suspension rheology and film structure, *Colloids Surf A Physicochem Eng Asp* 225:49–61, 2003.

Haderlein M, Segets D, Gröschel M, Pflug L, Leugering G, Peukert W: FIMOR: an efficient simulation for ZnO quantum dot ripening applied to the optimization of nanoparticle synthesis, *Chem Eng J* 260:706–715, 2015.

Halbig CE, Nacken TJ, Walter J, Damm C, Eigler S, Peukert W: Quantitative investigation of the fragmentation process and defect density evolution of oxo-functionalized graphene due to ultrasonication and milling, *Carbon* 96:897–903, 2016.

Hartge EU, Pogodda M, Reimers C, Schwier D, Gruhn C, Werther J: SolidSim—a tool for the flowsheet simulation of solids processes, *AT Miner Process* 47:42–51, 2006.

Hartig MAJ, Jacobsen N, Peukert W: Multi-component and multi-phase population balance model: the case of georgeite formation as methanol catalyst precursor phase, *Chem Eng Sci* 109:158–170, 2014.

Hartig MAJ, Peukert W, Jacobsen N, Leuthold A: A model-based precipitation study of copper-based catalysts, *AICHE J* 61:2104–2116, 2015.

Haslinger J, Kočvara M, Leugering G, Stingl M: Multidisciplinary free material optimization, *SIAM J Appl Math* 70:2709–2728, 2010.

Hesselmann A, Görling A: Efficient exact-exchange time-dependent density-functional theory methods and their relation to time-dependent Hartree–Fock, *J Chem Phys* 134:034120, 2011.

Hill M: Product and process design for structured products, *AICHE J* 50:1656–1661, 2004.
Hiptmair R, Paganini A: Shape optimization by pursuing diffeomorphisms, *CMAM* 15:291–305, 2015.
Hirschberger CB, Sukumar N, Steinmann P: Computational homogenization of material layers with micromorphic mesostructure, *Philos Mag* 88:3603–3631, 2008.
Israelachvili JN: *Intermolecular and surface forces,* San Diego, USA, 2011, Academic Press.
Iveson SM, Litster JD, Hapgood K, Ennis BJ: Nucleation, growth and breakage phenomena in agitated wet granulation processes: a review, *Powder Technol* 117:3–39, 2001.
Kaewsaneha C, Tangboriboonrat P, Polpanich D, Eissa M, Elaissari A: Janus colloidal particles: preparation, properties, and biomedical applications, *ACS Appl Mater Interfaces* 5:1857–1869, 2013.
Kayrak-Talay D, Dale S, Wassgren C, Litster J: Quality by design for wet granulation in pharmaceutical processing: assessing models for a priori design and scaling, *Powder Technol* 240:7–18, 2013.
Khalili S, Lin Y, Armaou A, Matsoukas T: Constant number Monte Carlo simulation of population balances with multiple growth mechanisms, *AICHE J* 56:3137–3145, 2010.
Kind M: Product engineering, *Chem Eng Process* 38:405–410, 1999.
Kirchhof MJ, Förster H, Schmid H-J, Peukert W: Sintering kinetics and mechanism of vitreous nanoparticles, *J Aerosol Sci* 45:26–39, 2012.
Klaumünzer M, Distaso M, Hübner J, et al: ZnO superstructures via oriented aggregation initiated in a block copolymer melt, *CrystEngComm* 16:1502–1513, 2014.
Knieke C, Sommer M, Peukert W: Identifying the apparent and true grinding limit, *Powder Technol* 195:25–30, 2009.
Knieke C, Romeis S, Peukert W: Influence of process parameters on breakage kinetics and grinding limit at the nanoscale, *AICHE J* 57:1751–1758, 2011.
Körmer R, Schmid H-J, Peukert W: Aerosol synthesis of silicon nanoparticles with narrow size distribution—part 2: theoretical analysis of the formation mechanism, *J Aerosol Sci* 41:1008–1019, 2010a.
Körmer R, Jank MPM, Ryssel H, Schmid H-J, Peukert W: Aerosol synthesis of silicon nanoparticles with narrow size distribution—part 1: experimental investigations, *J Aerosol Sci* 41:998–1007, 2010b.
Krasnyk M, Mangold M, Ganesan S, Tobiska L: Numerical reduction of a crystallizer model with internal and external coordinates by proper orthogonal decomposition, *Chem Eng Sci* 70:77–86, 2012.
Krekel J, Polke R: Qualitätssicherung bei der Verfahrensentwicklung, *Chem Ing Technol* 64:528–535, 1992.
Kruis FE, Kusters KA, Pratsinis SE, Scarlett B: A simple model for the evolution of the characteristics of aggregate particles undergoing coagulation and sintering, *Aerosol Sci Technol* 19:514–526, 1993.
Kwade A: Wet comminution in stirred media mills—research and its practical application, *Powder Technol* 105:14–20, 1999.
Kwon JSI, Nayhouse M, Christofides PD, Orkoulas G: Modeling and control of crystal shape in continuous protein crystallization, *Chem Eng Sci* 107:47–57, 2014.
Lal S, Link S, Halas NJ: Nano-optics from sensing to waveguiding, *Nat Photonics* 1:641–648, 2007.
La Mer V, Dinegar RH: Theory, prediction and mechanism of formation of monodispersed hydrosols, *J Am Chem Soc* 72:4847–4854, 1950.
Landau DP, Binder K: *A guide to Monte Carlo simulations in statistical physics,* ed 3, Cambridge, UK, 2009, Cambridge University Press.
Lee KW, Liu BYH: Theoretical study of aerosol filtration by fibrous filters, *Aerosol Sci Technol* 1:147–161, 1982.
Li J: Mesoscales: the path to transdisciplinarity, *Chem Eng J* 277:112–115, 2015.

Lin MY, Lindsay HM, Weitz DA, Ball RC, Klein R, Meakin P: Universality in colloidal aggregation, *Nature* 339:360–362, 1989.
Lin W, Walter J, Burger A, et al: A general approach to study the thermodynamics of ligand adsorption to colloidal surfaces demonstrated by means of catechols binding to zinc oxide quantum dots, *Chem Mater* 27:358–369, 2015.
Lipovskii A, Kolobkova E, Petrikov V, et al: Synthesis and characterization of PbSe quantum dots in phosphate glass, *Appl Phys Lett* 71:3406–3408, 1997.
Liu Y, Gibbs M, Puthussery J, et al: Dependence of carrier mobility on nanocrystal size and ligand length in PbSe nanocrystal solids, *Nano Lett* 10:1960–1969, 2010.
Lu W, Lieber C: Nanoelectronics from the bottom up, *Nat Mater* 6:841–850, 2007.
Lu L, Xu J, Ge W, Yue Y, Liu X, Li J: EMMS-based discrete particle method (EMMS-DPM) for simulation of gas–solid flows, *Chem Eng Sci* 120:67–87, 2014.
Madadi-Kandjani E, Passalacqua A: An extended quadrature-based moment method with log-normal kernel density functions, *Chem Eng Sci* 131:323–339, 2015.
Målqvist A: Multiscale methods for elliptic problems, *Multiscale Model Simul* 9:1064–1086, 2011.
Mann S, Ozin GA: Synthesis of inorganic materials with complex form, *Nature* 382:313–318, 1996.
Marchisio DL, Fox RO: *Computational models for polydisperse and particulate and multiphase systems,* New York, 2013, Cambridge University Press.
Marchisio DL, Pikturna JT, Fox RO, Vigil D: Quadrature method of moments for population-balance equations, *AICHE J* 49:1266–1276, 2003.
McMillan J, Shaffer F, Gopalan B, et al: Particle cluster dynamics during fluidization, *Chem Eng Sci* 100:39–51, 2013.
Meakin P: A historical introduction to computer models for fractal aggregates, *J Sol–Gel Sci Technol* 15:97–117, 1999.
Mehringer C, Wagner R, Jakuttis T, Butz B, Spiecker E, Peukert W: Gas phase synthesis of anisotropic silicon germanium hybrid nanoparticles, *J Aerosol Sci* 67:119–130, 2014.
Meier M, John E, Wieckhusen D, Wirth W, Peukert W: Generally applicable breakage functions derived from single particle comminution data, *Powder Technol* 194:33–41, 2009a.
Meier M, John E, Wieckhusen D, Wirth W, Peukert W: Influence of mechanical properties on impact fracture: prediction of the milling behaviour of pharmaceutical powders by nanoindentation, *Powder Technol* 188:301–313, 2009b.
Menz WJ, Shekar S, Brownbridge GPE, et al: Aerosol synthesis of silicon nanoparticles with narrow size distribution: a theoretical study, *J Aerosol Sci* 44:46–61, 2011.
Mersmann A: *Crystallization technology handbook,* New York, Basel, 2001, Marcel Dekker, Inc.
Messing GL, Zhang S-C, Jayanthi GV: Ceramic powder synthesis by spray pyrolysis, *J Am Ceram Soc* 76:2707–2726, 1993.
Michalewicz Z, Schoenauer M: Evolutionary algorithms for constrained parameter optimization problems, *Evol Comput* 4:1–32, 1996.
Miesbauer O, Götzinger M, Peukert W: Molecular dynamics simulations of the contact between two NaCl nano-crystals: adhesion, jump to contact an indentation, *Nanotechnology* 14:371–376, 2003.
Milek T, Zahn D: Molecular simulation of Ag nanoparticle nucleation from solution: redox-reactions direct the evolution of shape and structure, *Nano Lett* 14:4913–4917, 2014.
Miller RE, Tadmor EB: The quasicontinuum method: overview, applications and current directions, *J Computer-Aided Mater Des* 9:203–239, 2002.
Miller RE, Tadmor EB: A unified framework and performance benchmark of fourteen multiscale atomistic/continuum coupling methods, *Model Simul Mater Sci Eng* 17:053001, 2009.

Min Y, Akbulut M, Kristiansen K, Golan Y, Israelachvili J: The role of interparticle and external forces in nanoparticle assembly, *Nat Mater* 7:527–538, 2008.
Moreels I, Lambert K, De Muynck D, et al: Composition and size-dependent extinction coefficient of colloidal PbSe quantum dots, *Chem Mater* 19:6101–6106, 2007.
Mothes H, Löffler F: Prediction of particle removal in cyclone separators, *Int Chem Eng* 28:231–240, 1988.
Negishi E: Magical power of transition metals: past, present, and future, *Angew Chem Int Ed Engl* 50:6738–6764, 2011.
Nelder JA, Mead R: A simplex method for function minimization, *Comput J* 7:308–313, 1965.
Okuyama K, Lenggoro IW: Preparation of nanoparticles via spray route, *Chem Eng Sci* 58:537–547, 2003.
Özpinar GA, Peukert W, Clark T: An improved generalized AMBER force field (GAFF) for urea, *J Mol Model* 16:1427–1440, 2010.
Palma C-A, Cecchini M, Samori P: Predicting self-assembly: from empiricism to determinism, *Chem Soc Rev* 41:3713–3730, 2012.
Park J, Joo J, Kwon SG, Jang Y, Hyeon T: Synthesis of monodisperse spherical nanocrystals, *Angew Chem Int Ed Engl* 46:4630–4660, 2007.
Parteli EJR, Schmidt J, Blümel C, Wirth K-E, Peukert W, Pöschel T: Attractive particle interaction forces and packing density of fine glass powders, *Sci Rep* 4:6227, 2014.
Patruno LE, Dorao CA, Svendsen HF, Jakobsen HA: Analysis of breakage kernels for population balance modelling, *Chem Eng Sci* 64:501–508, 2009.
Paul J, Romeis S, Tomas J, Peukert W: A review of models for single compression and their application to silica microspheres, *Adv Powder Technol* 25:136–153, 2014.
Paul J, Romeis S, Mačković M, et al: In situ cracking of silica beads in the SEM and TEM—effect of particle size on structure–property relations, *Powder Technol* 270:337–347, 2015.
Paulick M, Morgeneyer M, Kwade A: A new method for the determination of particle contact stiffness, *Granul Matter* 17:83–93, 2014.
Peev G, Zambov L, Shanov V, Tserovski L: Kinetics of polysilicon film growth by thermal decomposition of silane, *Semicond Sci Technol* 6:281–286, 1991.
Penn RL, Banfield JF: Imperfect oriented attachment: dislocation generation in defect-free nanocrystals, *Science* 281:969–971, 1998.
Peters W, Eypasch M, Frank T, et al: Efficient hydrogen release from perhydro-N-ethylcarbazole using catalyst-coated metallic structures produced by selective electron beam melting, *Energy Environ Sci* 8:641–649, 2015.
Petersen EL, Crofton MW: Measurements of high-temperature silane pyrolysis using $SiH_4$ IR emission and $SiH_2$ laser absorption, *J Phys Chem A* 107:10988–10995, 2003.
Peukert W: Material properties of fine grinding, *Int J Miner Process* 74:S3–S17, 2004.
Peukert W, Schwarzer H-C, Götzinger M, et al: Control of particle interfaces—the critical issue in nanoparticle technology, *Adv Powder Technol* 14:411–426, 2003.
Peukert W, Schwarzer HC, Stenger F: Control of aggregation in production and handling of nanoparticles, *Chem Eng Process* 44:245–252, 2005.
Peukert W, Segets D: Funktionale Partikelsysteme interdisziplinär erforschen, *Nanotechnologie aktuell* 9:1–9, 2015.
Pietryga JM, Schaller RD, Werder D, et al: Pushing the band gap envelope: mid-infrared emitting colloidal PbSe quantum dots, *J Am Chem Soc* 126:11752–11753, 2004.
Portillo PM, Ierapetritou MG, Muzzio FJ: Characterization of continuous convective powder mixing processes, *Powder Technol* 182:368–378, 2008.
Pratsinis SE: Aerosol-based technologies in nanoscale manufacturing: from functional materials to devices through core chemical engineering, *AICHE J* 56:3028–3035, 2010.
Rahmanian N, Naji A, Ghadiri M: Effects of process parameters on granules properties produced in a high shear granulator, *Chem Eng Res Des* 89:512–518, 2011.

Ramkrishna D: *Population balances: theory and applications to particulate systems in engineering*, San Diego, 2000, Academic Press.

Reiss P, Protière M, Li L: Core/shell semiconductor nanocrystals, *Small* 5:154–168, 2009.

Renze P, Buffo A, Marchisio DL, Vanni M: Simulation of coalescence, breakup, and mass transfer in polydisperse multiphase flows, *Chem Ing Technol* 86:1088–1098, 2014.

Reynolds GK, Fu JS, Cheong YS, Hounslow MJ, Salman AD: Breakage in granulation: a review, *Chem Eng Sci* 60:3969–3992, 2005.

Rogach AL, Talapin DV, Shevchenko EV, Kornowski A, Haase M, Weller H: Organization of matter on different size scales: monodisperse nanocrystals and their superstructures, *Adv Funct Mater* 12:653–664, 2002.

Roke S, Gonella G: Nonlinear light scattering and spectroscopy of particles and droplets in liquids, *Annu Rev Phys Chem* 63:353–378, 2012.

Romeis S: From single particle compression testing to chemical influences of the solvent on wet grinding, *Doktorarbeit* 2015.

Romeis S, Paul J, Ziener M, Peukert W: A novel apparatus for in situ compression of submicron structures and particles in a high resolution SEM, *Rev Sci Instrum* 83:095105, 2012.

Romeis S, Paul J, Hanisch M, et al: Correlation of enhanced strength and internal structure for heat-treated submicron Stöber silica particles, *Part Part Syst Charact* 31:664–674, 2014.

Romeis S, Paul J, Herre P, De Ligny D, Schmidt J, Peukert W: Local densification of a single micron sized silica by uniaxial compression, *Scr Mater* 108:84–87, 2015.

Rumpf H: Über die Eigenschaften von Nutzstäuben, *Staub-Reinhalt der Luft* 27:3–13, 1967.

Rupich SM, Shevchenko EV, Bodnarchuk MI, Lee B, Talapin DV: Size-dependent multiple twinning in nanocrystal superlattices, *J Am Chem Soc* 132:289–296, 2010.

Saad T, Abboud AW, Smith ST, Ring TA: A class of exact solutions for population balances with arbitrary internal coordinates, *AICHE J* 61:1691–1698, 2015.

Sanimi A, Hassanpour A, Ghadiri M: Single and bulk compressions of soft granules: experimental study and DEM evaluation, *Chem Eng Sci* 60:3993–4004, 2005.

Sashchiuk A, Langof L, Chaim R, et al: Synthesis and characterization of PbSe and PbSe/PbS core-shell colloidal nanocrystals, *J Cryst Growth* 240:431–438, 2001.

Sauerbeck C, Braunschweig B, Peukert W: Surface charging and interfacial water structure of amphoteric colloidal particles, *J Phys Chem C* 118:10033–10042, 2014.

Schäfer S, Srikantharajah R, Klaumünzer M, Lobaz V, Voigt M, Peukert W: Self-alignment of zinc oxide nanorods into a 3D-smectic phase, *Thin Solid Films* 562:659–667, 2014.

Shevchenko EV, Talapin DV, Kotov NA, O'Brien S, Murray CB: Structural diversity in binary nanoparticle superlattices, *Nature* 55–59, 2006.

Schindler T, et al: A combined SAXS/SANS study for the in situ characterization of ligand shells on small nanoparticles—the case of ZnO, *Langmuir* 31:10130–10136, 2015a.

Schindler, T., Walter, J., Peukert, W., Segets, D., Unruh, T: In situ study on the evolution of the multimodal particle size distribution of ZnO quantum dots: some general growth rules. J. Phys. Chem. B. 2015b. http://dx.doi.org/10.1021/acs.jpcb.5b08005. In press.

Schlauch E, Ernst M, Seto R, Briesen H, Sommer M, Behr M: Comparison of three simulation methods for colloidal aggregates in Stokes flow: finite elements, lattice Boltzmann and Stokesian dynamics, *Comput Fluids* 86:199–209, 2013.

Schlundt S, Kuzmanich G, Spänig F, et al: Dendritic porphyrin-fullerene conjugates: efficient light-harvesting and charge-transfer events, *Chem Eur J* 15:12223–12233, 2009.

Schmidt J, Plata M, Tröger S, Peukert W: Production of polymer particles below 5 μm by wet grinding, *Powder Technol* 228:84–90, 2012.

Schmidt J, Sachs M, Blümel C, et al: A novel process route for the production of spherical LBM polymer powders with small size and good flowability, *Powder Technol* 261:78–86, 2014.

Schröder-Turk GE, et al: Minkowski tensor shape analysis of cellular granular and porous structures, *Adv Mater* 23:2535–2553, 2011.

Schürer B, Wunderlich S, Sauerbeck C, Peschel U, Peukert W: Probing colloidal interfaces by angle-resolved second harmonic light scattering, *Phys Rev B* 82:241401, 2010.

Schürer B, Elser MJ, Sternig A, Peukert W, Diwald O: Delamination and dissolution of titanate nanowires: a combined structure and in situ second harmonic generation study, *J Phys Chem C* 115:12381–12387, 2011.

Schuchmann H, Schubert H: Emulsification in high-pressure homogenizers, *Chem Eng Technol* 24:151–157, 2001.

Schwager T, Pöschel T: Coefficient of restitution and linear-dashpot model revisited, *Granul Matter* 9:465–469, 2007.

Schwarzer HC, Peukert W: Experimental investigation into the influence of mixing on nanoparticle precipitation, *Chem Eng Technol* 25:657–661, 2002.

Segets D, Gradl J, Klupp Taylor R, Vassilev V, Peukert W: Analysis of optical absorbance spectra for the determination of ZnO nanoparticle size distribution, solubility and surface energy, *ACS Nano* 3:1703–1710, 2009.

Segets D, Hartig MAJ, Gradl J, Peukert W: A population balance model of quantum dot formation: oriented growth and ripening of ZnO, *Chem Eng Sci* 70:4–13, 2012a.

Segets D, Lucas JM, Klupp Taylor RN, et al: Determination of the quantum dot bandgap dependence on particle size from optical absorbance and transmission electron microscopical measurements, *ACS Nano* 6:9021–9032, 2012b.

Segets D, Lutz C, Yamamoto K, et al: Classification of zinc sulfide quantum dots by size: insights into the particle surface-solvent interaction of colloids, *J Phys Chem C* 119:4009–4022, 2015.

Segets D, Marczak R, Schäfer S, et al: Experimental and theoretical studies of the colloidal stability of nanoparticles—a general interpretation based on stability maps, *ACS Nano* 5:4658–4669, 2011.

Semmler J, Pflug L, Leugering G: Shape optimization in electromagnetic applications. In Leugering G, Pratelli A, editors: *New trends in shape optimization*, Basel, 2015, Birkhäuser.

Shekar S, Sander M, Riehl RC, Smith AJ, Braumann A, Kraft M: Modelling the flame synthesis of silica nanoparticles from tetraethoxysilane, *Chem Eng Sci* 70:54–66, 2012.

Shi D, El-Farra NH, Li M, Mhaskar P, Christofides PD: Predictive control of particle size distribution in particulate processes, *Chem Eng Sci* 61:268–281, 2006.

Sommer K: *Sampling of powders and bulk materials,* Heidelberg, 1986, Springer-Verlag Berlin.

Sommer M, Stenger F, Peukert W, Wagner NJ: Agglomeration and breakage of nanoparticles in stirred media mills—a comparison of different methods and models, *Chem Eng Sci* 61:135–148, 2006.

Song R-Q, Cölfen H: Mesocrystals—ordered nanoparticle superstructures, *Adv Mater* 22:1301–1330, 2010.

Sontag ED: *Mathematical control theory: deterministic finite dimensional systems,* New York, 1998, Springer.

Soos M, Sefcik J, Morbidelli M: Investigation of aggregation, breakage and restructuring kinetics of colloidal dispersions in turbulent flows by population balance modeling and static light scattering, *Chem Eng Sci* 61:2349–2363, 2006.

Sorensen CM: The mobility of fractal aggregates: a review, *Aerosol Sci Technol* 45:755–769, 2011.

Sotiriou GA, Hirt AM, Lozach P-Y, Teleki A, Krumeich F, Pratsinis SE: Hybrid, silica coated, Janus-like plasmonic–magnetic nanoparticles, *Chem Mater* 23:1985–1992, 2011.

Steckel JS, Coe-Sullivan S, Bulovic V, et al: 1.3 μm to 1.55 μm tunable electroluminescence from PbSe quantum dots embedded within an organic device, *Adv Mater* 15:1862–1866, 2003.

Steed JW, Atwood JL: *Supramolecular chemistry,* 2009, John Wiley and Sons. online text. ISBN:0470740892, 9780470740897.

Stenger F, Peukert W: The role of particle interactions on suspension rheology—application to submicron grinding in stirred ball mills, *Chem Eng Technol* 26:177–183, 2003.

Stöber W, Fink A, Bohn E: Controlled growth of monodisperse silica particles in the micron size range, *J Colloid Interface Sci* 26:62–69, 1968.

Sugimoto T: *Monodispersed particles,* ed 1, Amsterdam, 2001, Elsevier Science B.V.

Sutkar VS, Deen NG, Padding JT, et al: A novel approach to determine wet restitution coefficients through a unified correlation and energy analysis, *AICHE J* 61:769–779, 2015.

Suzuki A: Cross-coupling reactions of organoboranes: an easy way to construct C–C bonds (nobel lecture), *Angew Chem Int Ed Engl* 50:6723–6737, 2011.

Talapin DV, Lee J-S, Kovalenko MV, Shevchenko EV: Prospects of colloidal nanocrystals for electronic and optoelectronic applications, *Chem Rev* 110:389–458, 2010.

Tao A, Habas S, Yang P: Shape control of colloidal metal nanocrystals, *Small* 4:310–325, 2008.

Taylor RK, Seifrt F, Zhuromskyy O, Peschel U, Leugering G, Peukert W: Painting by numbers: nanoparticle-based colorants in the post-empirical age, *Adv Mater* 23:2554–2570, 2011.

Tenne R: Inorganic nanotubes and fullerene-like nanoparticles, *Nat Nanotechnol* 1:103–111, 2006.

Thekale A: *Trust-region methods for simulation based nonlinear optimization,* Doktorarbeit, 2011, Friedrich-Alexander-Universität Erlangen-Nürnberg (FAU).

Tomas J: Product design of cohesive powders—mechanical properties, compression and flow behavior, *Chem Eng Technnol* 27:605–618, 2004.

Toneva P, Wirth K-E, Peukert W: Grinding in an air classifier mill—part II: characterisation of the two-phase flow, *Powder Technol* 211:28–37, 2011a.

Toneva P, Epple P, Breuer M, Peukert W, Wirth K-E: Grinding in an air classifier mill—part I: characterisation of the one-phase flow, *Powder Technol* 211:19–27, 2011b.

Torczon V, Trosset MW: From evolutionary operation to parallel direct search: pattern search algorithms for numerical optimization, *Comput Sci Stat* 29:396–401, 1998.

Torquato S, editor: *Random heterogeneous materials: microstructure and macroscopic properties,* New York, 2002, Springer-Verlag.

Vanarase AU, Alcalà M, Jerez Rozo JI, Muzzio FJ, Romañach RJ: Real-time monitoring of drug concentration in a continuous powder mixing process using NIR spectroscopy, *Chem Eng Sci* 65:5728–5733, 2010.

Van Rijssel J, Peters VFD, Meeldijk JD, et al: Size-dependent second virial coefficients of quantum dots from quantitative cryogenic electron microscopy, *J Phys Chem B* 118:11000–11005, 2014.

Vassilev V, Gröschel M, Schmid H-J, Peukert W, Leugering G: Interfacial energy estimation in a precipitation reaction using the flatness based control of the moment trajectories, *Chem Eng Sci* 65:2183–2189, 2010.

Vieweg BF, Butz B, Peukert W, Klupp Taylor RN, Spiecker E: TEM preparation method for site- and orientation-specific sectioning of individual anisotropic nanoparticles based on shadow-FIB geometry, *Ultramicroscopy* 113:165–170, 2012.

Viswanatha R, Sapra S, Satpati B, Satyam PV, Dev BN, Sarma DD: Understanding the quantum size effects in ZnO nanocrystals, *J Mater Chem* 14:661–668, 2004.

Vogel L, Peukert W: Breakage behaviour of different materials—construction of a master curve for the breakage probability, *Powder Technol* 129:101–110, 2003.

Vogel N, Retsch M, Fustin C-A, Del Campo A, Jonas U: Advances in colloidal assembly: the design of structure and hierarchy in two and three dimensions, *Chem Rev* 115:6265–6311, 2015.

Vollmer U, Raisch J: Control of batch crystallization—a system inversion approach, *Chem Eng Process: Process Intensif* 45:874–885, 2006.
Vozdecky P, Roosen A, Knieke C, Peukert W: Direct tape casting of nanosized $Al_2O_3$ slurries derived from autogeneous nanomilling, *J Am Ceram Soc* 93:1313–1319, 2010.
Walker SW: *The shapes of things: a practical guide to differential geometry and the shape derivative,* Philadelphia, 2015, SIAM.
Walter J, Löhr K, Karabudak E, et al: Multidimensional analysis of nanoparticles with highly disperse properties using multiwavelength analytical ultracentrifugation, *ACS Nano* 8:8871–8886, 2014.
Walter J, Thajudeen T, Süß S, Segets D, Peukert W: New possibilities of accurate particle characterisation by applying direct boundary models to analytical centrifugation, *Nanoscale* 7:6574–6587, 2015a.
Walter J, Nacken TJ, Thajudeen T, Eigler S, Peukert W: Determination of the lateral dimension of graphene oxide nanosheets using analytical ultracentrifugation, *Small* 11:814–825, 2015b.
Walter J, Sherwood PJ, Lin W, Segets D, Stafford WF, Peukert W: Simultaneous analysis of hydrodynamic and optical properties using analytical ultracentrifugation equipped with multiwavelength detection, *Anal Chem* 87:3396–3403, 2015c.
Walther A, Müller AHE: Janus particles: synthesis, self-assembly, physical properties, and applications, *Chem Rev* 113:5194–5261, 2013.
Walzel P: Zerstäuben von Flüssigkeiten, *Chem Ing Tech* 62:983–994, 1990.
Wang J, Chen DR, Pui DYH: Modeling of filtration efficiency of nanoparticles in standard filter media, *J Nanopart Res* 9:109–115, 2007.
Wang J, Kim SC, Pui DYH: Investigation of the figure of merit for filters with a single nanofiber layer on a substrate, *J Aerosol Sci* 34:323–334, 2008.
Wang W, Lu B, Zhang N, Shi Z, Li J: A review of multiscale CFD for gas–solid CFB modeling, *Int J Multiphase Flow* 36:109–118, 2010.
Wasserscheid P, Keim W: Ionic liquids—new 'solutions' for transition metal catalysis, *Angew Chem Int Ed Engl* 39:3773–3789, 2000.
Weidmann MC, Beck ME, Hoffmann RS, Prins F, Tisdale WA: Monodisperse, air-stable PbS nanocrystals via precursor stoichiometry control, *ACS Nano* 8:6363–6371, 2014.
Weinan E, Engquist B, Li X, Ren W, Vanden-Eijnden E: Heterogeneous multiscale methods: a review, *Comput Phys Commun* 2:367–450, 2007.
Whitesides GM, Boncheva M: Beyond molecules: self assembly of mesoscopic and macroscopic components, *Proc Natl Acad Sci USA* 99:4769–4774, 2002.
Witt W, Köhler U, List J: Hochgeschwindigkeits-Bildanalyse zur Charakterisierung von Partikelgröße und -form, *Produktgestaltung in der Partikeltechnologie* 6:213–239, 2013.
Wunderlich S, Schürer B, Sauerbeck C, Peukert W, Peschel U: Molecular Mie model for second harmonic generation and sum frequency generation, *Phys Rev B* 84:235403, 2011.
Xiong Q, et al: Large-scale DNS of gas–solid flows on Mole-8.5, *Chem Eng Sci* 71:422–430, 2012.
Yow HN, Pitt MJ, Salman AD: Drag correlations for particles of regular shape, *Adv Powder Technol* 16:363–372, 2005.
Zhang Z, Ghadiri M: Impact attrition of particulate solids. Part 2: experimental work, *Chem Eng Sci* 57:3671–3686, 2002.
Zheng J, Papajak E, Truhlar DG: Phase space prediction of product branching ratios: canonical competitive nonstatistical model, *J Am Chem Soc* 131:15754–15760, 2009.
Zhu HP, Yang RY, Yu AB: Discrete particle simulation of particulate systems: a review of major applications and findings, *Chem Eng Sci* 63:5728–5770, 2008.

CHAPTER TWO

# Multiscale Analysis of a Coating Process in a Wurster Fluidized Bed Apparatus

**Stefan Heinrich*[,1], Maksym Dosta*, Sergiy Antonyuk[†]**
*Institute of Solids Process Engineering and Particle Technology, Hamburg University of Technology, Hamburg, Germany
[†]Chair of Particle Process Engineering, University of Kaiserslautern, Kaiserslautern, Germany
[1]Corresponding author: e-mail address: stefan.heinrich@tuhh.de

## Contents

1. Introduction 85
2. Multiscale Simulation Framework 88
   2.1 Macroscale 90
   2.2 Mesoscale 92
   2.3 Microscale 98
   2.4 Submicroscale 101
3. Experimental Estimation of the Material Parameters 103
   3.1 Studied Materials 103
   3.2 Compression Test of Particles 104
   3.3 Shear Test of Particles 109
   3.4 Collision Tests of Particles 111
4. Analysis of the Process Behavior 113
   4.1 Apparatus Geometry 113
   4.2 Process and Material Parameters 115
   4.3 Results of the Microscale Simulation 117
   4.4 Mesoscale Model 127
   4.5 Macroscale Calculations 130
5. Conclusions 131
Acknowledgments 132
References 133

## Abstract

This contribution outlines a multiscale simulation approach for analysis of a Wurster coating process occurring in a fluidized bed. The processes occurring in the apparatus are described on four different time and length scales: The Discrete Element Method coupled with Computational Fluid Dynamics, where each particle is considered as a separate entity and its motion in fluid field is calculated, play a central role in the modeling framework. On the macroscale, the Population Balance Model describes the particle

*Advances in Chemical Engineering*, Volume 46
ISSN 0065-2377
http://dx.doi.org/10.1016/bs.ache.2015.10.012

growth. The nozzle zone model calculates the wetting and overspray rate on the mesoscale. The model on the lowest scale predicts the impact behavior of wetted particles. The influence of key processing parameters, such as gap distance or atomization air flow, on the particle-fluid dynamics during coating process is analyzed. In order to get material parameters, such as coefficient of restitution and friction, a set of experimental investigations of a model material is performed.

## ABBREVIATIONS

***Ar*** Archimedes number (–)
***A*** surface ($m^2$)
$\boldsymbol{B_{nuc}}$, $\boldsymbol{B_{agg}}$, $\boldsymbol{B_{break}}$ particle birth rate due to nucleation, aggregation, or breakage (1/s)
$\boldsymbol{D_{nuc}}$, $\boldsymbol{D_{agg}}$, $\boldsymbol{D_{break}}$ particle death rate due to nucleation, aggregation, or breakage (1/s)
***D*** diameter (m)
***e*** restitution coefficient (–)
***E*** modulus of elasticity (Pa)
***F*** force (N)
***G*** shear modulus (Pa)
**Gr** growth rate (m/s)
***h*** height (m)
$\boldsymbol{K_{dust}}$ part of overspray, which remains in apparatus (–)
$\boldsymbol{K_{os}}$ overspray part of suspension (–)
$\boldsymbol{L_{film}}$ thickness of liquid film (m)
***m*** mass (kg)
$\boldsymbol{\dot{M}_{sol}}$ solution mass flow (kg/s)
***R*** particle radius (m)
***Re*** Reynolds number (–)
***Sc*** Schmidt number (–)
***Sh*** Sherwood number (–)
***V*** volume ($m^3$)
***Y*** humidity (kg/kg)
$\boldsymbol{x_w}$ water part of suspension (–)
$\boldsymbol{\Sigma}$ surface tension (N/m)
$\boldsymbol{\mu}$ dynamic viscosity (Pa s)
$\boldsymbol{v}$ velocity (m/s)
$\boldsymbol{v}$ kinematic viscosity of gas ($m^2$/s)
$\boldsymbol{\rho}$ density (kg/$m^3$)

## SUBSCRIPTS

**cap** capillary
**D** drag
**drop** droplet
**g** gas
**lb** liquid bridge
**n** normal direction

**p** particle
**sat** saturation
**sol** solution
**t** tangential direction
**tot** total
**vap** vapor
**vis** viscous

## 1. INTRODUCTION

The fluidized bed (FB) process allows the integration of diverse subprocesses like particle formulation, wetting, homogenization, separation, or drying into one single step. This technology is widely used for many applications, such as encapsulation, coating, granulation, or agglomeration (Mörl et al., 2007). In an FB coating process, a coating solution or suspension is injected into the apparatus and the forming liquid droplets are deposited on solid particles such as pellets or mini-tablets. Further drying of the solids containing liquid film leads to layer-wise particle growth. The main advantages of FB technology are: liquid-like behavior, which leads to easy handling and particles transport; intensive solids mixing and thus intensive heat, mass, and momentum transfer between all phases, which results in the uniform temperature distribution; large solid-gas exchange area by virtue of small solids grain size. However, some disadvantages are related to equipment required because of solids entrainment by fluidizing gas, complexity with the scale-up, attrition of solid material and erosion of internals and apparatus walls (Werther, 2007). Two more challenges regarding the usage of the FB is the nonuniform residence time of solids, which may occur especially in the continuous operation mode and difficult maintaining of stable fluidization of cohesive powders, which usually show channeling.

The existing FB apparatuses, which are used for coating of particles, can be classified into four main categories, such as: bubbling bed, rotated bed, spouted bed, and Wurster. According to the position of the spraying nozzle, the systems are classified in top, bottom, and side (tangential) spray. The Wurster-coater apparatus, which is investigated in this contribution, is commonly used in the pharmaceutical industry for coating of particles smaller than 1 mm and tablets. In order to create a circulating flow pattern, Wurster (1966) has proposed to install a vertical draft tube into the apparatus, which enables operation with a large bed height and small particles. The perforated plate has larger orifices in the tube region than in the annulus. Due a larger flow velocity in the tube as in the outer bed range, the particles are moved into the tube by the Venturi effect and will be transported upwards forming a spout. During upward motion of particles within the tube, they pass through the nozzle zone and are covered with atomized droplets which coalescence to form a liquid film. In the expansion area of the apparatus, which typically has a conical form, the particle velocity is decreased and solid material is circulates back to the coating zone.

Despite the relatively simple principle of the coating process, efficient operation of an industrial FB apparatus is not a trivial task. Regardless of the particle size distribution (PSD), the apparatus should provide uniformly coated particles with desired coating thickness, which can be in the order of 10 μm for some controlled release formulations. There exist several main pitfalls, which can arise during coating in an FB:

- *Disruption of fluidization behavior*: on the one hand, the gas flow should be large enough to bring particles into a fluid-like state and to provide particle recirculation. On the other hand, if the flow velocity is too large, it can disrupt particle recirculation.
- *Formation of stagnation zones*: in the case of the incorrect particle dynamics, stagnation zones can be formed in the bottom part of the annulus, which can lead to nonuniform coating.
- *Undesired particle agglomeration*: the high thickness and viscosity of the liquid film on the surface of contacted particles and the small collision velocities and sizes of particles can lead to the their sticking and formation of agglomerates due to the total loss of the impact energy by capillary and viscous forces.
- *Large overspray ratio*: a part of the liquid droplets are dried before contact with solid particles. These droplets can elutriate from the apparatus as dust or can serve as source for new nuclei.

In recent years, various techniques have been employed to experimentally investigate particle dynamics in FBs. A nonintrusive optical technique to follow single particles in the fountain region of Wurster coater was developed by Karlsson et al. (2006). As it will be shown in the next sections, there is a good agreement between experimental results (Karlsson et al., 2006) and numerical results obtained in this contribution. Perpar et al. have shown that the analysis of pressure drop fluctuations in the draft tube can be applied to control quality of the final product (Perpar et al., 2013). To analyze spatially distributed impact rates of particles inside the draft tube, Luštrik et al. have used a piezoelectric probe and have found that the gap size has an influence on height-dependent impact frequencies (Luštrik et al., 2013). The technique based on positron emission particle tracking (PEPT) has been used by Li et al. (2015a,b) to determine the cycle and residence times of particles in different regions of a Wurster FB. Their experimental results show that with increasing batch size cycle time decreases and that in binary mixtures the smaller particles have a shorter cycle time. Another experimental technique, such as electrical capacitance tomography, has been applied to measure particle concentrations in a Wurster FB (Ge et al., 2014).

However, despite such a wide distribution of different experimental techniques as listed above, the experimental investigations are cost intensive and in many cases do not give a full overview of processes occurring in the apparatus. Therefore, a lot of research work has been done in the area of modeling the Wurster coating process. Toschkoff and Khinast (2013) have proposed to classify all simulation methods applied for modeling of a coating process into four main groups:

- General analytical modeling
- Multiphase Computational Fluid Dynamics (CFD) simulations
- Discrete Particle Modeling (DPM): coupling between Discrete Element Method (DEM) and CFD.
- Population Balance Modeling (PBM) and stochastic simulations.

The PBM is one of the widely used approaches to predict particle formulation in FBs (Dosta, 2013; Heinrich et al., 2003; Litster et al., 1995; Ramkrishna, 2000). The main challenge of the usage of the PBM is the determination of parameters, which define the process kinetics, such as size-dependent growth rate, aggregation or breakage kernel, nucleation rate. Nevertheless, the PBM does not allow an apriori exact prediction of the rate processes on the investigated parameter domain. Contrary to this, the DPM is an effective deterministic approach to predict particle dynamics in laboratory-scale apparatuses (Fries et al., 2011b; Li et al., 2015a,b). In DPM, each particle is tracked individually and the material parameters can be considered properly. However, due to the high-computational effort, caused by the usage of the DEM, only a relatively small amount of particles can be simulated on a short time interval. From the computational point of view, the multiphase CFD calculations allow much faster simulations compared to DPM. Karlsson et al. (2009) have extended CFD to include heat and mass transfer between particles and consider liquid injection in a Wurster coater. Rajniak et al. (2009) have combined CFD and PBM to predict agglomeration of particles. Nonetheless, in CFD calculations the Eulerian multiphase model cannot describe interparticle friction and their interactions correctly.

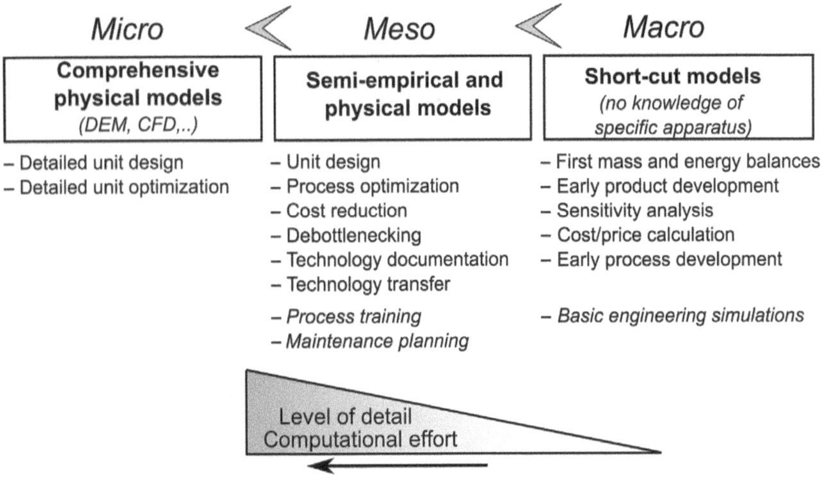


**Figure 1** Models classification according to application area.

Another class of methods, which is not listed in the classification proposed by Toschkoff and Khinast (2013) and which can be effectively applied for modeling of industrial scale FBs, is a multiscale simulation approach. This methodology implies a combination of submodels and computational methods applied for process simulation on different time and length scales. According to Werther et al. (2011), the models of different apparatuses and processes in the solids industry can be distinguished by detailing levels and application purposes, whereby each level can only be applied to some specific application of modeling Fig. 1.

On the one hand, the combination between several submodels allows simulations of large-scale apparatuses, on the other hand, material microproperties and specific apparatus geometry can be considered properly. Many attempts have been done to apply this strategy onto granulation processes (Barasso et al., 2015; Bouffard et al., 2012; Dosta et al., 2012; Ingram and Cameron, 2005). In this contribution we want build on this topic by applying it to multiscale simulation and optimization of a coating process occurring in an FB apparatus with installed draft tube.

## 2. MULTISCALE SIMULATION FRAMEWORK

For a detailed description of the complex FB process, the particle dynamics on the micro- and macrolevel needs to be well understood and predicted. For this purpose, a multiscale simulation concept has been proposed. According to this concept, the FB coating in the Wurster coater has been modeled on four different scales with different resolution of the physical mechanisms. In Fig. 2, the general architecture and coupling between submodels with main transferred parameters are schematically shown.

On the macroscale, the one-dimensional PBM is used for calculation of the time-dependent PSD and for description of the corresponding particle growth in the apparatus. According to the results obtained from the macroscale and other process parameters micro- and mesoscales are generated. Following parameters are transferred from macroscale to the microscale: fluidization and atomization gas properties (flow, temperature), Sauter mean diameter of PSD and bed mass.

The FB process is simulated using a DPM. This model couples the DEM (Cundall and Strack, 1979), which describes the motion and interactions of every individual particle by equations of motion, and the CFD, which calculates the process gas dynamics in the apparatus by volume-averaged Navier–Stokes equations. The interactions between particles, particles and apparatus walls as well as between particles and gas are described.

Due to the high-computational effort, these calculations are performed only for a relatively short period of real time (in order of seconds). The particle trajectories and the fluid profile obtained by this modeling are transferred to the mesoscale, where the simulations of particle wetting and drying are performed. On the mesoscale two different models are employed. The first model describes droplets as separate objects solving Newtonian equations of motion for each droplet and analyzing collisions between particles and droplets. These calculations allow to predict the overspray rate $K_{os}$. The second model is used to calculate the drying of the liquid from particle

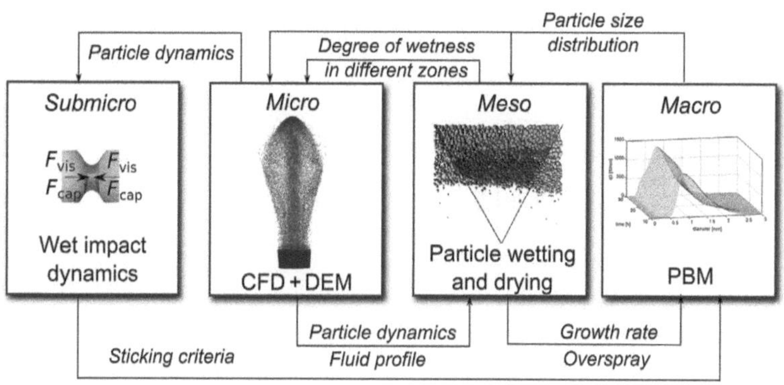


**Figure 2** General representation of the multiscale architecture.

surfaces and gives a possibility to analyze the degree of wetness in different zones of the Wurster coater.

The model on the submicroscale is developed to calculate the collision of single particles considering the influence of the liquid layers on their surfaces by solving of the force balance, which includes the capillary and viscous forces. As a result from this model, the particle sticking criterion is estimated. Due to a combination of this criterion with characteristics obtained from microscale and mesoscales, the agglomeration rate in the apparatus can be predicted.

In this study, the DPM calculations on the microscale have been performed neglecting the influence of the liquid on the particle dynamics. Such simplification can be made only when the fraction of particles covered with liquid film is relative small. To consider the influence of liquid, the submicro- and mesoscale models can be employed. On the submicroscale, detailed simulation of particle impacts with different liquid amount can be performed to predict energy dissipation. The mesoscale model can be used to approximate wetted surface fractions of particles in different zones. Transferring the data from both scales to the DPM calculations gives a possibility to consider liquid.

## 2.1 Macroscale

The PBM is an efficient method (Ramkrishna, 2000) to describe time-dependent changes of the multidimensional particle property distribution due to various effects occurring in the system. The one-dimensional PBM with the particle diameter as property coordinate was used in this contribution. In the case of FB spray granulation five main mechanisms, which may lead to changes of PSD can be distinguished, namely: breakage, nucleation, agglomeration, continuous growth, and attrition (Uhlemann and Mörl, 2000).

A general form of the one-dimensional PBM for an ideally mixed granulation system can be expressed as (Dosta, 2013; Heinrich, 2001):

$$\frac{\partial n(t,v)}{\partial t} = B_{\mathrm{nuc}}(t,v) - \frac{\partial[(\mathrm{Gr}(t,v) - \mathrm{Attr}(t,v)) \cdot n(t,v)]}{\partial v} + B_{\mathrm{agg}}(t,v) - D_{\mathrm{agg}}(t,v) + B_{\mathrm{break}}(t,v) - D_{\mathrm{break}}(t,v) \quad (1)$$

where $v$ is the size of the particles, the terms $B(t,v)$ and $D(t,v)$ describe particle birth and death, respectively. The abbreviations nuc, break, agg are used to distinguish between nucleation, breakage, and aggregation mechanisms. The continuous growth is considered by the term $\mathrm{Gr}(t,v)$ and the attrition rate by $\mathrm{Attr}(t,v)$.

The usage of the PBM is combined with two main challenges. On the one hand, since the PBM is described with a complex partial integro-differential equation for the agglomeration and breakage terms, even in the simplified case for pure convection growth the numerical solution of it is a great challenge. There exist various numerical techniques, such as fixed-pivot approach (Kumar and Ramkrishna, 1996) or cell-average technique (Kumar et al., 2006), which can be used to minimize numerical error. On the other hand, the determination of the parameters which defines the process kinetics, such as breakage function, coalescence kernel, or size-dependent growth rate, is not a trivial task. In the literature various correlations, which have been experimentally obtained for specific process parameters, can be found. However, the application of these correlations to the new process conditions is a challenging and in many cases impossible task. Therefore, to get process-related PBM parameters, the models on additional scales were used in this contribution.

With help of mesoscale model described in the next section, the overspray rate $K_{os}$ is calculated. This is the amount of the sprayed solution which is dried before contact with particles. Overspray can be elutriated from the apparatus as dust or it may serve as a source of new solid nuclei. The $K_{dust}$ parameter defines the amount of solution, which is elutriated from the apparatus as dust. In Fig. 3, the main mass flows in the coater are shown.

In the case of size-independent (uniform) growth, the particle growth rate can be obtained as (Mörl et al., 2007):

$$\mathrm{Gr} = \frac{2 \cdot \dot{M}_{sol} \cdot (1 - x_w)}{\rho_{sol} \cdot A_{tot}} \tag{2}$$

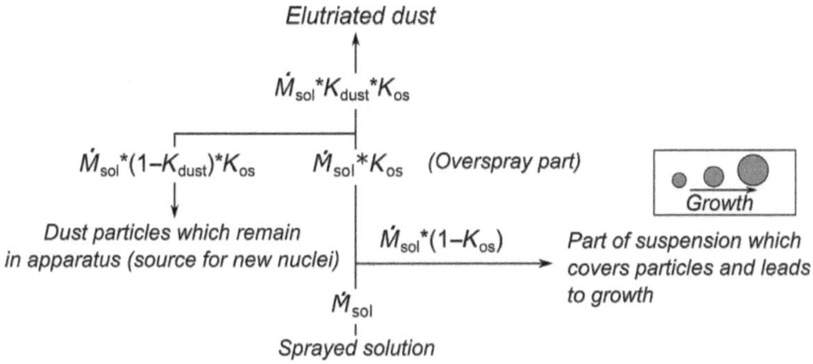


**Figure 3** Main mass flows in apparatus.

where $\dot{M}_{sol}$ is the suspension or solution mass flow rate, $A_{tot}$ is the total surface of all particles in the apparatus, $\rho_{sol}$ is the suspension or solution density and $x_w$ is the water content in the suspension or solution.

By the PBM modeling of granulation, a simplification has been made in this work that particles are ideally mixed in the apparatus. However, in the reality this is not always the case as, for example, the partial segregation of particles can result in the size-dependent growth rate. In this case, particles of different size fractions have various residence times in wetting zone and as a consequence they have different growth rates (Zank et al., 2001). In the case of the Wurster coater, two main reasons can lead to the size-dependent growth rate: formation of the stagnation zones, where particles of specified size classes are situated; difference in the periods of circulation cycles for different particles.

## 2.2 Mesoscale

### *2.2.1 Modeling of Droplets as Discrete Elements*

In order to calculate the wetting of the granules or pellets with injected liquid, the mesoscale model of the nozzle zone has been used. The implemented model is based on the DEM coupled with CFD calculations and allows the approximation of the overspray part of the suspension (see Fig. 3).

On the mesoscale, not only particles have been treated as separate objects, but also the liquid droplets have been represented as discrete elements and their translational and rotational motion in the fluid field has been calculated. To perform modeling of the nozzle zone, following a set of simplifications and assumptions have been carried out:

- *spherical shape*: droplets are treated as ideally spherical objects and they do not deform during motion in the fluid field;
- *no aggregation*: there is no collision between droplets and there is no droplet coalescence;
- *no breakage or size decrease*: droplets have a constant diameter, which does not change due to the breakage or liquid evaporation. The break-up of liquid in a gas stream occurs when the dynamic pressure of the gas exceeds the pressure inside the droplet (Walzel, 1993). In this contribution, the drying of the droplets and dust formation is not calculated on the mesoscale;
- *simplified coalescence model between droplets, particles, and apparatus walls*: it is assumed that if a droplet has a contact with the wall or with granules,

then the surface wetting occurs and the droplet is removed from the simulation domain.

The trajectories of all particles and the fluid profile are transferred as input parameters from DPM calculation to the mesoscale model. On the micro- and mesoscale, irregular and Cartesian mesh are used accordingly. To transfer parameters from one scale to another additional remeshing is performed (Dosta, 2013).

The main calculation algorithm is shown in Fig. 4. In the first three steps, the import and transformation of particle trajectories and the fluid profile are performed. Afterwards the iterative part of the algorithm is started. In step 4, new liquid droplets are generated in the system. Number of droplets $N_{drop}$, which are generated per one time step $\Delta t$, was calculated as:

$$N_{drop} = \Delta t \cdot \frac{6 \cdot \dot{M}_{sol}}{\pi \cdot d_{drop}^3 \cdot \rho_{sol}}, \tag{3}$$

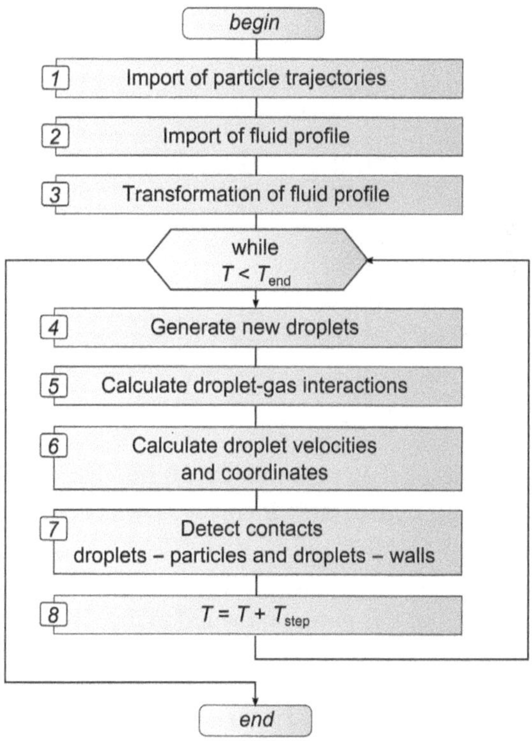


**Figure 4** Calculation algorithm of mesoscale model with discrete droplets.

where $d_{\mathrm{drop}}$ is the droplet diameter. In the literature, several correlations to describe size of droplets can be found (Mueller and Kleinebudde, 2007; Teunou and Poncelet, 2002; Toschkoff et al., 2013; Varga et al., 2013). For example, Mueller and Kleinebudde (2007) have proposed to calculate the droplet diameter according to:

$$d_{\mathrm{drop}} = \frac{585 \times 10^3}{v_{\mathrm{rel}}} \sqrt{\frac{\sigma}{\rho_{\mathrm{sol}}}} + 1683 \cdot \mu_{\mathrm{sol}}^{0.45} \cdot (\sigma \cdot \rho_{\mathrm{sol}})^{-0.225} \cdot \left(\frac{1000}{J}\right)^{1.5} \tag{4}$$

where $v_{\mathrm{rel}}$ is the air velocity relative to the liquid velocity at the nozzle exit, $\mu_{\mathrm{sol}}$ is the solution viscosity, $J$ is the dimensionless air to liquid volume ratio, $\sigma$ is the liquid surface tension.

In step 5, the interactions between gas and droplets are calculated (Crown et al., 2011). For this purpose, one-way coupling is employed and the drag force $F_{\mathrm{D}}$, which acts on droplets, is calculated depending on the relative velocity between droplet and gas $v_{\mathrm{rel}}$, as it is shown in Eqs. (5–7). More precise results can be obtained with two-way coupling, where the influence of droplets on the gas phase is modeled.

$$Re = \frac{d_{\mathrm{drop}} \cdot v_{\mathrm{rel}}}{v_{\mathrm{g}}} \tag{5}$$

$$C_{\mathrm{D}} = \begin{cases} \dfrac{17}{Re^{0.6}}, & 10.1 < Re \\ \dfrac{27}{Re^{0.8}}, & 0.5 \leq Re \leq 10.1 \\ \dfrac{24}{Re}, & Re < 0.5 \end{cases} \tag{6}$$

$$F_{\mathrm{D}} = C_{\mathrm{D}} \cdot \frac{\pi \cdot d_{\mathrm{drop}}^2}{4} \cdot \frac{\rho_{\mathrm{g}}}{2} \cdot v_{\mathrm{rel}}^2, \tag{7}$$

where $\rho_{\mathrm{g}}$ is the gas density, $Re$ is the Reynolds number, $\nu_{\mathrm{g}}$ is the gas kinematic viscosity. The gas density and kinematic viscosity does not depends on liquid evaporation.

After calculation of forces, the new droplets velocities $v_{\mathrm{drop}}$ and positions $\mathrm{pos}_{\mathrm{drop}}$ are predicted. For this purpose, the following system of ordinary differential equations (ODEs) is solved:

$$\frac{\mathrm{d}v_{\mathrm{drop}}}{\mathrm{d}t} = \frac{F_{\mathrm{D}} + F_{\mathrm{G}} + F_{\mathrm{B}}}{m_{\mathrm{drop}}} \tag{8}$$

$$\frac{\mathrm{dpos_{drop}}}{\mathrm{d}t} = v_{\mathrm{drop}}, \tag{9}$$

where $m_{\mathrm{drop}}$ is the mass of a droplet, $F_{\mathrm{G}}$ is the gravity force, $F_{\mathrm{B}}$ is the buoyancy force.

In calculation of step 7 (Fig. 4), the contacts between droplets and particles and between droplets and apparatus walls are analyzed. If a droplet hits a solid material, then it is removed and the respective solid surface is assumed to be wetted. In reality, four different scenarios can occur when a spray impacts onto a solid surface (Panao and Moreira, 2004):

- *stick*: the droplet sticks to the surface and remains in nearly spherical form;
- *spread*: the droplet covers the surface as uniformly distributed liquid film;
- *splash*: the liquid film spreads out and a crown is formed owing kinematic discontinuity;
- *rebound*: the droplet does not stick to the solid surface.

In Fig. 5, an example of generated droplets is illustrated.

### 2.2.2 Uniform Wetting Zone

The uniform model of the wetting zone has been developed to predict the wetted surface fraction of pellets in the FB and how it is influenced by air temperature and suspension mass flow. For effective operation of the FB, it is important that particles are dried before they are coming into the annulus with high solids concentration. If particles with wet surfaces enter the dense region, it may lead to undesirable agglomeration.

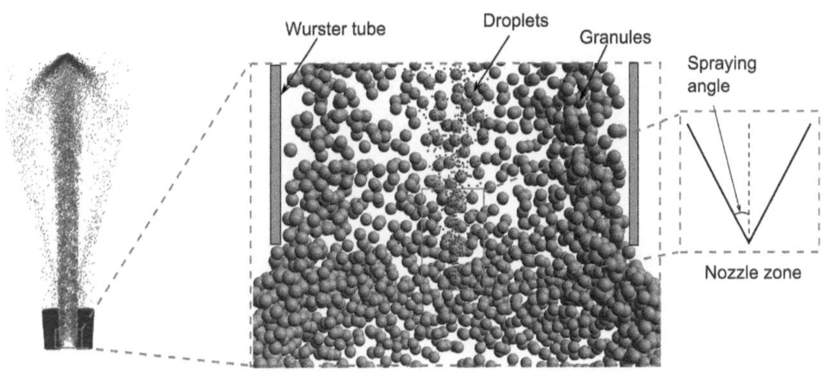


**Figure 5** Generation of discrete droplets (blue gray in the print version dots) on the mesoscale.

Contrary to the nozzle zone model described in the previous section, individual liquid droplets were not tracked here. Instead, the nozzle has been modeled as uniform wetting zone (Fries et al., 2011a). This wetting zone is described as a cone-shaped region with specified centre, angle, and spraying length. Link et al. (2007) analyzed the deposition of droplets on particles in a spout FB and found that 95% of the droplets were deposited within a limited cone-shaped geometry. Similarly, Palzer (2000) has described particle wetting in a rotating drum granulator.

In this simplified model, the solution mass flow $\dot{M}_{\mathrm{sol},i}$ onto a particle with index $i$, which is situated in the wetting region, is calculated as:

$$\dot{M}_{\mathrm{sol},i} = \frac{\dot{M}_{\mathrm{sol}} \cdot A_{\mathrm{p},i}}{A_{\mathrm{tot,p}}}, \tag{10}$$

where $A_{\mathrm{p},i}$ is the particle surface and $A_{\mathrm{tot}}$ is the total surface of all particles which are situated in the wetting region.

To calculate liquid drying, the mass transfer coefficient $\beta$ has been calculated for each particle as:

$$\beta = \frac{Sh \cdot D}{d_{\mathrm{p}}}, \tag{11}$$

where $Sh$ is the Sherwood number, $d_{\mathrm{p}}$ is the particle diameter, and $D$ is the diffusion coefficient of water vapor in air. The diffusion coefficient is dependent on fluidization gas temperature $\vartheta_{\mathrm{g}}$ and is calculated according to the following correlation (Schirmer, 1938):

$$D = 26 \times 10^{-6} \left( \frac{\vartheta_{\mathrm{g}} + 273.15}{273.15} \right)^{1.81} \tag{12}$$

The Sherwood number ($Sh$) depends on Reynolds number ($Re$) and Schmidt number ($Sc$) and is calculated with (Gnielinski, 1980):

$$Sh = 2 + \sqrt{Sh_{\mathrm{lam}}^2 + Sh_{\mathrm{turb}}^2} \tag{13}$$

$$Sh_{\mathrm{lam}} = 0.664 \cdot \sqrt{Re} \cdot \sqrt[3]{Sc} \tag{14}$$

$$Sh_{\mathrm{turb}} = \frac{0.037 \cdot Re^{0.8} \cdot Sc}{1 + 2.443 \cdot Re^{-0.1} \left( Sc^{0.67} - 1 \right)} \tag{15}$$

The Schmidt number depends on dynamic viscosity of gas $\nu_{\mathrm{g}}$ and is calculated as:

$$Sc = \frac{\nu_g}{D} \tag{16}$$

The obtained mass transfer coefficient $\beta$ afterwards is used to predict the mass flow of vapor $\dot{M}_{vap}$. It is assumed that the liquid film on the particle surface has a constant thickness $L_{film}$. Therefore, the particles are only partially covered with a liquid film. The particles surface covered with liquid $A_{wet}$ depends on the mass of the liquid on the particle surface $M_l$, density of liquid $\rho_l$ and is calculated as:

$$A_{wet} = \frac{M_l}{\rho_l \cdot L_{film}} \tag{17}$$

Finally, the mass flow of vapor from the surface of the particles is calculated as:

$$\dot{M}_{vap} = \rho_g \cdot \beta \cdot A_{wet} \cdot (Y_{sat} - Y_g) \tag{18}$$

where $Y_g$ and $Y_{sat}$ are the average and saturation humidity of the air. In Fig. 6, exemplary application of the mesoscale model on the Wurster coater apparatus is shown. Here, the particles are colored according to the mass of the liquid deposited on the pellet surface.

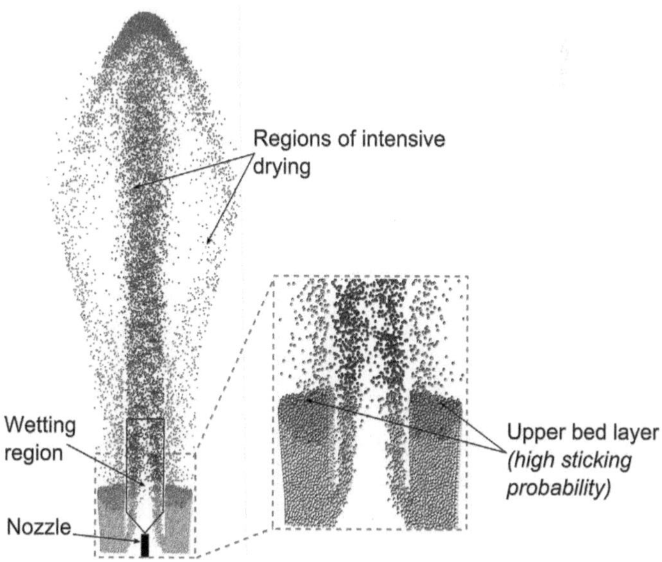


**Figure 6** Application of the mesoscale model on Wurster coater.

## 2.3 Microscale

### 2.3.1 Discrete Particle Model

On the microscale, the DPM has been used to describe the particle dynamics in the apparatus (Deen et al., 2007). In the DPM, the motion of individual particles is predicted using Newton's and Euler's laws:

$$m_\mathrm{p}\frac{\mathrm{d}v_\mathrm{p}}{\mathrm{d}t} = -V_\mathrm{p}\cdot\nabla\mathrm{p} + \frac{V_\mathrm{p}\cdot\beta_{\mathrm{g-p}}}{1-\varepsilon}\left(u_\mathrm{g}-v_\mathrm{p}\right) + m_\mathrm{p}\cdot g + F_\mathrm{c}, \tag{19}$$

$$I_\mathrm{p}\frac{\mathrm{d}\omega_\mathrm{p}}{\mathrm{d}t} = \sum_{i=1}^{n} M_i, \tag{20}$$

where $m_\mathrm{p}$ and $V_\mathrm{p}$ is the mass and volume of particle accordingly. The force balance on the right side of Eq. (19) consists of the force due to pressure gradient, fluid drag, gravity, and contact forces by particle–particle and particle–wall collisions, respectively. $I_\mathrm{p}$ and $\omega_\mathrm{p}$ in Eq. (20) are the moment of inertia and angular velocity of the particle. $M_\mathrm{i}$ are the moments generated from tangential contact forces acting on the particle.

In the drag force, the influence of the gas on the particle is described with interphase momentum transfer coefficient $\beta_{\mathrm{g\text{-}p}}$ which is calculated depending on the porosity of the solid phase $\varepsilon$. For the dense regimes, when porosity is smaller than 0.8, the coefficient is calculated according to the Ergun correlation (Ergun, 1952):

$$\beta_{\mathrm{g-p}} = \frac{\eta_\mathrm{g}}{d_\mathrm{p}^2}\left(150\frac{(1-\varepsilon)^2}{\varepsilon} + 1.75\cdot(1-\varepsilon)\cdot Re\right) \tag{21}$$

At the porosity $\varepsilon \geq 0.8$ (dilute regime) the Wen and Yu (1966) model is applied:

$$\beta_{\mathrm{g-p}} = \frac{3}{4}c_\mathrm{D}\cdot Re\frac{(1-\varepsilon)}{\varepsilon^{2.65}}\cdot\frac{\eta_\mathrm{g}}{d_\mathrm{p}} \tag{22}$$

$$C_\mathrm{D} = \begin{cases} 24\cdot\left(\dfrac{1+0.15Re^{0.687}}{Re}\right) & Re < 1000 \\ 0.44 & Re \geq 1000 \end{cases} \tag{23}$$

It should be mentioned that the correlation described above can be effectively applied only for FBs with monodisperse PSD. In the case when solid particles have wide PSD, other correlations should be used, as for example correlation obtained with Lattice–Boltzmann simulations for polydisperse systems (Beetstra et al., 2007).

The contact forces between particles and between particles and apparatus walls were calculated by a viscoelastic contact model proposed by Tsuji et al. (1992, 1993). The normal and tangential components of the contact force are defined as follows:

$$\vec{F}_{\mathrm{c,n}} = -\vec{F}_{\mathrm{c,n,el}} - \vec{F}_{\mathrm{c,n,d}} = -\frac{2}{3} E^* \sqrt{R^* s_{\mathrm{n}}^3}\ \vec{n} - \eta_{\mathrm{n}} \vec{v}_{\mathrm{n}} \tag{24}$$

$$\begin{aligned}\vec{F}_{\mathrm{c,t}} &= -\vec{F}_{\mathrm{c,t,el}} - \vec{F}_{\mathrm{c,t,d}} \\ &= \left| \begin{array}{ll} -4\, G^* \sqrt{R^* s_{\mathrm{n}}}\ \vec{t} - \eta_{\mathrm{t}} \vec{v}_{\mathrm{t}} & \text{for} \left|\vec{F}_{\mathrm{c,t}}\right| \le \mu \left|\vec{F}_{\mathrm{c,n}}\right| \\ -\mu \left|\vec{F}_{\mathrm{t}}\right| \vec{t} & \text{for} \left|\vec{F}_{\mathrm{c,t}}\right| = \mu \left|\vec{F}_{\mathrm{c,n}}\right| \end{array} \right|\end{aligned} \tag{25}$$

where $s_{\mathrm{n}}$ is the normal displacement in the contact and $\mu$ is the dynamic friction coefficient.

The elastic contribution of the impact energy, absorbed during the compression, is released during the restitution phase of the impact and leads to the force that separates the contact partners. The elastic part of the contact force $F_{\mathrm{el}}$ is described by the approach based on the Hertz theory (Hertz, 1882). The elastic force is proportional to the equivalent modulus of elasticity $E^*$ and the equivalent shear modulus $G^*$ for normal and shear direction respectively, which are given as:

$$E^* = 2\left(\frac{1-\nu_1^2}{E_1} + \frac{1-\nu_2^2}{E_2}\right)^{-1} \tag{26}$$

$$G^* = 2\left(\frac{2-\nu_1}{G_1} + \frac{2-\nu_2}{G_2}\right)^{-1} \tag{27}$$

where indexes 1 and 2 mark the parameters of two contact partners. $\nu$ is the Poisson ratio. The median radius $R^*$ of the contact partners is a characteristic radius of the contact surface curvature, as defined in Eq. (28).

$$R^* = \left(\frac{1}{R_1} + \frac{1}{R_2}\right)^{-1} \tag{28}$$

The absorption of kinetic energy during the impact is described by viscous part of the contact force $F_{\mathrm{d}}$. To account for viscoelastic material properties that cause energy dissipation, a damping factor $\eta$ related to the coefficient of restitution $e$ is included into the DEM model, as proposed by Tsuji et al. (1993):

$$\eta_{n/t} = -2\sqrt{\frac{5}{6}}\alpha\sqrt{s_{n/t}\,m^*} \tag{29}$$

$$\alpha = \begin{cases} \dfrac{\ln e}{\sqrt{\pi^2 + \ln^2 e}} & \text{if } 0 < e \leq 1 \\ 1 & \text{if } e = 0 \end{cases} \tag{30}$$

where $m^*$ is the effective mass of the contact partners given as:

$$m^* = \left(\frac{1}{m_1} + \frac{1}{m_2}\right)^{-1} \tag{31}$$

### 2.3.2 Scaled Discrete Particle Model

The real coating process in the studied Wurster coater apparatus with the bed mass of 3 kg contains about $21.8\times10^6$ particles with the size of 550 μm. Unfortunately, the numerical effort for the calculation of the DPM model increases with increasing the number of simulated particles. The DPM model is unable to represent this number of particles, at least with the actually available computing power. However, the number of particles can be reduced by conservation of the particle and fluid dynamics in the simulated apparatus and its real geometry. In this work a scaling approach proposed by Link et al. (2009) and extended by Sutkar et al. (2013) has been used, in which the scaling of the particle size was carried out. Due to the size increase, the adequate properties of solid and gas phase have been adapted to keep the dimensionless numbers Archimedes (*Ar*) and Reynolds (*Re*) and the velocities of minimal fluidization and elutriation constant.

The Archimedes number is calculated as:

$$Ar = \frac{g\cdot d_p^3}{\nu_g^2}\,\frac{\rho_p - \rho_g}{\rho_g} \tag{32}$$

Reynolds number and velocity of elutriation are given as

$$Re_{elu} = \sqrt{\frac{4}{3}Ar} \tag{33}$$

$$u_{elu} = \frac{Re_{elu}\nu_g}{d_p} \tag{34}$$

The Reynolds number corresponding to the point of the minimal fluidization depends on porosity $\varepsilon_{mf}$ and is given as:

$$Re_{mf} = 42.9 \cdot (1 - \varepsilon_{mf}) \sqrt{1 + \frac{\varepsilon_{mf}^3}{(1 - \varepsilon_{mf})^2} \frac{Ar}{3214}} - 1 \tag{35}$$

The parameter $k$ describes the ratio of both particle sizes (index 1, experiment; index 2, scaled DPM simulation):

$$d_{p,2} = k \cdot d_{p,1} \tag{36}$$

At first, we assume that the velocity of minimal fluidization and accordingly the Reynolds number are constant.

$$u_{mf,1} = u_{mf,2} = \frac{Re_{mf,1} \cdot \nu_{g,1}}{d_{p,1}} = \frac{Re_{mf,2} \cdot \nu_{g,2}}{d_{p,2}} \tag{37}$$

Thus the dynamic viscosity of the gas for DPM is given as:

$$\nu_{g,2} = \frac{d_{p,2} \cdot \nu_{g,1}}{d_{p,1}} = k \cdot \nu_{g,1} \tag{38}$$

We assume that the gas density in the DPM simulation remains constant and is equal to that of the experiment. To calculate the particle density the Archimedes number $Ar_1 = Ar_2$ can be applied:

$$\frac{g \cdot d_{p,1} \cdot \left(\rho_{p,1} - \rho_{g,1}\right)}{\nu_{g,1}^2 \cdot \rho_{g,1}} = \frac{g \cdot d_{p,2} \cdot \left(\rho_{p,2} - \rho_{g,2}\right)}{\nu_{g,2}^2 \cdot \rho_{g,2}} \tag{39}$$

After simplification and introducing the ratio $k$, this equation yields

$$\rho_{p,2} = \frac{\rho_{g,2}}{k} \frac{\rho_{p,1} - \rho_{g,1}}{\rho_{g,1}} + \rho_{g,2} \tag{40}$$

## 2.4 Submicroscale

This model was developed to investigate collision dynamics of individual wet particles under different impact conditions (velocity, size of contact partners, impact angle) and material properties (particle density, suspension viscosity, etc.). The submicroscale model is a stochastic model, where a set of simulations is performed on specific parameters domain to obtain average collision characteristics. In every simulation study, two contact partners are generated in the three-dimensional space and their impact behavior is modeled with the DEM. Afterwards, the restitution coefficient has been analyzed as main criteria to characterize impact dynamics and to obtain agglomeration probability. If the restitution coefficient equals to zero, this indicates that particles stick to each other and agglomeration occurs. Otherwise rupture of liquid bridge between granules occurs.

The described above Hertz–Mindlin discrete element model developed for dry materials has been extended with the model to calculate forces acting in the liquid bridge (Antonyuk et al., 2009). Due to the large amount of material parameters, which cannot be directly measured, additional experiments are often needed to validate the model. For the model validation, the free-fall experimental setup can be effectively used (Crüger et al., 2015).

In this contribution, the simplification has been made and it was assumed that liquid film on particles has a constant thickness (Heinrich, 2001). To consider liquid film on particle surface correctly, the radius for contact detection of DEM has been increased by the thickness of liquid film. To detect collision between two wet particles with radii $R_1$ and $R_2$ the following criteria has been used:

$$|\mathrm{Pos}_1 - \mathrm{Pos}_2| < R_1 + R_2 + 2 \cdot L_{\mathrm{film}}, \tag{41}$$

where $\mathrm{Pos}_1$ and $\mathrm{Pos}_2$ are positions of both particles accordingly and $L_{\mathrm{film}}$ is the thickness of the liquid film.

To calculate the capillary $F_{\mathrm{cap}}$ and viscous forces in normal $F_{\mathrm{vis,n}}$ and tangential $F_{\mathrm{vis,t}}$ direction, Eq. (42) (Pitois et al., 2000) and Eqs. (43) and (44) (Popov, 2011) have been used accordingly.

$$F_{\mathrm{cap}} = 2\pi \cdot \bar{R} \cdot \sigma \cdot \cos(\theta) \left[ 1 - \frac{1}{\sqrt{1 + \frac{2V_{\mathrm{lb}}}{\pi R h^2}}} \right] \tag{42}$$

$$F_{\mathrm{vis}}^{\mathrm{n}} = 6\pi \cdot \mu_{\mathrm{l}} \cdot v_{\mathrm{rel,n}} \frac{\bar{R}^2}{h} \tag{43}$$

$$F_{\mathrm{vis}}^{\mathrm{t}} = 2 \cdot \pi \cdot \mu_{\mathrm{l}} \cdot \bar{R} \cdot v_{\mathrm{rel,t}} \cdot \ln\left(1 + \frac{\bar{R}}{2h}\right) \tag{44}$$

where $V_{\mathrm{lb}}$ is the volume of a liquid bridge, $v_{\mathrm{rel,n}}$ and $v_{\mathrm{rel,t}}$ are relative velocities of particles in normal and shear direction accordingly, $\mu_{\mathrm{l}}$ is dynamic liquid viscosity, $h$ is the distance between particle surfaces, $\overline{R}$ is the equivalent radius of contact partners.

Another important parameter by modeling of liquid bridge is a rupture criterion, i.e., the critical distance between particles $h_{\mathrm{rupt}}$ when the rupture of the bridge occurs. To detect rupture of the liquid bridge the criteria proposed by Lian et al. (1993) was used in this contribution. However, this correlation can be effectively applied for contact partners of same diameter. To calculate rupture distance between spheres of different size another correlation can be used (Willett et al., 2007).

$$h_{\mathrm{rupt}} > (1 + 0.5 \cdot \theta)\sqrt[3]{V_{\mathrm{lb}}} \tag{45}$$

# 3. EXPERIMENTAL ESTIMATION OF THE MATERIAL PARAMETERS

## 3.1 Studied Materials

The sugar pellets "SUGLETS" produced by Colorcon (UK) are processed in the simulated Wurster coater. These particles are studied in the compression, shear, and impact tests to estimate the material parameters for the Hertz–Mindlin contact model of particles used in DEM on the microscale of the simulation.

The PSD of the investigated particles (Fig. 7) has been determined by the image processing with Camsizer® XT device. The maximum Feret diameter of captured projection areas was used as a characteristic particle size. The particles size is distributed in a range of 430–900 μm with a mean size of 555 μm.

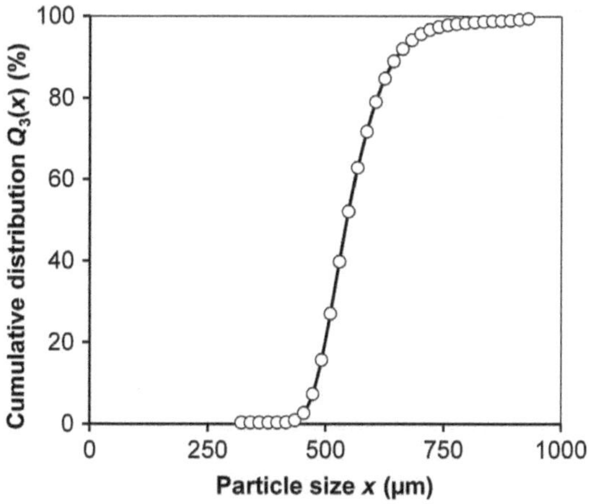


**Figure 7** Volume-based particle size distribution of studied sugar pellets measured by Camsizer® XT device.

## 3.2 Compression Test of Particles

### 3.2.1 Experimental Method

The device "Texture Analyser" from the company "Stable Micro Systems Ltd" has been used for the investigation of deformation and breakage behavior of sugar pellets during single particle compression test (Fig. 8). It serves for the determination of important model material parameters for the simulations, such as stiffness, modulus of elasticity, breakage force, and compressive strength of particles. The particle is deposited on the stationary test steel bench by means of vacuum tweezers. The cylindrical-shaped steel stamp with a diameter of 5 mm moves downwards until particle contact. In this moment, the height of the particle was measured, and the recording of the force–displacement curve starts until breakage is reached. The compression tests have been conducted at a constant deformation rate of 0.4 mm/s. The experiments have been repeated 105 times to increase the confidence level.

### 3.2.2 Description of Elastic Contact Deformation with the Hertz Theory

During uniaxial loading of a comparatively soft spherical elastic pellet with a smooth stiff wall (flat surface), the contact area of the particle deforms as a circle. The model of Hertz (Hertz, 1882) describes the force and internal pressure distribution as well as the contact radius depending on the particle radius and moduli of elasticity behavior of the two contacting materials.

The nonlinear relation between elastic contact force and displacement in normal direction, found by Hertz, is given as:

$$F_{\mathrm{el}} = \frac{2}{3} E^* \sqrt{R^*}\, s_{\mathrm{n}}^{3}, \tag{46}$$

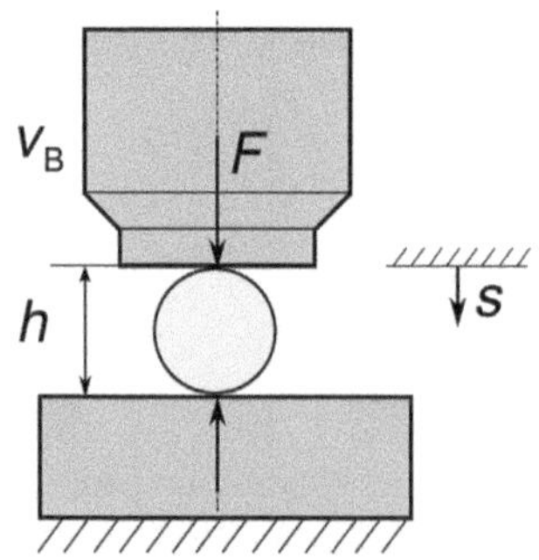


**Figure 8** Schematic representation of the compression test of a single particle.

where $s_n$ is the full elastic displacement composed of the contact and of the whole spherical particle. Note that we use index $p$ for the pellet and index $w$ for the wall. The equivalent radius of contact partners $R^*$ is a characteristic radius of contact surface curvature:

$$R^* = \left(\frac{1}{R_p} + \frac{1}{R_w}\right)^{-1} \approx R_p \tag{47}$$

The effective modulus of elasticity $E^*$ of contact partner ($E_w \gg E_p$) is given as:

$$E^* = 2\left(\frac{1-\nu_p^2}{E_p} + \frac{1-\nu_w^2}{E_w}\right)^{-1} \approx \frac{2\,E_p}{1-\nu_p^2} \tag{48}$$

where $\nu$ is Poisson ratio.

Due to the parabolic curvature of the force–displacement curve $F_{el}(s_n)$, the equivalent contact stiffness in normal direction increases with increasing displacement $s_n$ and pellet radius $R_p$:

$$k^*_{n,el} = \frac{dF_{el}}{ds} = E^*\sqrt{R^*\,s_n} \approx \frac{E_p}{1-\nu_p^2}\sqrt{R_p\,s_n} \tag{49}$$

During compression between two walls, two contact areas are formed. In this case the force–displacement relationship is given as:

$$F_{el,w-p-w} = \frac{1}{3}\frac{E_p}{\left(1-\nu_p^2\right)}\sqrt{2\,R_p\,s_n^3} \tag{50}$$

Here, the displacement $s$ is the sum of the displacements of both contacts and of the whole pellet.

The average elastic stiffness of this wall-pellet-wall-systems results from sequence contacts (Antonyuk et al., 2010). Hence, the equivalent contact stiffness ($k^*_{n,el,p}$), usually measured by a compression test, equals half the stiffness of a pellet ($k_{n,el,p}$):

$$k^*_{n,el,w-p-w} = \frac{dF_{el,w-p-w}}{ds_n} = \frac{1}{2\left(1-\nu_p^2\right)}E_p\sqrt{2\,R_p\,s_n} \tag{51}$$

### 3.2.3 Results of the Compression Tests

The particle deformation and breakage behavior can be illustrated by means of the measured force–displacement curves. A typical force–displacement curve for the investigated sugar pellets is shown in Fig. 9. The force $F$ characterizes the contact force between stamp and particle. The distance $s$ shows the corresponding displacement of the particle. All compression tests were carried out up to a compressive strain of 20%.

The measured force–displacement curves include two sections:

1. Contact deformation of the particle, beginning with the first contact with the stamp. The slope of this curve quantifies the particle contact stiffness, which depends on the deformation behavior. A slight force increase in this area means plastic (soft) material behavior and a strong increase means stiff material behavior. The force drops immediately at the primary breakage point $B_p$, where the resistance of the particle disappears. The enclosed area of the force–displacement curve in this first section is equal to the primary breakage energy $W_P$, which is required for inducing the first breakage event of the particle.
2. After the primary breakage event, the particle is further compressed up to 20% strain. Fragments formed during the first breakage event are broken down further. These secondary breakage events can be seen as loading and unloading behavior in the curve ($B_1 \cdots B_n$).

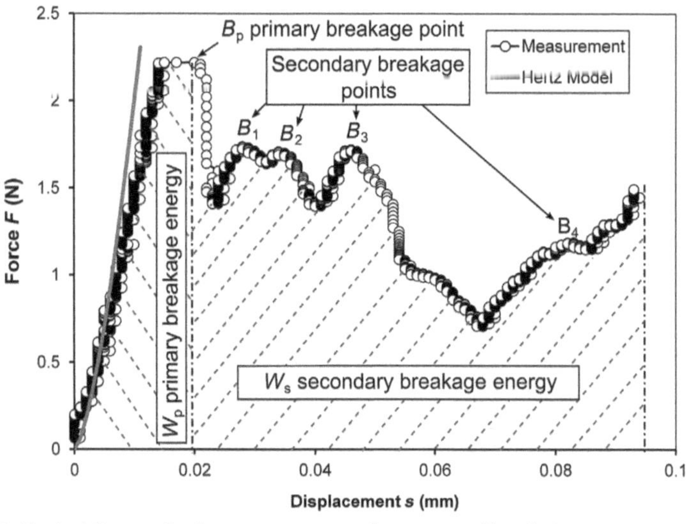


**Figure 9** Typical force–displacement curve of a sugar pellet during compression.

The typical ranges of elastic and elastic-plastic deformation of the tested sugar pellets are shown in Fig. 10 on a force–displacement curve. At the beginning of the punch–particle contact the elastic contact deformation of the pellet takes place. The elastic force was well described by Hertz in Eq. (50). Young's modulus and the stiffness of the pellet during elastic deformation were calculated with Eqs. (48) and (51) correspondingly. For that the Poisson ratio for the sugar particles was assumed to be 0.3. The shear modulus of the particles could be determined from the Young's modulus by

$$G_p = E_p / \left(2 \cdot \left(1 + \nu_p\right)\right).$$

Due to the parabolic curvature of the function $F(s)$, the contact stiffness $k_{n,el,p}$ increases with increasing displacement $s$ and reaches the maximum value in the yield point $F$ in Fig. 10. In this point the plastic deformation begins. This is confirmed by the increasing deviation of the experimental curve $F-B$ from the theoretical Hertz curve (Fig. 10).

The yield strength $p_F$ can be characterized at the beginning of the plastic deformation $s_F$ with the contact force $F_F$ as (Thornton and Ning, 1998; Tomas, 2000).

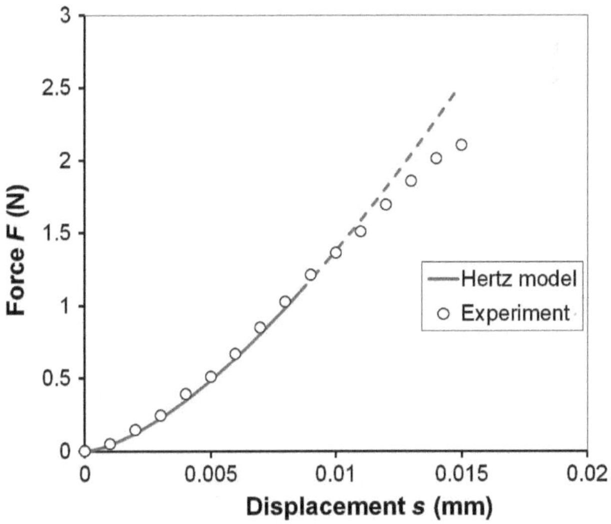


**Figure 10** Force–displacement curve of a sugar pellet during compression.

**Table 1** Mean Mechanical Material Parameters of the Analyzed Sugar Pellets

| Material Parameters | Value |
|---|---|
| Young's modulus $E_p$ (GPa) | 5.4±1.7 |
| Shear modulus $G_P$ (GPa) | 2.1±0.6 |
| Yield point: | |
| | 1.42±0.27 |
| – Yield force $F_F$ ($N$) | 11.5±3.1 |
| – Displacement until yield point $s_F$ (μm) | 2.5±0.03 |
| – Compression until yield point $\varepsilon_F$ (%) | 27.2±2.1 |
| – Yield pressure $\rho_F$ (MPa) | |
| Contact stiffness $k_{n,el,p}$ at yield point $F$ (N/mm) | 427±80 |

$$p_F = \frac{E^*}{\pi}\sqrt{\frac{2\,s_F}{d}} \tag{52}$$

$$s_F = \frac{\pi^2 d\, p_F^2}{2\,E^*} \tag{53}$$

The measured and calculated mean mechanical material parameters for the particles are presented in Table 1. The standard deviation of the determined material parameters are marked with the sign "±." The individual force–displacement curves show a large scattering in both the force and the displacement values, which is to be expected for an inhomogeneous material. The particles differ in size and shape, which has a considerable strong influence on this statistical data distribution.

Compared to the Young's modulus of typical construction materials (Fig. 11) like metals and ceramics (Lama and Vutukuri, 1978) the mean Young's modulus of the sugar pellets indicates that the sugar is an elastic soft material. The Young's modulus of sugar lies in the range of polymers and wood and is comparable to the Young's modulus of maize and microcrystalline cellulose.

The following breakage properties were determined from force–displacement curves:

1. Breakage force $F_B$ and breakage displacement $s_B$ characterize the primary breakage point of the particle.
2. Compression strength, defined as quotient of breakage force $F_B$ and breakage surface:

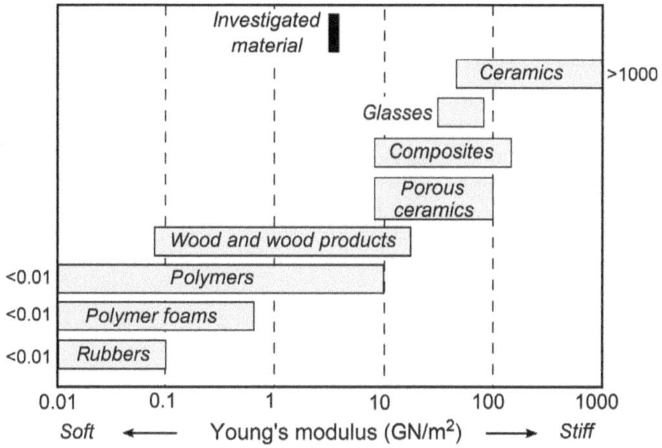


**Figure 11** Mean Young's modulus of the sugar particles compared to various other materials.

$$\sigma = \frac{4\,F_B}{\pi\,d^2} \tag{54}$$

where $d$ is the diameter of the particle. The compression strength characterizes the breakage stress, which corresponds to the maximum displacement or breakage strain $\varepsilon_B$:

$$\varepsilon_B = \frac{s_B}{d} \tag{55}$$

3. The primary breakage energy $W_p$ represents the minimum energy required for the primary breakage event of the particle. It is defined as the work applied to the particle during compression until the first breakage event appears.

$$W_p = \int_0^{s_B} F ds \tag{56}$$

The mechanical properties of the sugar pellets were determined from force–displacement curves and are summarized in Table 2.

## 3.3 Shear Test of Particles

The friction of the particles can have a significant effect on the particle dynamics in FB, on the velocities and rotation of particles (Salikov et al., 2015). To estimate the particle–particle and particle–wall friction coefficients, the flow behavior of the sugar particles was measured with the Jenike

**Table 2** Mean Values of the Mechanical Properties of the Sugar Pellets

| **Primary Breakage Limit** | | | | |
|---|---|---|---|---|
| **Breakage Force** | **Breakage Displacement** | **Compressive Strength** | **Failure Strain** | **Primary Breakage Energy** |
| $F_B$ **(N)** | $s_B$ **(μm)** | $\sigma$ **(N/mm$^2$)** | $\varepsilon_B$ **(%)** | $W_p$ **(mJ)** |
| 2.43±0.64 | 17±7 | 14.4±3.9 | 3.62±1.41 | 0.113±0.057 |

±, Confidence interval 95% statistical certainty.

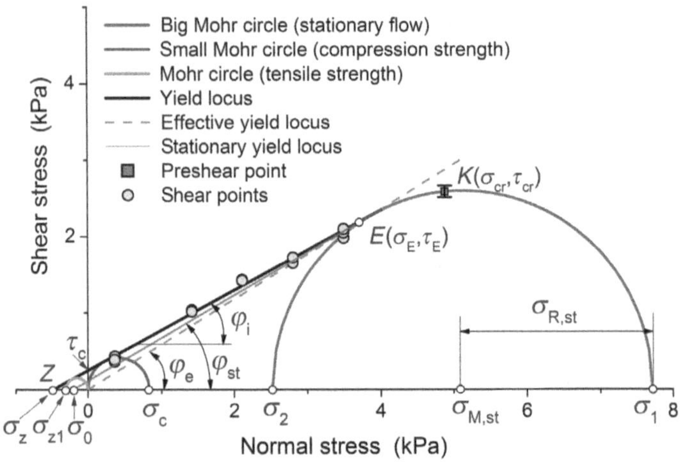


**Figure 12** Yield locus of tested sugar pellets with corresponding Mohr circles and parameters: $\sigma_1$, consolidation stress; $\sigma_c$, uniaxial compression strength; $\sigma_Z$, isostatical tensile strength; $\tau_c$, cohesion, $\varphi_i$, internal friction angle; $\varphi_e$, effective friction angle.

shear cell. Which is often used for examination of the inner friction and the strength of bulk materials under conditions of consolidation (Schulze, 2008).

Figure 12 shows the obtained yield points and their linear approximations for the internal friction, i.e., friction between sugar particles. The mean wall yield locus measured using the stainless steel wall element is presented in Fig. 13. The obtained parameters are listed in Table 3. The investigated material has a negligible cohesion and a high friction. The mean values of the measured friction coefficients between particles is $\mu_{p\text{-}p}=0.86\pm0.07$. The obtained wall friction coefficient is $\mu_{p\text{-steel}}=0.482\pm0.021$. The value of flow function at $ff_c=34$ shows that nearly spherical sugar pellets can be characterized as free flowing.

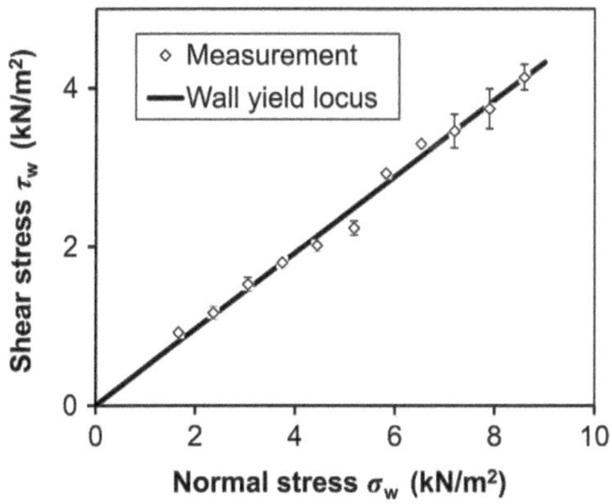


**Figure 13** Wall friction yield locus of spherical sugar pellets with steel wall.

**Table 3** Measurement Results of Shear Tests with Sugar Pellets

| Parameter | Value |
|---|---|
| Critical shear stress $\tau_{cr}$ (kN/m$^2$) | 3.92 |
| Critical normal stress $\sigma_{cr}$ (kN/m$^2$) | 4.88 |
| Cohesion $\tau_c$ (kN/m$^2$) | 0.07 |
| Internal friction angle $\varphi_i$ (°) | 40.8 |
| Effective friction angle $\varphi_e$ (°) | 41.4 |
| Uniaxial compression strength $\sigma_c$ (kN/m$^2$) | 0.30 |
| Isostatical tensile strength $\sigma_z$ (kN/m$^2$) | -0.08 |
| Midpoint of big Mohr circle $\sigma_{M.st}$ (kN/m$^2$) | 6.30 |
| Radius big Mohr circle $\sigma_{R.st}$ (kN/m$^2$) | 4.17 |
| Consolidation stress $\sigma_1$ (kN/m$^2$) | 10.47 |
| Flow function $ff_c$ | 34.33 |
| Wall friction angle $\varphi_{p\text{-steel}}$ (°) | 25.7 |

## 3.4 Collision Tests of Particles

The restitution coefficient is an essential parameter, which is needed to describe the energy absorption and damping force in the contact model. The coefficient of restitution is a ratio of impulse during the restitution phase of impact ($t_C \leq t \leq t_R$) to that during the compression phase ($0 \leq t \leq t_C$) (Eq. 57). Moreover, the impulse ratio in Eq. (57) gives the square root of

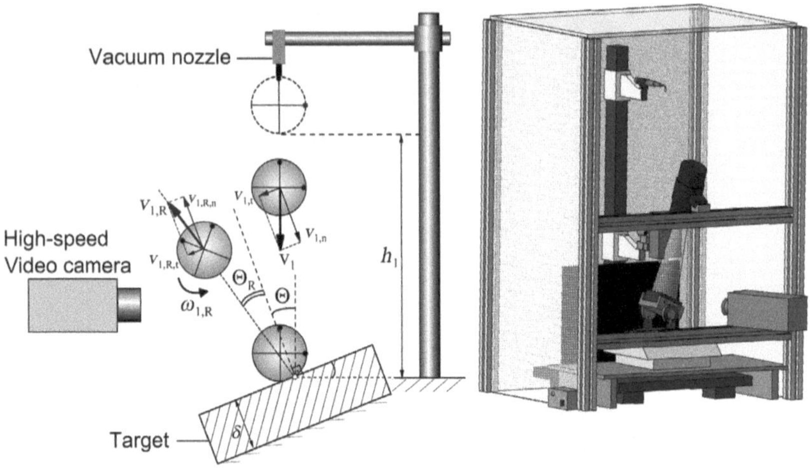


**Figure 14** Schematic representation of the particle impact device.

the ratio of elastic strain energy $W_{\text{kin,R}}$ released during the restitution to the impact energy, i.e., initial kinetic energy $W_{\text{kin}}$.

$$e = \frac{\int_{t_c}^{t_R} F\mathrm{d}t}{\int_{0}^{t_c} F\mathrm{d}t} = \frac{\sqrt{W_{\text{kin},R}}}{\sqrt{W_{\text{kin}}}} = \frac{v_{1,\mathrm{R}}}{v_1} \approx \sqrt{\frac{h_{1,\mathrm{R}}}{h_1}} \tag{57}$$

Figure 14 shows the experimental setup for the free-fall experiments. Before it is dropped, a pellet is held at a height $h_1$ above the target (flat plane) with the aid of vacuum tweezers. The pellet falls freely onto a target and reaches a rebound height $h_{1,\mathrm{R}}$ after the impact. The movement of the pellet near the contact point before and after the impact is recorded by a high-speed video camera with a frequency of 8.000 frames per second and a resolution of $1200 \times 1200$ pixels.

From the ratio of energies in Eq. (57), it follows that the coefficient of restitution is a ratio of relative rebound velocity $v_{1,\mathrm{R}}$ to that before impact $v_1$. These velocities were determined from captured impacts in free-fall experiments with help of a software tool developed on the basis of Matlab.

Two different tests were performed:

- Particle—Wurster granulator wall, which made of stainless steel.
- Particle—Elastic glass plate impact, which mimics the impact between two particles, due to a stiff elastic material of the used plate.

To detect the relatively small pellets, a black background was installed. For accurate measurement of the normal restitution coefficient normal impact and rebound should be ensured. During our experiments, some pellets without initial spin falling at normal impact rebounded of the horizontal plate under an angle (0–8°). This can be explained by the nonideal spherical shape of the pellets. During data processing no influence on the mean restitution coefficient was observed for rebound angles below 8°. Therefore, we determined the coefficient of restitution using free-fall tests, in which the rebound took place in this small angle range. The free-fall test was repeated for each model material until 28 nearly vertically rebounds (angle below 8°) were captured. The number of performed experiments for each material (steel and glass plates) was 116, i.e., the number of valid measurements was about 24%.

The coefficients of normal restitution of the sugar pellets was obtained at an impact velocity of $1.28 \pm 0.03$ m/s for particle–steel apparatus wall impact with $e_{\text{p-w}} = 0.54 \pm 0.08$ and for particle–glass wall impact with $e_{\text{p-p}} = 0.59 \pm 0.11$.

# 4. ANALYSIS OF THE PROCESS BEHAVIOR

## 4.1 Apparatus Geometry

In this contribution, the coating process in the Wurster coater apparatus produced by Glatt® (GPCG 3/5, Wurster 7) has been investigated. In Fig. 15, the geometry of the apparatus is illustrated.

The diameters of the inlet and outlet of the apparatus are equal to 180 and 500 mm, respectively. The Wurster tube has a diameter of 89 mm and a height of 330 mm.

The liquid is injected with a bottom-spray nozzle, which is situated in the center inside the Wurster tube. In the Wurster coater, air enters the equipment through the bottom via a segmented distributor plate that consists of a perforated plate, as illustrated in Fig. 16.

To perform DPM calculations, the internal volume of the Wurster coater was discretized in a mesh consisting of almost $10^6$ cells using the grid generator Gambit 2.4.6 (Ansys, Inc.). The geometry was divided into tetrahedral mesh cells according to the TGrid algorithm. To avoid local extrema of the solids concentration, the gas flow field is solved on a relatively coarse grid compared to the particle size.

To describe the effects of turbulent fluctuations of velocities on the pressure drop in the apparatus a $k-\varepsilon$ model (with the turbulence intensity of 5%)

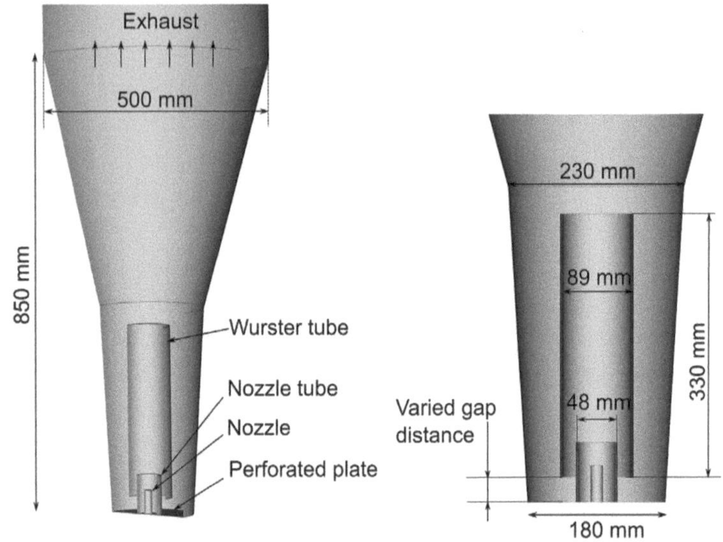


**Figure 15** Cross-cut view of the studied Wurster coater apparatus.

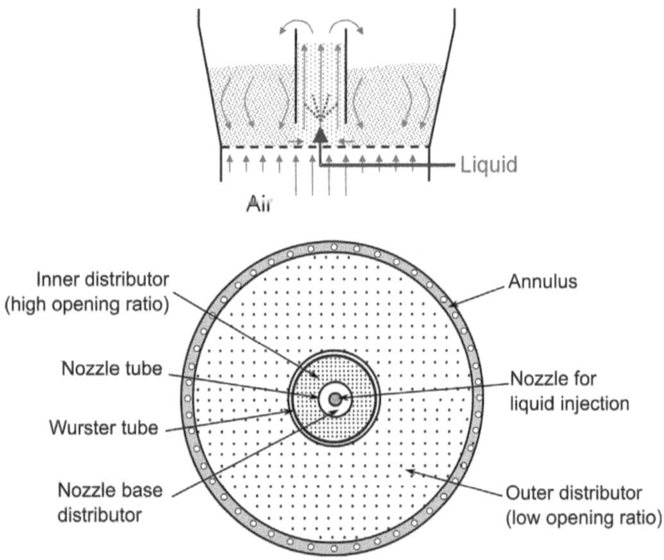


**Figure 16** Segmented gas distributor plate of the Wurster coater, with four homogeneous zones used in the simulations.

was applied for the calculation. This model showed a good agreement with the experiment for the spouted bed in the work of Gryczka et al. (2009). The influence of the wall presence on the turbulent flow has been calculated with standard wall function according to the proposal of Launder and Spalding (1974).

The gas injection by the nozzle is included as a free jet. The nozzle tip is a circular area (diameter 3 mm), where the gas velocity is prescribed by the boundary condition (velocity inlet). The gas volume around the nozzle tip has a refined mesh compared to the rest of the volume of the coater (see Fig. 17).

## 4.2 Process and Material Parameters

For CFD calculations, the ANSYS Fluent system was used. In order to specify correct boundary conditions, four different zones have been distinguished on the perforated plate and for each zone the separate gas inlet velocities were specified as boundary conditions. The diameter and number

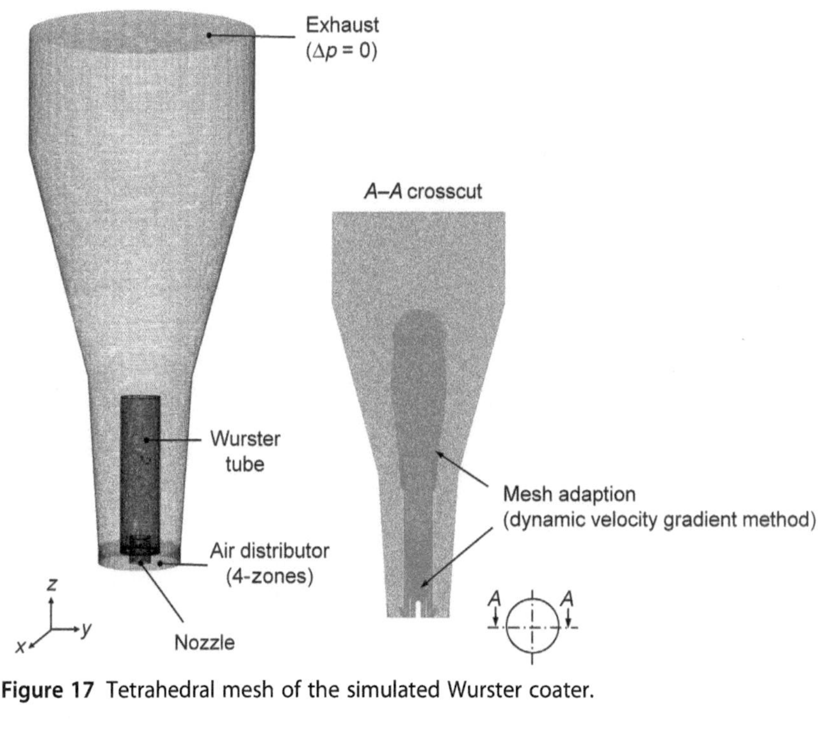


**Figure 17** Tetrahedral mesh of the simulated Wurster coater.

of orifices is different for different zones of the perforated plate. Therefore, based on the equal pressure drop for all zones, the gas velocity for each zone has been approximated. For this purpose, the correlation proposed by Hunt et al. (1955) and McAllister et al. (1958) was used:

$$\Delta p_{\text{distr}} = K \cdot \frac{\rho_{\text{g}}}{2}\left(\frac{\dot{V}_{\text{g}}}{\Psi}\right)^2 \cdot \left(5 - 4\Psi + (1-\Psi)^2\right) \tag{58}$$

where $K$ is the outflow coefficient and $\Psi$ is the open area ratio (Mörl et al., 2007). As outlet pressure condition the atmospheric pressure was chosen.

According to the scaling laws, described above, the material parameters for a simulation have been calculated. In Table 4, the scaled DPM parameters are compared with the real data. As it can be seen here, all dimensionless numbers and velocities are the same in the real and simulated process. The obtained mechanical parameters of particles (Table 5) described in the previous section were used in the contact model.

Firstly, three different case studies of particle dynamics under dry conditions (without injection of the binder) have been performed until steady-state (time interval 3–4 s). The values of main parameters, which have been varied, are listed in Table 6. The first case study has been used as a basic parameter set. In the second study, the influence of atomization gas flow rate has been investigated. In the third study, the gap distance below the Wurster tube has been increased from 22 to 42 mm.

**Table 4** Comparison of the Scaled DPM Parameters and Real Experimental Parameters

| Parameters | Experiment | Simulation |
|---|---|---|
| Particle diameter $d_{50,3}$ (mm) | 0.55 | 2.96 |
| Bed mass (kg) | 3.00 | 0.56 |
| Number of particles (–) | 21,796,034 | 140,000 |
| Particle density (kg/m$^3$) | 1580 | 295 |
| Minimal fluidization velocity (m/s) | 0.14 | 0.14 |
| $Re_{\text{mf}}$ (–) | 4.2 | 4.2 |
| $Ar$ (–) | 9219 | 9219 |
| Gas flow (m$^3$/h) | 180 | 180 |
| Gas density (kg/m$^3$) | 1.1894 | 1.1894 |
| Gas dynamic viscosity (Pa s) | 1.823°×°10$^{-5}$ | 9.809°×°10$^{-5}$ |

**Table 5** Mechanical Properties of Particles Parameters Used in the Contact Model

| Parameters | Value |
|---|---|
| **Restitution coefficient** | |
| Between dry particles | 0.54 |
| Between dry particle and dry steel wall | 0.59 |
| Young's modulus | 5.4 GPa |
| **Coefficient of the friction** | |
| Between particles | 0.86 |
| Between particles and steel wall | 0.48 |

**Table 6** Parameters of DPM Simulation for Three Case Studies

| Case Study | Case 1 | Case 2 | Case 3 |
|---|---|---|---|
| Atomization air flow velocity (m/s) | 80 | **120** | 80 |
| Gap distance (mm) | 22 | 22 | **42** |
| Fluidization gas flow rate ($m^3$/h) | 180 | 180 | 180 |

## 4.3 Results of the Microscale Simulation

### *4.3.1 Particle and Fluid Dynamics in the Wurster Tube and Annulus*

First simulations were performed for the empty apparatus without the solid particles. The atomizer gas flow velocity was set to 80 m/s. The Wurster gap distance was 22 mm. Figure 18 illustrates the simulated steady-state flow field in the empty apparatus. Figure 19 shows the time-averaged steady-state flow profile obtained with CFD simulation of the empty apparatus at a volume flow of 143 $m^3$/h. In the vicinity of the atomizer, the gas velocity reaches its maximum. Figure 19 shows the distribution of the gas velocity depending on the distance to the bottom plate. The flow becomes wider and slower with the apparatus height. At the small distances to the bottom the distribution of the air velocity is bimodal due to high differences in the air velocity in different zones of the bottom plate. At the height $H=820$ mm the velocity distribution shows only a small wide peak.

In Fig. 20, the instantaneous flow profile for a basic case study with particles colored according to their absolute velocity is shown. The particle velocity inside the tube increases with the height because of the acceleration of the particles due to process gas flow and atomizer gas flow. The gas

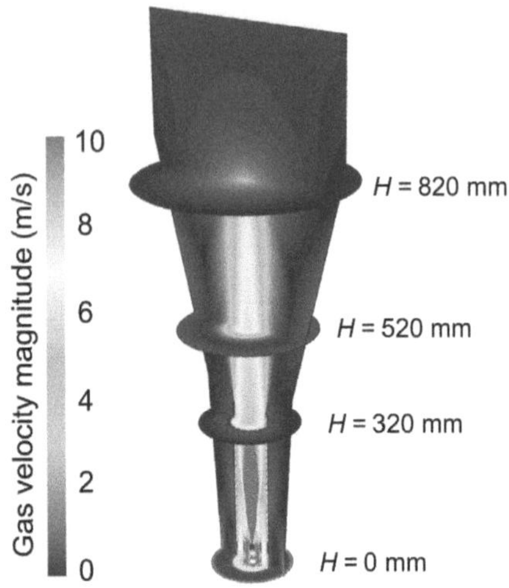


**Figure 18** Simulated steady-state flow field in the empty Wurster coater apparatus.

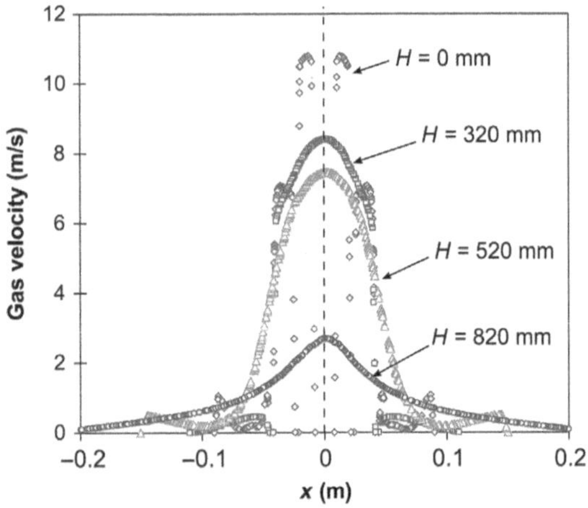


**Figure 19** The simulated velocity distribution on the different distances from the bottom plate in the empty Wurster coater.

velocity decreases with the apparatus height and the decelerating particles deviate from vertical air flow moving downward to the area of outside the Wurster tube where they are picked up again by the flow into the Wurster tube. Outside the Wurster tube, the particles are distributed over the full width of the expansion chamber before falling back down to the bottom zone. This results in low particle concentrations in the radial profile, which are decreasing toward the outer walls of the equipment.

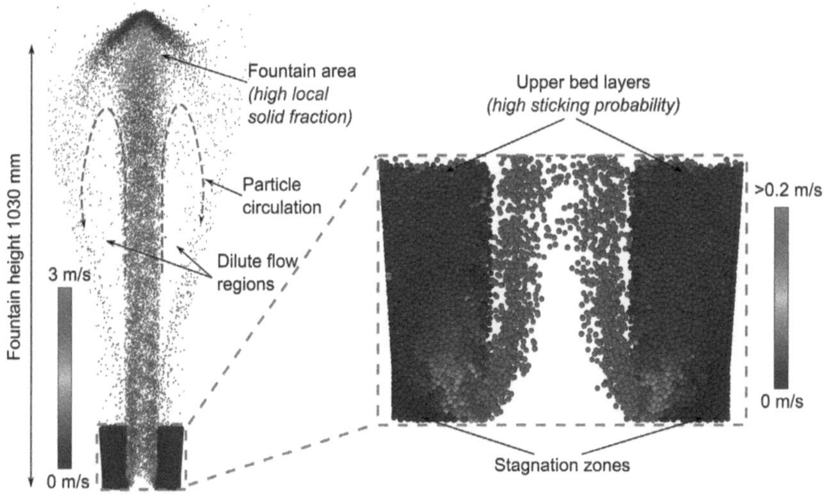


**Figure 20** Instantaneous particle velocity flow profile for basic case study.

In the left part of Fig. 20, the bottom zone of apparatus is shown. Here, the particles are sucked into Wurster tube and their recirculation occurs. To analyze the efficiency of the process, it is important to analyze the dynamics of particles in the bottom zone of apparatus. Here, the stagnation zones, where particles building almost fixed bed regions, can appear. As a consequence, this can lead to the nonuniform coating of material or to undesired agglomeration.

The particle dynamics in the apparatus was analyzed inside the Wurster tube (coating zone) and annulus (drying zone) and compared in some horizontal cross-sectional areas of the apparatus. The first cross-sectional area is situated at the height of $H = 22$ mm from the bottom plate and cuts the nozzle under its tip and the bed in the dense bottom zone. The second cross-sectional area is above at the upper end of the nozzle ($H = 100$ mm) and the third is at the upper end of the Wurster tube ($H = 320$ mm).

Figure 21 shows the radial distribution of the time-averaged particle velocities in the Wurster coater for at different heights from the bottom. Inside the Wurster tube the flow velocities are significantly higher than outside the tube. The particle velocities inside the tube clearly increase with the height because of the acceleration of the particles due to process gas flow and atomizer gas flow.

The particle movement is directed upwards inside the tube and downwards in the annulus as it can be seen in Fig. 20. This indicates that the circulating regime is intact over a wide fluidization range. From a height of

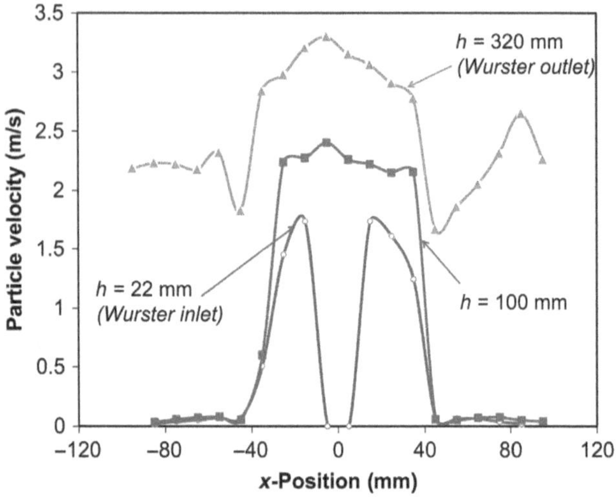


**Figure 21** Time-averaged radial distribution (absolute values) of the particles in the different zones of the Wurster coater for case study 1.

about 600 mm above the bottom of the apparatus a deceleration of the particles can be observed. The particles are moving downwards outside the tube, which indicates the increase the particle velocity at the height of 320 mm at $x > 42$ mm in Fig. 21. The peak of the particle velocity distribution inside the tube is shifted to a higher value in comparison to the entire apparatus in Fig. 22.

Figure 23 compares the particle–particle collision velocity distribution for the entire apparatus and inside the Wurster tube. Due to higher absolute particle velocities inside the tube the particles collide here at higher velocities than in the annulus. The average values of the parameters characterizing particle dynamics inside the Wurster tube are given in Table 7. The average particle–particle impact velocity is about 7 × smaller than the absolute particle velocity. The particles collide inside the tube mainly in normal direction. Therefore the normal component of the impact force is on the average 3.9 × higher than its tangential component.

Inside the Wurster tube, the particles show a high rotation (average rotational velocity equals 253.2 rad/s for the study case 1), which is very desirable for the homogenous distribution of the spray droplets on the particle surface to produce a homogenous and closed film. However in a moment only 3.48% of all particles are situated inside the tube. This results in a smaller number of collisions in comparison with the dense bed in the annulus

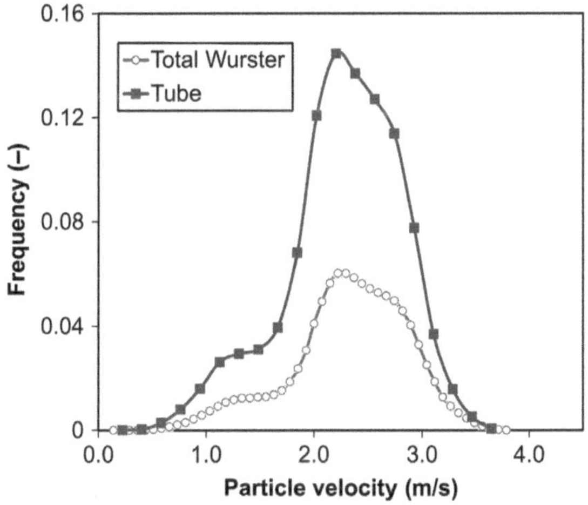


**Figure 22** Time-averaged distribution of the particle velocities in the whole apparatus (all particles) and inside the tube ($h \leq 320$ mm) for the case study 1.

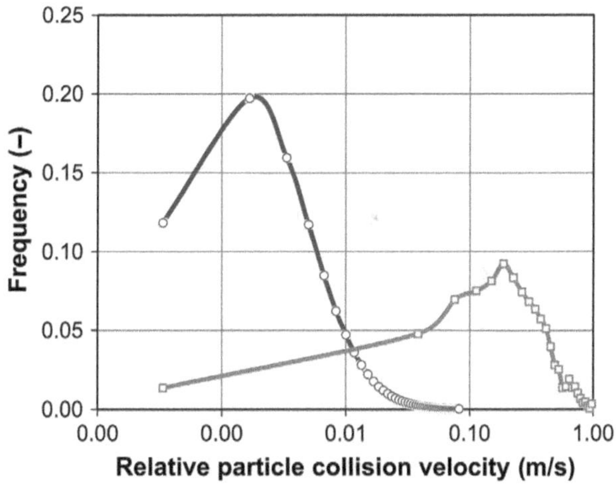


**Figure 23** Time-averaged particle–particle collision velocity distribution for entire apparatus and inside the Wurster tube (for the case study 1).

(Fig. 24). Generally, the higher collision velocity of wet particles inside the Wurster tube and the smaller number of collision lead to the smaller agglomeration rate and improves the coating process.

A quantitative comparison of the particle dynamics in the Wurster tube for all three case studies is presented in Table 8. The time-averaged distributions of the vertical component of translational velocity and angular velocity of particles inside the tube are shown in Figs. 25 and 26, respectively.

**Table 7** The Parameters Characterizing Particle Dynamics Inside the Wurster Tube for the Case Study 1

| Parameter | Value |
|---|---|
| Average particle fraction (%) | 3.48 |
| Average particle velocity (m/s) | 2.34 |
| Collisions per pair of particles ($s^{-1}$) | 102 |
| Average relative collision velocity ($p-p$) (m/s) | 0.33 |
| Average contact time ($p-p$) (μs) | 29 |
| Average collision force normal ($p-p$) (mN) | 3.61 |
| Average collision force tangential ($p-p$) (mN) | 0.93 |
| Average particle residence time (s) | 0.15 |

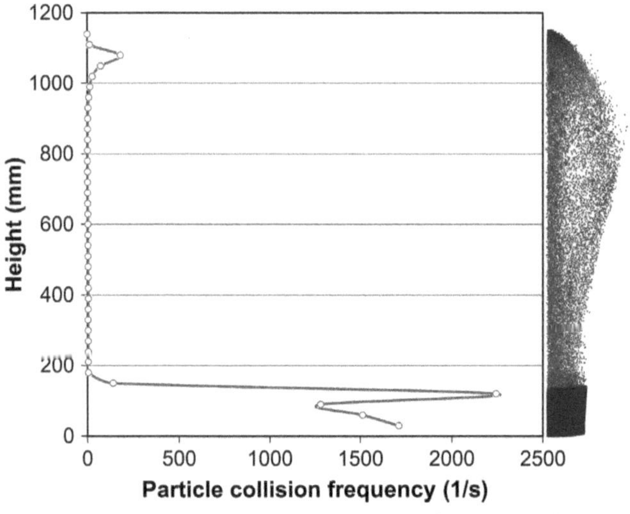


**Figure 24** Time-averaged particle–particle collision frequency depending on the height (case study 1).

By adjusting the gap distance below the Wurster tube (cases 1 and 3), the number of particles that are dragged into the tube can be controlled. It can be seen in Table 8 that larger gap sizes lead to increase of the particle concentration in the Wurster tube. It is found that the average solids volume fraction inside the draft tube increases by 19% from Case 1 to Case 3. At the same time, the average particle velocity is reduced by 6.8%, which leads to a small increase of their residence time in the tube. This is mainly related to the

**Table 8** Comparison of the Parameters Characterizing Particle Dynamics Inside the Wurster Tube

| Parameter | Case 1 | Case 2 | Case 3 |
|---|---|---|---|
| Average particle velocity in vertical direction (m/s) | 2.34 | 2.32 | 2.18 |
| Average particle velocity (m/s) | 2.36 | 2.33 | 2.2 |
| Average angular velocity (rad/s) | 125.23 | 126.82 | 97.15 |
| Average particle number in tube zone | 4960 | 4667 | 5881 |
| Average particle fraction (%) | 3.48 | 3.28 | 4.14 |
| Average particle residence time (s) | 0.15 | 0.15 | 0.16 |

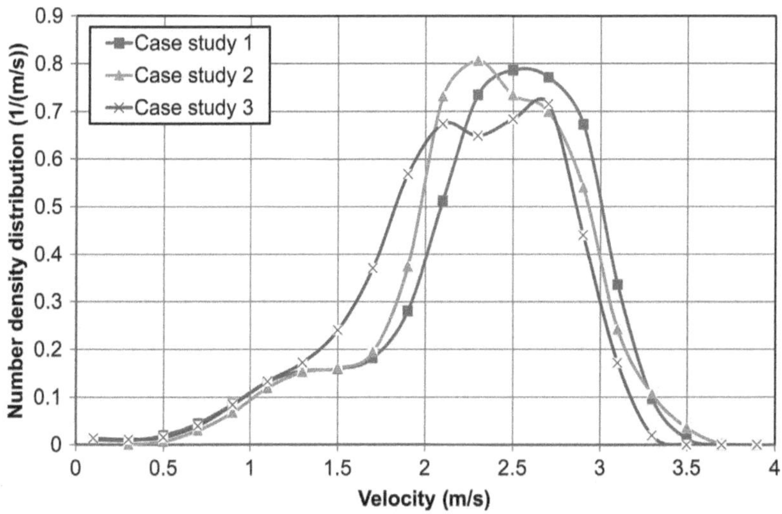


**Figure 25** Time-averaged distribution of the vertical velocity of the particles inside the tube.

increase of the pressure drop in the tube and to the formation of particle strands at the wall of the tube. In these regions of high solids volume fraction, the frictional energy dissipation is increased, leading to lower average particle velocities.

The radial distribution of the particle velocities depends significantly on the height, as it can be seen in Fig. 27. Here, the time-averaged total velocities and their vertical components are presented for different gap distances. The time-averaged total velocity of particles $\bar{v}$, which are moved on a radial distance $r$ from the central axis of the Wurster-coater apparatus, is given as:

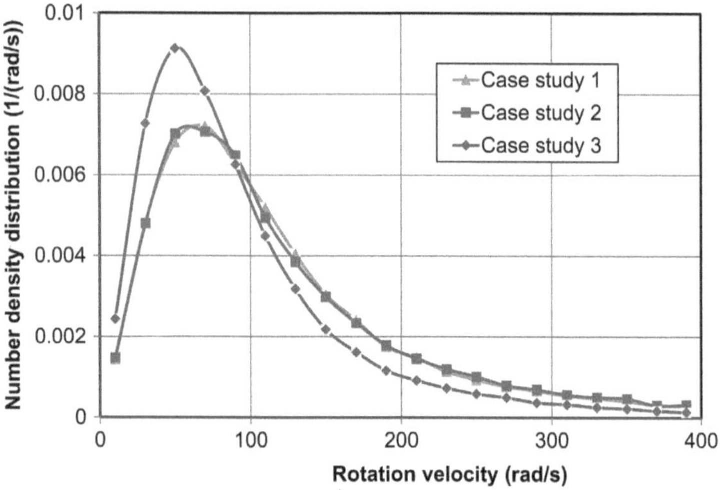


**Figure 26** Time-averaged distribution of the angular velocity of the particles inside the tube.

$$\bar{v}(r) = |\bar{v}(r)| = \frac{1}{N(r)} \sum_{i}^{N(r)} \sqrt{v_{x,i}^2(r) + v_{y,i}^2(r) + v_{z,i}^2(r)} \tag{59}$$

where $v_x$, $v_y$, $v_z$ are projections of the total velocity vector on the axis $x$, $y$, and $z$, respectively. Positive values of the vertical projection $v_z$ in Fig. 27 mean that it is directed upwards. The total number of particles $N(r)$, which are moved during the full simulation time on the radial distance $r$, was considered for the calculation of the distribution. In the case for a gap distance of 44 mm (case 3), the Wurster tube was lifted 22 mm upwards. For the comparison with the case 1, the velocities were evaluated also in the cross-sectional areas, which are 22 mm higher than in the case 1, i.e., 44, 124, and 344 mm.

In the zone of the tube inlet (at $r < r_{\text{tube}} = 0.0395$ m and at the height $H = 22$ mm and $H = 44$ mm for the cases 1 and 4, respectively) higher position of the tube resulting in increasing mean particle velocity in 14%; however, the vertical component of velocity decreases with the height. In the case of the higher position of the tube, particles move more in the radial direction to the centre of the tube. Figure 28 shows the instantaneous particle positions in these zones. It can be seen that the particle flow from the annulus into the tube begins at different heights for different gap distances. At the higher heights (200/220 and 320/344 mm), the increase of the gap distance leads to the decreasing average particle velocity inside the tube

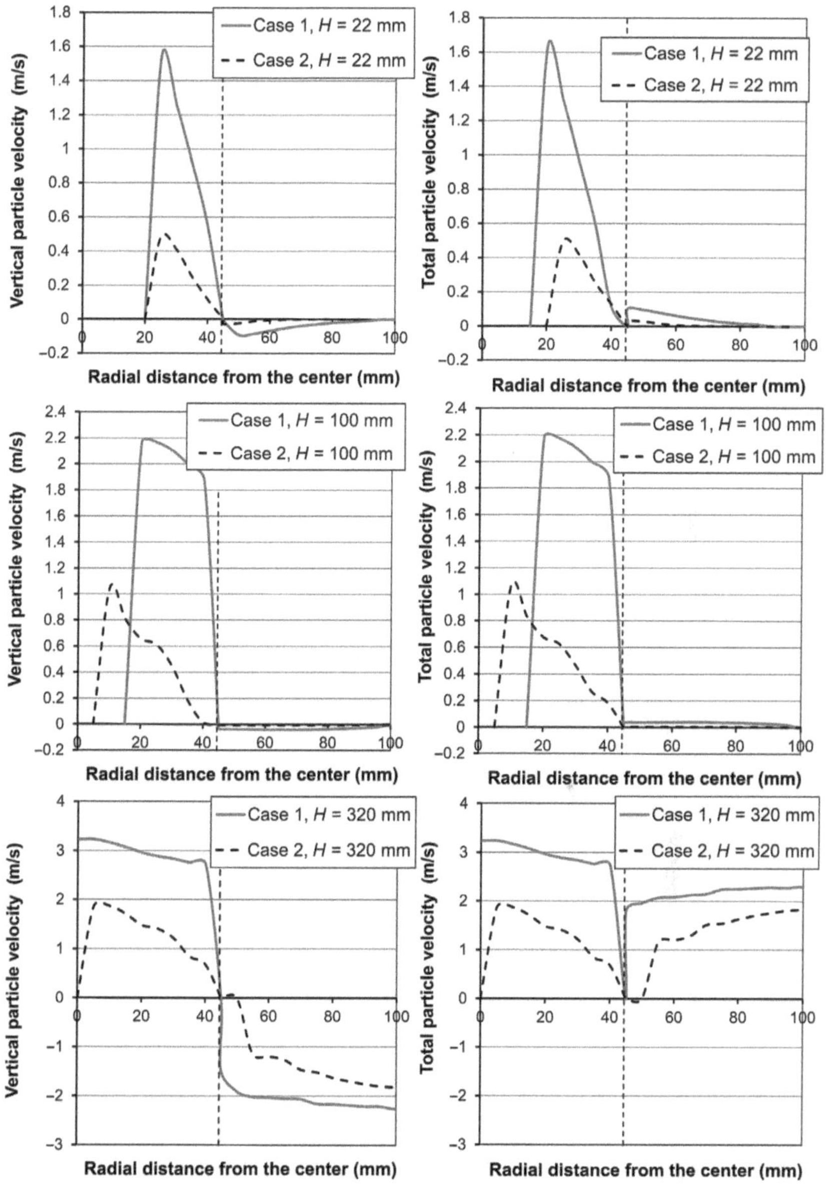


**Figure 27** Time-averaged radial distribution of vertical and total velocities of the particles in horizontal cross-sections at different distances from the bottom plate for case study 1 and 3.

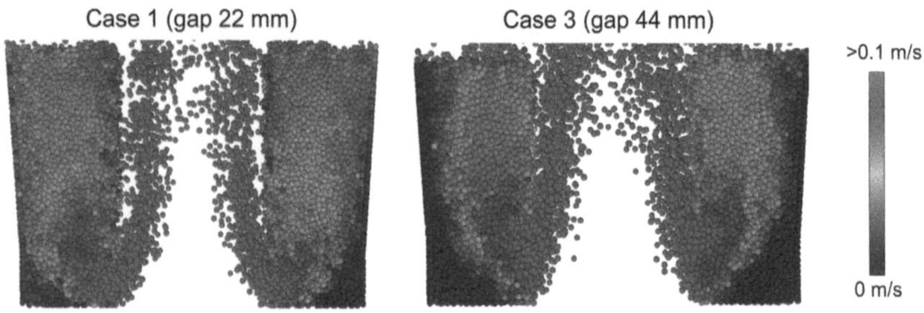


**Figure 28** Instantaneous particle positions and particle velocities in the lower part of the tube for two different gap distances: 22 mm (case 1) and 44 mm (case 3). The walls are not shown here.

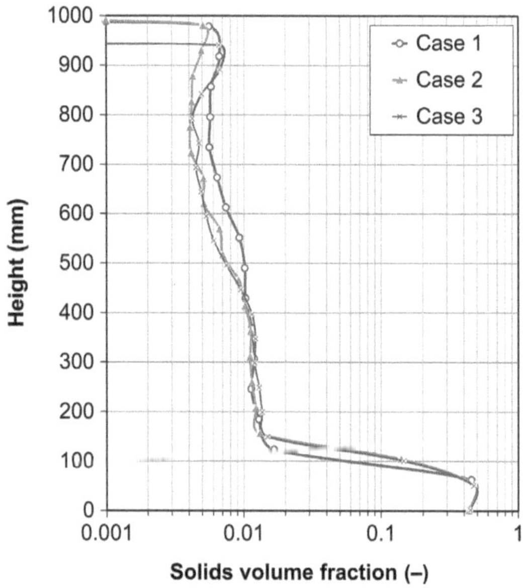


**Figure 29** Time-averaged solids volume fraction along the apparatus height.

(both the total and vertical velocities) in 1–5%. Due to the higher gap distance and smaller average velocities of the particles, the average solids volume fraction inside the tube in the study case 3 is higher than in the study case 1 (Table 8). The solids volume fraction Fig. 29 decreases with increase of the height. With increasing gap distance, the jet height decreases, as it can be seen in Fig. 29. The increase of the gap distance leads to a significant decrease of the particle mobility in the annulus (other bottom zone). From obtained results it can be seen that increase of atomization air flow leads to the higher particle trajectories. Same dependencies have been experimentally obtained by Karlsson et al. (2006) in their work, where the fountain area of the Wurster FB has been analyzed with high-speed video camera.

DPM simulation results summarized in Table 8 show that the jet injection via the nozzle has no significant influence on the particle dynamics in the granulator. The distribution of the particle velocities inside the tube is approximately identical for different atomization flow rates.

## 4.4 Mesoscale Model

In order to predict the sticking criteria of investigated granules, the submicroscale model has been used. In Table 9, the main material parameters used for calculations are listed.

In the three-dimensional dependency between restitution coefficient, impact velocity and liquid viscosity is illustrated. From obtained results it can be seen that during collisions with small impact velocities the restitution coefficient equals to zero and this indicates that particles stick to each other. Increasing of the liquid viscosity leads to an increased amount of energy dissipated in contact and as a consequence to the larger values of the minimal rebound velocity (Fig. 30).

In order to investigate the overspray rate for all case studies and to analyze the dependency of overspray on various parameters such as droplets size, initial droplets velocity and spraying angle a set of mesoscale calculations has been performed.

On the mesoscale, the liquid drying process was not simulated, that is why the droplets generation rate was not dependent on suspension mass flow and was equal to $8 \times 10^5$ droplets per second. With such a large droplet generation rate numerical error caused by random generation of droplets is avoided. To predict the size of droplets it was simply assumed that droplets have a monomodal distribution with diameter equals to 80 μm.

In Fig. 31, the dependency of overspray rate ($K_{os}$) on spraying angle for different case studies is shown. From the obtained results the conclusion can

**Table 9** Material Parameters Used for Simulation on Microscale

| Parameter | Value |
|---|---|
| Wetting degree | 10% |
| Liquid film thickness | 50 μm |
| Surface tension of liquid film | 0.1 N/m |

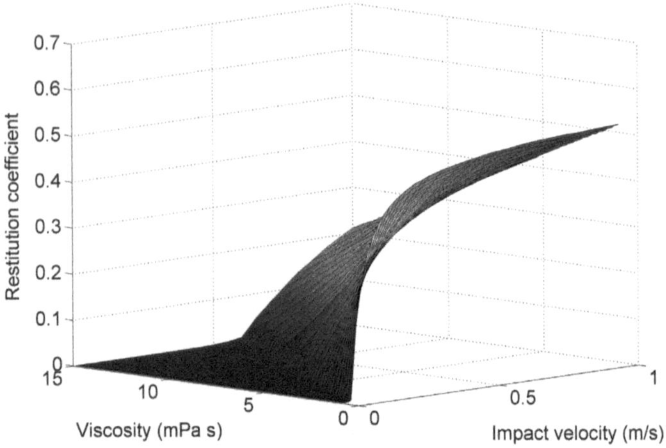


**Figure 30** Influence of impact velocity and liquid viscosity on restitution coefficient.

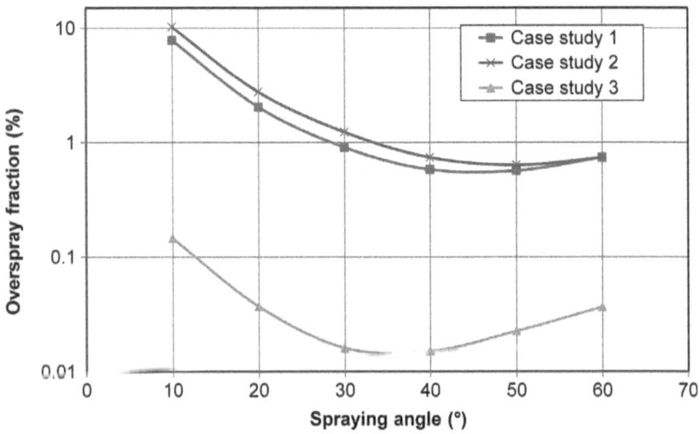


**Figure 31** Dependency of overspray on spraying angle for different case studies.

be drawn that in the case study 3, where the gap distance between bottom plate and Wurster tube was increased to 42 mm, the overspray rate is smaller than in all other studies. This is due to the larger amount of particles which are located in the tube zone of the apparatus. The comparison between case studies 1 and 2 shows that in the first case study the overspray rate is smaller than in the second. This happens due to the larger nozzle gas flow that leads to larger porosity in the vicinity of the nozzle orifice and faster motion of droplets in the vertical direction.

The variation of the spraying angle shows a similar influence in all four case studies. The increase of the spraying angle from 10° up to approximately

35° leads to the decrease of the vertical component of droplets initial velocity and, as a consequence, to the continuous decrease of the overspray fraction. During further enlarging of spraying angle, a larger amount of droplets collides with the walls of the Wurster tube that leads to an increase of overspray fraction.

The main aim of the simulations with uniform wetting zone was to obtain the wetted surface fraction of particles which, after being wetted in the tube zone, are recycled back to the upper layers of the bed (Fig. 6). This parameter can have a decisive influence on the growth dynamics. Due to the small relative velocity and large number of interparticle contacts, the high amount of the liquid on the particle surface in this region can lead to undesired product agglomeration.

The mesoscale simulations with uniform wetting zone have been performed for all case studies. In Fig. 32, the normalized particle distributions of three case studies are shown. These results have been obtained after process simulation for a time interval of 1.6 s with a gas temperature of 75 °C and suspension mass flow of 1 g/s. This is a sufficiently long interval for pellets to move from the wetting region through the fountain area and to fall down onto the bed surface.

From results represented in Fig. 32, it can be seen that the case study 3, when the gap distance between tube and gas distributor has been increased, has the narrowest distribution compared to the other studies. Therefore,

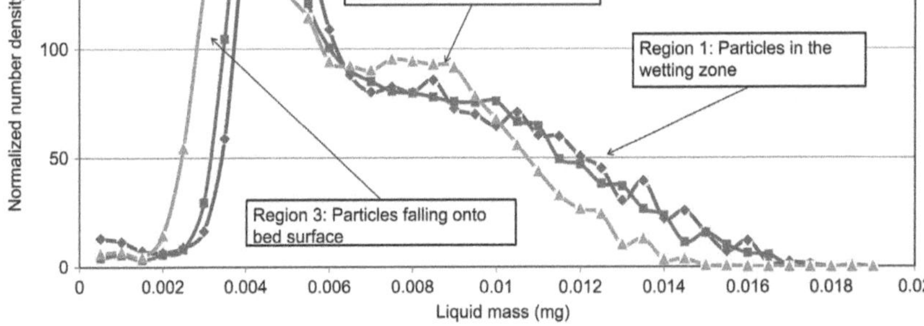


**Figure 32** Particle distribution according to amount of liquid film mass for different case studies.

increasing of the gap distance leads to a smaller amount of particles with large mass of liquid film. From the comparison between results obtained for the cases 1 and 2, the conclusion can be drawn that increasing of the atomization air flow has no significant influence on the liquid distribution of the pellets.

In the pellet distribution represented in Fig. 32, three different regions can be distinguished. The particles with the largest mass of liquid film are situated in the region 1. These particles are leaving the wetting zone. Afterwards, due to the drying the mass of the liquid on the particles is decreased and particles moved into the region 2. Finally, the particles shifted into the region 3. This region represents particles, which are falling down onto the bed. Here, the relative velocity between pellet and gas is very small that leads to the significant decrease of drying intensity.

## 4.5 Macroscale Calculations

To simulate particle growth dynamics, the one-dimensional PBM with parameters, which are listed in Table 10, have been used. The PSD of nuclei appearing due to the overspray and initial distribution of granules in apparatus have been distributed according to the normal (Gaussian) distribution.

In order to find a numerical solution of the PBM the finite volume method has been applied. The investigated size domain has been discretized with equal-sized grid into 1000 intervals and partial differential equation has been represented as a system of ODE's. In Fig. 33, the time progressions of Sauter mean diameter of PSD for three different overspray rates are shown.

**Table 10** Process Parameters Used in the PBM

| Parameter | Value/-s |
|---|---|
| Initial bed mass (kg) | 3 |
| Solution flow (kg/h) | 2 |
| Solid fraction (wt%) | 14 |
| Initial PSD median (mm)/deviation (mm) | 0.55/0.1 |
| Nuclei PSD median (mm)/deviation (mm) | 0.12/0.05 |
| $K_{dust}$ (Vol.-%) | 50 |
| $K_{os}$ (Vol.-%) | 4 |

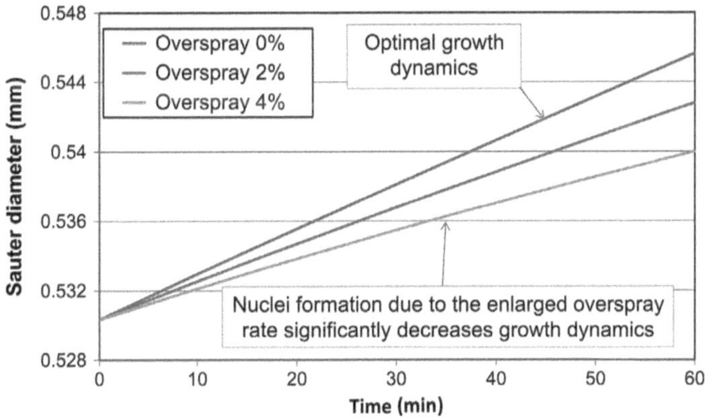


**Figure 33** Time progression of PSD.

## 5. CONCLUSIONS

In this work, a numerical study of the particle and fluid dynamics in the Wurster coater apparatus (Glatt® GPCG 3/5, Wurster 7) for coating of sugar pellets has been performed. The process was modeled using a multi-scale approach. On the microscale the FB process has been simulated using DPM. This model couples the DEM, which describes the motion and interaction of every individual particle, and the CFD, which calculates the process gas dynamics in the apparatus. To reduce the high-computational effort of the simulation, the DPM model was scaled. The relevant properties of solid and gas phase have been adapted to save fluid and particles dynamics of the investigated apparatus, which is characterized by the similarity of the relevant dimensionless numbers, i.e., Archimedes and Reynolds numbers.

To predict the real interactions of the particles in the FB, their material properties were obtained experimentally and used as parameters of DEM contact model. The deformation and breakage behavior of pellets were obtained with the help of single particle compression tests. The energy loss during the collisions was obtained by means of free-fall impact tests.

Three different case studies have been performed using DPM and the influence of such parameters as fluidization gas flow rate, atomization gas flow rate, and gap distance between tube and perforated bottom plate has been described in terms of the particle and collision dynamics.

The results of the microscale simulations showed that the gap distance between Wurster tube and bottom plate strongly influences the particle dynamics in the Wurster coater as it controls the recirculation of particles into the spout. The gap distance has a significant effect on the particle circulation and on the particle concentration in the spray zone. The larger gap distance increases the particle concentration in the Wurster tube and reduces the particle translational and rotational velocities, and the collision velocity as well as the fountain height. The increase of the gap distance leads to a significant decrease of the particle mobility in the annulus.

The atomization jet flow rates in the studied range do not have a significant influence on the particle dynamics. Only in the inlet and outlet areas of the Wurster tube, a small increase of particle velocity with increasing atomization air flow was obtained.

In the submodel on the mesoscale simulation, the overspray part of suspension and its influence on the coating layer growth dynamics have been investigated. The droplets have been considered as discrete elements and the Newton–Euler equations of motions have been solved for all droplets. The dynamics of solid pellets and fluid profile were taken from the previously obtained DPM results. To describe the particle growth the one-dimensional PBM has been used. From the analysis of the obtained results, the following main conclusions can be drawn:

- Increase of the gap distance between Wurster tube and perforated bottom plate leads to a decrease of the overspray fraction. Therefore a smaller amount of suspension droplets hits the apparatus walls and leaves the FB as dust. This happens because an increase of gap distance leads to a larger volume fraction of particles in the Wurster tube. However, such an increase of the gap distance leads to larger stagnation zones in the corners of the apparatus. As a consequence, nonuniform coating of pellets can occur.
- An increase of the atomization gas flow leads to a small increase of the overspray fraction. In the case of the larger atomization flow, a smaller amount of pellets is situated near the nozzle and as a consequence a larger amount of droplets dried before they contact particle. However, it should be mentioned that the influence of atomization gas flow on the droplet diameter was not considered in this study.

## ACKNOWLEDGMENTS

The authors gratefully acknowledge Merck Sharp and Dohme for financial support of this work and appreciate for the many helpful discussions with Pavithra Sundararajan and Gerard Byrne from Merck Sharp and Dohme Limited.

## REFERENCES


Antonyuk S, Heinrich S, Dosta M: Energetic aspects of granule impact: influence of the liquid layers. In *Powders and grains 2009: proceedings of the 6th international conference on micromechanics of granular media*, vol 1145, 2009, pp 796–799.

Antonyuk S, Heinrich S, Tomas J, Deen NG, van Buijtenen MS, Kuipers JAM: Energy absorption during compression and impact of dry elastic-plastic spherical granules, *Granul Matter* 1(12):15–47, 2010.

Barasso D, Eppinger T, Pereira FE, et al: A multi-scale, mechanistic model of a wet granulation process using a novel bi-directional PBM-DEM coupling algorithm, *Chem Eng Sci* 123:500–513, 2015.

Beetstra R, van der Hoef MA, Kuipers JAM: Drag force of intermediate Reynolds number flow past mono- and bidisperse arrays of spheres, *AIChE J* 53:489–501, 2007.

Bouffard J, Bertrand F, Chaouki J: A multiscale model for the simulation of granulation in rotor-based equipment, *Chem Eng Sci* 81:106–117, 2012.

Crown C, Schwarzkopf JD, Sommerfeld M, Tsuji Y: *Multiphase flows with droplets and particles,* ed 2, Boca Raton, 2011, CRC Press, p. 509.

Crüger B, Salikov V, Heinrich S, et al: Coefficient of restitution for particles impacting on wet surfaces: an improved experimental approach, *Particuology*, 2015. http://www.sciencedirect.com/science/article/pii/S1674200115001157.

Cundall PA, Strack ODL: A discrete numerical model for granular assemblies, *Geotechnique* 29:47–65, 1979.

Deen NG, van Sint Annaland M, van der Hoef MA, Kuipers JAM: Review of discrete particle modeling of fluidized beds, *Chem Eng Sci* 62(1–2):28–44, 2007.

Dosta M: *Dynamic flowsheet simulation of solids processes and its application to fluidized bed spray granulation,* Göttingen, 2013, Cuvillier Verlag, p. 142.

Dosta M, Antonyuk S, Heinrich S: Multiscale simulation of the fluidized bed granulation process, *Chem Eng Technol* 35:1373–1380, 2012.

Ergun S: Fluid flow through packed columns, *Chem Eng Prog* 48(2):89–94, 1952.

Fries L, Dosta M, Antonyuk S, Heinrich S, Palzer S: Moisture distribution in fluidized beds with liquid injection, *Chem Eng Technol* 34:1076–1084, 2011a.

Fries L, Antonyuk S, Heinrich S, Palzer S: DEM-CFD modeling of a fluidized bed spray granulator, *Chem Eng Sci* 66:2340–2355, 2011b.

Ge R, Ye J, Wang H, Yang W: Measurement of particle concentration in a Wurster fluidized bed by electrical tomography sensors, *AIChE J* 60:4051–4064, 2014.

Gnielinski V: Wärme- und Stoffübertragung in Festbetten, *Chem Ing Tech* 52:228–236, 1980.

Gryczka O, Heinrich S, Deen NG, et al: Characterization and CFD-modeling of the hydrodynamics of a prismatic spouted bed apparatus, *Chem Eng Sci* 64(14):3352–3375, 2009.

Heinrich S: *Modellierung des Wärme- und Stoffübergangs sowie der Partikelpopulationen bei der Wirbelschicht Sprühgranulation,* Düsseldorf, 2001, VDI-Verlag.

Heinrich S, Peglow M, Ihlow M, Mörl L: Particle population modeling in fluidized bed-spray granulation—analysis of the steady state and unsteady behaviour, *Powder Technol* 138:154–161, 2003.

Hertz H: Über die Berührung fester elastischer Körper, *J Reine Angew Math* 92:156–171, 1882.

Hunt CD'A, Hanson DN, Wilke CR: Capacity factors in the performance of perforated-plate columns, *AIChE J* 1:4, 1955.

Ingram GD, Cameron IT: Formulation and comparison of alternative multiscale models for drum granulation. In *Proceedings of ESCAPE-15.* 2005, pp 481–486.

Karlsson S, Björn IN, Folestad S, Rasmuson A: Measurement of the particle movement in the fountain region of a Wurster type bed, *Powder Technol* 165:22–29, 2006.

Karlsson S, Rasmuson A, van Wachem B, Björn IN: CFD modeling of the Wurster bed coater, *AIChE J* 55:2578–2590, 2009.

Kumar J, Peglow M, Warnecke G, Heinrich S, Mörl L: Improved accuracy and convergence of discretized population balance for aggregation: the cell average technique, *Chem Eng Sci* 61:3327–3342, 2006.

Kumar S, Ramkrishna D: On the solution of population balance equations by discretization —I, *A fixed pivot technique. Chem Eng Sci* 51:1311–1332, 1996.

Lama RD, Vutukuri VS: *Series on rock and solid mechanics, Bd. II: handbook on mechanical properties of rocks – testing techniques and results*, Clausthal, 1978, Trans Tech Publications.

Launder BE, Spalding DB: The numerical computation of turbulent flows, *Comput Methods Appl Mech Eng* 3:269–289, 1974.

Li L, Rasmuson A, Ingram A, et al: PEPT study of particle cycle and residence time distributions in a Wurster fluid bed, *AIChE J* 61:756–768, 2015a.

Li L, Remmelgas J, van Wachem BGM, et al: Residence time distributions of different size particles in the spray zone of a Wurster fluid bed studied using DEM-CFD, *Powder Technol* 280:124–134, 2015b.

Lian G, Thornton C, Adams MJ: A theoretical study of the liquid bridge force between rigid spherical bodies, *J Colloid Interface Sci* 161:138–147, 1993.

Link JM, Godlieb W, Deen NG, Kuipers JAM: Discrete element study of granulation in a spout-fluidized bed, *Chem Eng Sci* 62:195–207, 2007.

Link JM, Godlieb W, Tripp P, et al: Comparison of fibre optical measurements and discrete element simulations for the study of granulation in a spout fluidised bed, *Powder Technol* 189:202–217, 2009.

Litster JD, Smit DJ, Hounslow MJ: Adjustable discretization balance for growth and aggregation, *AIChE J* 41:591–603, 1995.

Luśtrik M, Sibanc R, Srćic S, Perpar M, Žun I, Dreu R: Characteristics of pellet flow in a Wurster coater draft tube utilizing piezoelectric probe, *Powder Technol* 235:640–651, 2013.

McAllister RA, McGinnis PH Jr., Plank CA: Perforated-plate performance, *Chem Eng Sci* 9:25–35, 1958.

Mörl L, Heinrich S, Peglow M: Fluidized bed spray granulation. In Salman AD, Hounslow MJ, Seville JPK, editors: *Handbook of powder technology*, vol. 11, Amsterdam, 2007, Elsevier Science, pp 121–188.

Mueller R, Kleinebudde P: Comparison of a laboratory and a production coating spray gun with respect to scale-up, *AAPS PharmSciTech* 8:1–11, 2007.

Palzer S: *Anreichern und Benetzen von pulverförmigen Lebensmitteln mit Flüssigkeiten in diskontinuierlichen Mischaggregaten*, 2000. Ph.D. Thesis, Technical University of Munich.

Panao MRO, Moreira ALN: Experimental study of the flow regimes resulting from the impact of an intermittent gasoline spray, *Exp Fluids* 37:834–855, 2004.

Perpar M, Luśtrik M, Dreu R, Srćic S, Žun I: Estimating coating quality parameters on the basis of pressure drop measurements in a Wurster draft tube, *Powder Technol* 246:41–50, 2013.

Pitois O, Moucheront P, Chateau X: Liquid bridge between two mobbing spheres: an experimental study on viscosity effects, *J Colloid Interface Sci* 231:26–31, 2000.

Popov V: *Kontaktmechanik und Reibung,* Berlin, 2011, Springer Verlag, p. 374.

Rajniak P, Stepanek F, Dhanasekharan K, Fan R, Mancinelli C, Chern RT: A combined experimental and computational study of wet granulation in a Wurster fluid bed granulator, *Powder Technol* 189:190–201, 2009.

Ramkrishna D: *Population balances: theory and applications to particulate systems in engineering,* ed 1, New York, NY, 2000, Academic Press.

Salikov V, Antonyuk S, Heinrich S, Sutkar VS, Deen NG, Kuipers JAM: Characterization and CFD-DEM modelling of a prismatic spouted bed, *Powder Technol* 270(Part B):622–636, 2015.

Schirmer R: Die Diffusionszahl von Wasserdampf-Luftgemischen und die Verdampfungsgeschwindigkeit, *VDI Beiheft Verfahrenstechnik* 170:170–177, 1938.

Schulze D: *Powders and bulk solids. Behavior, characterization, storage and flow,* Berlin, 2008, Springer Verlag. 512 pp.

Sutkar VS, Deen NG, Mohan M, et al: Numerical investigations of a pseudo-2D spout fluidized bed with draft plates using a scaled discrete particle model, *Chem Eng Sci* 104:790–807, 2013.

Teunou E, Poncelet D: Batch and continuous fluid bed coating—review and state of the art, *J Food Eng* 53:325–340, 2002.

Thornton C, Ning Z: A theoretical model for the stick/bounce behaviour of adhesive, elastic-plastic spheres, *Powder Technol* 99:154–162, 1998.

Tomas J: Particle adhesion fundamentals and bulk powder consolidation, *KONA Powder Particle* 18:157–169, 2000.

Toschkoff G, Khinast JG: Mathematical modeling of the coating process, *Int J Pharm* 457:407–422, 2013.

Toschkoff G, Just S, Funke A, et al: Spray models for discrete element simulations of particle coating process, *Chem Eng Sci* 101:603–614, 2013.

Tsuji Y, Tanaka T, Ishida T: Lagrangian numerical simulation of plug flow of cohesionless particles in a horizontal pipe, *Powder Technol* 71:239–250, 1992.

Tsuji Y, Kawaguchi T, Tanaka T: Discrete particle simulation of two-dimensional fluidized bed, *Powder Technol* 77(1):79–87, 1993.

Uhlemann H, Mörl L: *Wirbelschicht-Sprühgranulation,* Berlin, 2000, Springer Verlag, p. 617.

Varga CM, Lasheras JC, Hopfinger EJ: Initial breakup of a small-diameter liquid jet by a high-speed gas stream, *Journal of Fluid Mechanics* 497:405–434, 2013.

Walzel P: Spraying and atomizing of liquids. In *Ullmann's encyclopedia of industrial chemistry*, vol. 34, Weinheim, 1993, Wiley-VCH, pp 79–98.

Wen CY, Yu YH: A generalized method for predicting the minimum fluidization velocity, *AIChE J* 12(3):610–612, 1966.

Werther J: Fluidized bed reactors. In *Ullmann's encyclopedia of industrial chemistry*, Weinheim, 2007, Wiley-VCH.

Werther J, Heinrich S, Dosta M, Hartge E-U: The ultimate goal of modeling—simulation of system and plant performance, *Particuology* 9:320–329, 2011.

Willett C, Johnson S, Adams M: Pendular capillary bridges. In Salman AD, Hounslow MJ, Seville JPK, editors: *Handbook of powder technology: granulation*, Amsterdam, 2007, Elsevier Science (Chapter 28).

Wurster DE: Particle coating proces. US Patent No. 3.253.944, Wisconsin Alumni Research Foundation, 1966.

Zank J, Kind M, Schlünder E-U: Particle growth in a contniuously operated fluidized bed granulator, *Dry Technol* 19:1755–1772, 2001.

CHAPTER THREE

# Euler–Lagrange Modeling of the Hydrodynamics of Dense Multiphase Flows

**Johan T. Padding, Niels G. Deen, E.A.J.F. (Frank) Peters, J.A.M. (Hans) Kuipers[1]**

Department of Chemical Engineering and Chemistry, Eindhoven University of Technology, Eindhoven, The Netherlands

[1]Corresponding author: e-mail address: j.a.m.kuipers@tue.nl

## Contents

1. Introduction 138
2. Governing Equations for the Continuous and Discrete Phases 141
   2.1 Balance Equations for the Continuous Phase 141
   2.2 Equations of Motion for the Discrete Phase 145
   2.3 Interphase Momentum Transfer 145
   2.4 From the Eulerian Grid to Lagrangian Positions and Back 148
   2.5 Numerical Techniques: Deterministic Approach 151
   2.6 Numerical Techniques: Stochastic Approach 153
3. Dimensionless Numbers for Multiphase Flows 156
4. Flows with Particles: Discrete Particle Models (CFD–DEM) 159
   4.1 Particle–Fluid Interactions 159
   4.2 Particle–Particle Interaction 165
   4.3 Application: Gas–Solid Fluidized Beds 167
5. Flows with Bubbles: DBMs 167
   5.1 Bubble Interactions 167
   5.2 Application: Bubble Column Reactors 170
6. Flows with Droplets: DSMC 172
   6.1 Binary Droplet–Droplet Interactions 172
   6.2 Application: Droplet Sprays for Spray Drying 176
7. Summary and Outlook 180
Acknowledgments 182
Appendix A. Derivation of the Filtered Fluid Equation 182
Appendix B. Direct Numerical Simulation of Droplet Collision Outcomes 186
References 188

*Advances in Chemical Engineering*, Volume 46
ISSN 0065-2377
http://dx.doi.org/10.1016/bs.ache.2015.10.005

## Abstract

The large-scale hydrodynamic behavior of relatively dense dispersed multiphase flows, such as encountered in fluidized beds, bubbly flows, and liquid sprays, can be predicted efficiently by use of Euler–Lagrange models. In these models, grid-averaged equations for the continuous-phase flow field are solved, where the grid size is larger than the discrete phase size, while the discrete phase is explicitly tracked and experiencing forces in a Lagrangian fashion. In this chapter, we provide a summary of our own efforts in this field, including details which we deem necessary for a novice to be aware of. We start with a theoretical introduction to Euler–Lagrange models, emphasizing the importance of the availability of high-quality correlations for the interphase momentum transfer and the outcome of binary interactions between members of the discrete phase. Then, in three topical sections, we discuss implementations of the methods which are used intensely in our group: the computational fluid dynamics/discrete element method (CFD–DEM), discrete bubble method (DBM), and direct simulation Monte Carlo (DSMC). CFD–DEM is most suitable for solid particles moving in a gas. The interplay between hydrodynamic flow and dissipative collisions between these particles leads to inhomogeneities at meso- and larger scale. DBM applies to bubbly flows, where the additional complication of coalescence and splitting of bubbles needs to be taken into account accurately. DSMC is suitable for not-too-dense systems of particles or droplets in a gas (dispersed volume fraction less than 10%). Collisions between the discrete phase elements are detected stochastically from the local number density, relative velocities, and sizes of neighboring dispersed elements, leading to a considerable saving of computer time. We end with an outlook into directions of research which would lead to an even more comprehensive use of Euler–Lagrange models in the future.

## 1. INTRODUCTION

In this chapter, we focus on our efforts to model dispersed multiphase flows in which a discrete phase (consisting of solid particles, gas bubbles, or liquid droplets) is moving through, or is moved by, a continuous Newtonian fluid phase. Such flows appear frequently in process equipment in the chemical, metallurgical, pharmaceutical, and food industries. Examples include fluidized bed reactors, spouted bed reactors, pneumatic conveyors, bubble column reactors, slurry reactors, and spray driers. Figure 1 shows a schematic overview of typical dispersed multiphase systems.

In most of the above-mentioned systems, inhomogeneous mesoscale structures appear in the spatial distribution of the discrete phase, caused by an interplay of hydrodynamic flow and enhanced energy dissipation associated with mutual particle, bubble, or droplet collisions (Li and Kuipers, 2005; Sundaresan, 2000). In turn, these mesoscale structures influence the

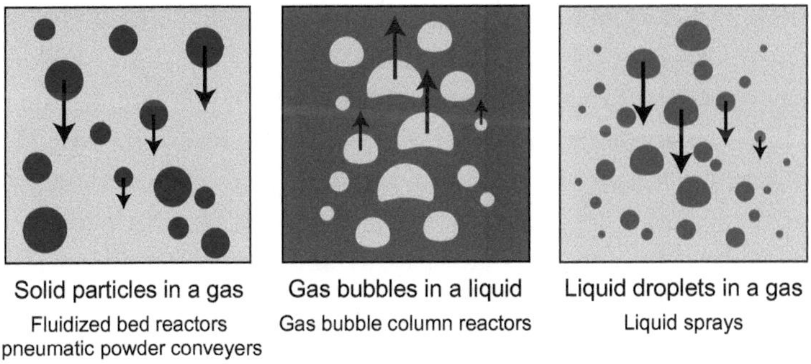


**Figure 1** Schematic examples of multiphase systems in which a discrete dispersed phase is moving through, or moved by, a continuous fluid phase. The discrete phase can be a solid (left), a gas (center), or a liquid (right). In many cases, inhomogeneous mesoscale structures appear in the spatial distribution of the discrete phase, caused by interplay of hydrodynamic flow and local energy dissipation. More complicated cases with three or more phases are also possible, such as encountered in slurry reactors (where solid particles are also present in the continuous liquid phase) or trickle bed reactors (where the droplets are sprayed on a packed bed of particles). To focus on the essentials, the topical sections will focus mostly on the two-phase examples depicted here.

mixing rates, heat transfer rates, and chemical species mass transfer rates. Therefore, it is very important to understand and to predict the (influence of) mesoscale structures appearing in multiphase systems. This can not only be achieved through careful experimentation but increasingly also through computer simulation. The rising interest in computer simulation is driven by the availability of ever faster computing facilities and by the availability of validated simulation models which can provide us with detailed insights that are very difficult, if not impossible, to obtain from experiments.

A number of methods exist to simulate dispersed multiphase flows. When choosing a particular simulation method, it is important to consider first the relevant length scales. The most obvious length scales are, from large to small, the dimensions of the confinement (equipment dimensions), the dimensions of the discrete elements (particles, bubbles, or droplets), and the mean free path of the molecules in the continuous fluid phase. The molecular mean free path ranges from less than a nanometer in a liquid to the order of 100 nm in a gas at ambient pressure. Discrete molecular effects such as Brownian forces and molecular slip conditions are therefore very important in nanofluidic and small microfluidic devices (Hadjiconstantinou, 2006). They are also very important for the dynamic behavior of nano(structured) particles in gas flows and colloidal particles suspended in a liquid. In these situations, particle-based simulation methods are ideally suited to investigate the dynamic processes (Padding and Louis, 2006; Zhang, 2011). Still, for the majority of current process equipment, the confining dimensions are much larger than a micrometer and may even reach up to several meters. The discrete molecular nature of the continuous fluid can then be safely neglected. This naturally leads to a computational approach in which the Navier–Stokes momentum equations are spatially and temporally discretized, i.e., computational fluid dynamics (CFD).

Within the realm of CFD, there are still a number of choices to make. The most accurate approach is to perform direct numerical simulations (DNSs) resolving all flow details, down to the Kolmogorov turbulent length scale and, when dealing with bubbles or droplets, including discrete phase surface deformations and internal flow (see, e.g., Gorokhovski and Herrmann, 2008; Tenneti and Subramaniam, 2014). However, because the dimensions of the process equipment are usually many orders of magnitude larger than the dimensions of the discrete phase, such a detailed approach is computationally far too expensive. Even with fully parallelized codes running on supercomputers, only a few thousand particles, bubbles, or droplets can be treated with such a high level of detail. It therefore makes sense to not try to resolve the detailed flow of the fluid but rather use correlations for the interphase forces (i.e., drag, lift, etc.) and subgrid scale turbulent effects. Using these correlations, equations can be solved for the averaged (filtered) continuous-phase flow field. This is called the Euler–Lagrange approach, because the continuous phase is treated in an Eulerian CFD fashion, while the discrete particles, bubbles, or droplets are treated explicitly in a Lagrangian fashion. It is important to realize that the continuous phase not only influences the dynamics of the discrete phase, but the discrete phase can also have a profound influence on the dynamics of the continuous fluid phase. Only if the volume fraction of the discrete phase is extremely low (much less than $10^{-5}$ for a typical gas–solid system) can this back-coupling to the continuous phase be neglected. In such extremely dilute systems, the interactions between the discrete particles themselves can also be neglected. We refer to Balachandar and Eaton (2010) for an excellent review focusing on the effects of turbulence in such dilute multiphase flows.

Here, we will explore Euler–Lagrange methods for the case of relatively dense multiphase flows where both two-way coupling between the discrete and continuous phase and interactions between individual members of the discrete phase are important. A number of reviews have already appeared in the past focusing on relatively dense solid particle–fluid flows (Deen et al, 2007; Portela and Oliemans, 2006; Silva et al, 2015; Subramaniam, 2013; Van der Hoef et al, 2008; Zhu et al, 2007), bubbly flows (Jakobsen et al, 1997, 2005), and flows with droplets (Jamaleddine and Ray, 2010; Jiang et al, 2010; Kolakaluri et al, 2014). The goal of this chapter is to provide an up-to-date overview of the recent activities of our own group, with a focus on exposing the similarities in modeling different multiphase systems. To this end, Sections 2 and 3 provide a general theoretical introduction to the governing equations, numerical methods, and dimensionless analysis, applicable to most multiphase systems. Subsequently, three topical sections follow, each treating a particular multiphase systems coupled to a particular Euler–Lagrange method. These topical sections serve as an illustration of different Euler–Lagrange methods, which may however also be applied to other or more complex multiphase systems. Section 4 treats the discrete particle model for solid particle flows, Section 5 treats the discrete bubble model (DBM) for bubbly flows, and Section 6 treats the direct simulation Monte Carlo (DSMC) model for flows with droplets. We end with a summary and outlook to future directions of research.

## 2. GOVERNING EQUATIONS FOR THE CONTINUOUS AND DISCRETE PHASES

### 2.1 Balance Equations for the Continuous Phase

The balance equations for the continuous fluid phase follow from the conservation of mass and momentum and are given by

$$\frac{\partial}{\partial t}(\varepsilon_f \rho_f) + \nabla \cdot (\varepsilon_f \rho_f \mathbf{u}) = 0 \tag{1}$$

$$\frac{\partial}{\partial t}(\varepsilon_f \rho_f \mathbf{u}) + \nabla \cdot (\varepsilon_f \rho_f \mathbf{u}\mathbf{u}) = -\nabla p - \nabla \cdot \boldsymbol{\tau}_f + \mathbf{S}_{p\to f} + \varepsilon_f \rho_f \mathbf{g} \tag{2}$$

where $\varepsilon_f$ is the fluid volume fraction, $\rho_f$ its mass density, $\mathbf{u}$ the fluid velocity, $p$ the fluid pressure, $\boldsymbol{\tau}_f$ the fluid stress tensor, $\mathbf{S}_{p\to f}$ the force density exerted by the discrete elements (particles) on the fluid, and $\mathbf{g}$ the gravitational acceleration. These equations are spatially filtered or smoothed equations. By construction, the velocity field of the fluid phase varies slowly over particle length scale. Variations on the particle scale are not resolved. This means that forces exerted by the fluid on discrete phase cannot be computed from the flow field but need to be obtained from a closure relation.

In the literature, there has been quite some discussion on the precise form of the governing equations. For example, Gidaspow (1994) distinguished models A and B in the setting of the two-fluid model (see also Bouillard et al, 1989). Zhou et al (2010) discuss the differences between different sets of governing equations in the setting of CFD–DEM (discrete element method). They define equation-sets I, II, and III. In their view, set I is the most basic equation, which gives rise to set II follows after splitting of the force due to a macroscopic pressure gradient (as we will do in Section 2.3). Their equation-set III follows after rather restrictive assumption. The model A of Gidaspow follows from set II after inserting simple closure relations and neglecting some relatively unimportant terms such as the viscous stress contributions, but model B is equivalent to set III after applying closures and simplifications. Our Eq. (2) is the same as set I of Zhou et al and thus closer to model A than B. We overall agree with Zhou et al's recommendations on the proper model definition and usage.

The following expressions relate the "micro"-quantities to the filtered or smoothed macro-quantities

$$\varepsilon_{\mathrm{f}}(\mathbf{x}) = \iiint_{V_{\mathrm{f}}} D(\mathbf{x}-\mathbf{y})\,\mathrm{d}V_y = 1 - \iiint_{V_{\mathrm{p}}} D(\mathbf{x}-\mathbf{y})\,\mathrm{d}V_y = 1 - \varepsilon_{\mathrm{p}}(\mathbf{x}) \tag{3}$$

$$\rho_{\mathrm{f}}(\mathbf{x}) = \frac{\iiint_{V_{\mathrm{f}}} D(\mathbf{x}-\mathbf{y})\rho_{\mathrm{f}}^{\mathrm{micro}}(\mathbf{y})\,\mathrm{d}V_y}{\iiint_{V_{\mathrm{f}}} D(\mathbf{x}-\mathbf{y})\,\mathrm{d}V_y} \tag{4}$$

$$\mathbf{u}(\mathbf{x}) = \frac{\iiint_{V_{\mathrm{f}}} D(\mathbf{x}-\mathbf{y})\rho_{\mathrm{f}}^{\mathrm{micro}}(\mathbf{y})\mathbf{u}^{\mathrm{micro}}(\mathbf{y})\,\mathrm{d}V_y}{\iiint_{V_{\mathrm{f}}} \rho_{\mathrm{f}}^{\mathrm{micro}}(\mathbf{y}) D(\mathbf{x}-\mathbf{y})\,\mathrm{d}V_y} \tag{5}$$

where the integration is performed over the fluid phase and the integration variable is $\mathbf{y}$. Here, a smoothing function or filter kernel, $D(\mathbf{x})$, is used. This function has a typical width of few times a particle diameter and its volume integral equals 1. Note that the velocity is a mass-averaged velocity. With these definitions, the smoothed continuity equation is exactly recovered from the resolved continuity equation.

In Appendix A, we provide a derivation of the macroscopic momentum balance equations from resolved ones by means of smoothing. The subtlety in deriving the basic equations lies in where factors $\varepsilon_\mathrm{f}$ appear in these equations, e.g., in or outside spatial derivatives. The explicit derivation is provided, because it illustrates that explicit use of the assumption of the separation of length scales is needed to get this correct. There is a microscale, namely, the size of the elements of the discrete phase. For now, we will talk about the elements of the discrete phase as "particles", but the discrete phase can also consist of e.g., droplets or bubbles. The filter kernel has a typical length scale of several times the particle size, $d_\mathrm{p}$. After filtering, the smoothed fluid fields are expected to vary significantly only over length scales much larger than this filter length scale. If this is the case, a range of filters with a width in between the micro and the macro length scale will all give the same equations for the fluid phase. Therefore, the fluid-phase equations are almost invariant to the precise choice of the filter. Also, the derivation provides more detailed results, such as a precise form of the smoothed viscous stress. It is, for example, illustrated that the properly smoothed viscous stress is different from the relation that is commonly assumed (which is the same as the microscopic relation between stress and deformation rate).

An interesting historic note concerns the proper prediction of the pressure drop in a fluidized bed. It is an empirical fact that in a stationary fluidized bed, the pressure drop (times area) is close to the particle plus fluid weight. From Eq. (2), this result can be easily obtained: Assume stationary state such that the left-hand side equals zero and assume that the contribution of the fluid stress is negligible. The two external forces acting on particles are the force applied by the liquid on the particles $\left(\mathbf{S}_{\mathrm{f}\to\mathrm{p}} = -\mathbf{S}_{\mathrm{p}\to\mathrm{f}}\right)$ and the gravitational pull on the particles. If we assume that the particle phase has no net acceleration, the two forces need to (nearly) balance, so $\mathbf{S}_{\mathrm{f}\to\mathrm{p}} = \varepsilon_\mathrm{p}\rho_\mathrm{p}\mathbf{g}$ and therefore the (average) pressure gradient is close to $\left(\varepsilon_\mathrm{p}\rho_\mathrm{p} + \varepsilon_\mathrm{f}\rho_\mathrm{f}\right)\mathbf{g}$. Some of the early models did not correctly reproduce this result as discussed by Xu and Yu (1998).

The smoothing of the convective term in the momentum balance gives extra contributions that can be incorporated into the stress term, i.e.,

$$\boldsymbol{\tau}_\mathrm{f}^{\mathrm{conv}} = \frac{1}{\varepsilon_\mathrm{f}} \iiint_{V_\mathrm{f}} D(\mathbf{x}-\mathbf{y})\rho_\mathrm{f}^{\mathrm{micro}}(\mathbf{y})\left(\mathbf{u}^{\mathrm{micro}}(\mathbf{y}) - \mathbf{u}(\mathbf{x})\right)\left(\mathbf{u}^{\mathrm{micro}}(\mathbf{y}) - \mathbf{u}(\mathbf{x})\right) \mathrm{d}V_y \tag{6}$$

This expression is known as the Reynolds stress in turbulence theory. However, also in laminar flow, due to the presence of the particles, there can be significant local subscale velocity variations that give rise to contributions to the smoothed stress. Because the particles and fluid are coupled through (slip, no slip, or partial slip) boundary conditions, the magnitude of the local subscale fluid velocity fluctuations is expected to scale with the fluctuations in the particle's velocities. Therefore, one might try to model the subscale fluid velocity variations as a type of pressure that is correlated with the local granular temperature and macroscopic slip velocity. Recently, a "pseudo-turbulent" Reynolds stress has been investigated using particle resolved flow simulations by Mehrabadi et al (2015). Note, however, that in that paper an ensemble-averaged Reynolds stress is considered, while for the stress needed is more of a subgrid stress similar to that is used in large eddy simulations of turbulence. Note that for high Reynolds numbers flow sometimes single-phase turbulence models are implemented. Well-validated closures for the pseudo-turbulence stress that include the effects of particles are currently not available.

Collecting all stress terms for the fluid up to second order, we find,

$$\boldsymbol{\tau}_{\mathrm{f}} = \boldsymbol{\tau}_{\mathrm{f}}^{\mathrm{Newt}} + \varepsilon_{\mathrm{f}} \boldsymbol{\tau}_{\mathrm{f}}^{\mathrm{conv}} + \sum_{\mathrm{i}} D(\mathbf{x} - \mathbf{x}_{\mathrm{i}}) \mathbf{s}_{\mathrm{i}} \tag{7}$$

The last term $\mathbf{s}_i$ is a contribution of individual particles. The symmetric part of these particle contributions are called stresslets. $\mathbf{s}_i$ also has antisymmetric contributions that are due to the torque of the particles applied to the fluid. In early models, none of these stress terms was accounted for. The general reasoning is that the force density corresponding to the divergence of the fluid stresses is small compared to the fluid–particle momentum exchange term. This is not necessarily the case in regions that only contain a low amount of particles. Nowadays, the viscous Newtonian stress term is usually modeled. The pseudo-turbulent fluid stress contribution is neglected except in turbulent flows where a single-phase turbulence model is sometimes implemented. Clearly, more accurate modeling of this term would be desirable, but that is currently not available. The (higher) order particle stresslet contribution is usually not included in the modeled macroscopic fluid stress. In the applications presented in Section 4, only the viscous Newtonian contribution is taken into account.

For the given derivation to be fully valid, a separation of length scales needs to be assumed. The macroscopic quantities should vary only over sizes that are large compared to the size of the used filter. In practice, this is often not the case. For example, in a bubbling fluidized bed, the fluid volume fraction changes within a few particle diameters from a value about 0.4 in the emulsion phase to near 0 in a bubble. These kinds of sudden macroscopic jumps are actually smoothed out by the use of a filter kernel. Very often, for this reason, the macroscopic solids volume fraction is not determined by means of a filter, but by locally determining the volume occupied by particles in a computational grid cell. This enables the method to deal with localized jumps better, but this also introduces a grid dependency introduced in the simulation method. At the local "discontinuities" of the macroscopic fields, the equations might break down and extra terms might be needed there that are grid dependent.

## 2.2 Equations of Motion for the Discrete Phase

The discrete phase is considered to be constituted by rigid bodies referred to as "particles" for now. This assumption is clearly an approximation if the discrete phase is in fact deformable, as is the case for droplets and bubbles. Rigid bodies can only translate and rotate. For the translational motion, the evolution of the center-of-mass velocity follows from Newton's equation, while for the rotations the change of angular velocity can be related to the torque (via the moment of inertia),

$$m_{\mathrm{i}}\frac{\mathrm{d}\mathbf{v}_{\mathrm{i}}}{\mathrm{d}t}=\mathbf{f}_{\mathrm{i}},\quad \frac{\mathrm{d}}{\mathrm{d}t}(\mathbf{I}_{\mathrm{i}}\cdot\boldsymbol{\omega}_{\mathrm{i}})=\mathbf{T}_{i} \tag{8}$$

Relevant forces are external volume forces like gravity, particle–particle interactions such as contact forces or electrostatic interactions, and the force exerted by the fluid on a particle. Some of these forces (that do not act on the center of mass) have a torque associated to it. The torque associated with the fluid–particle interaction follows from the antisymmetric part of the particle stress contributions. A detailed discussion of dominant forces in the case of gas–solid, liquid–bubble, and gas–droplet interactions will be provided in the topical sections (Sections 4–6).

## 2.3 Interphase Momentum Transfer

The force exerted by the fluid on a particle follows from integrating the total microscopic stress (pressure plus viscous stress) over the particle surface. Since the microscopic detail is lost from the description, closure relations need to be used that related the force to macroscopic quantities. It is customary to split the total fluid-to-particle force into different contributions. These individual forces are qualitatively different and have their own closure relations, which may also depend on the nature of the dispersed system (i.e., fluid–solid, gas–liquid, or liquid–gas). Details will be given in the topical sections (Sections 4–6), but generally speaking, the total force of the fluid exerted on particle i equals

$$\mathbf{f}_{\mathrm{f}\rightarrow\mathrm{i}} = -V_{\mathrm{i}}\nabla p + \mathbf{f}_{\mathrm{d,i}} + \mathbf{f}_{\mathrm{l,i}} + \mathbf{f}_{\mathrm{m,i}} \tag{9}$$

The first term corresponds to the buoyancy force. It is found when the stress contribution due to the macroscopic pressure is separated in the surface integral. The other force contributions are the drag force, $\mathbf{f}_{\mathrm{d,i}}$, the lift force, $\mathbf{f}_{\mathrm{l,i}}$, and added mass force, $\mathbf{f}_{\mathrm{m,i}}$. The drag force acts in the opposite direction of the relative velocity between particle and fluid. Dimensionless representations of the drag force are by means of a drag coefficient, $C_{\mathrm{D}}$, or a dimensionless Stokes coefficient, $F$,

$$\mathbf{f}_{\mathrm{d,i}} = -\frac{1}{2}C_{\mathrm{D}}A_{\mathrm{cs,i}}\rho_{\mathrm{f}}|\mathbf{v}_{\mathrm{i}} - \mathbf{u}|(\mathbf{v}_{\mathrm{i}} - \mathbf{u}) = -3\pi\mu_{\mathrm{f}}F d_{\mathrm{p,i}}(\mathbf{v}_{\mathrm{i}} - \mathbf{u}) \tag{10}$$

where $d_{\mathrm{p,i}}$ is the (characteristic) diameter of particle i, $A_{\mathrm{cs}}$ is the cross-sectional area ($A_{\mathrm{cs,i}} = (\pi/4)d_{\mathrm{p,i}}^2$ for spheres), and $\mu_{\mathrm{f}}$ is the fluid viscosity. For high Reynolds numbers, $C_{\mathrm{D}}$ is expected to be constant, while for low ones $F$ becomes constant. Typical correlations express $C_{\mathrm{D}}$ or $F$ in terms of local particle Reynolds numbers, $Re_{\mathrm{p,i}} = \rho_{\mathrm{f}}|\mathbf{v}_{\mathrm{i}} - \mathbf{u}|\varepsilon_{\mathrm{f}} d_{\mathrm{p,i}}/\mu_{\mathrm{f}}$, and fluid hold-up, $\varepsilon_{\mathrm{f}}$. Note that the most common way to define the particle Reynolds number is based on a superficial velocity. That is why the factor $\varepsilon_{\mathrm{f}}$ was included. The lift force is directed perpendicular to the direction of motion. It can be written as

$$\mathbf{f}_{\mathrm{l,i}} = -C_{\mathrm{L}}\rho_{\mathrm{f}} V_{\mathrm{i}}(\mathbf{v}_{\mathrm{i}} - \mathbf{u}) \times (\nabla \times \mathbf{u}) \tag{11}$$

For inviscid potential flow (using Bernoulli's equation to obtain the pressure) around a sphere, one finds that $C_{\mathrm{L}} = 1/2$. In this inviscid limit, the virtual or added mass force can also be computed using the instationary Bernoulli equation. In this case, for an accelerating spherical particle one finds

$$\mathbf{f}_{\mathrm{m,i}} = -C_{\mathrm{M}}\rho_{\mathrm{f}} V_{\mathrm{i}}\frac{D}{Dt}(\mathbf{v}_{\mathrm{i}} - \mathbf{u}) \tag{12}$$

with $C_{\mathrm{M}} = 1/2$. This force arises from the fact that if a particle accelerates also part of the surrounding fluid needs to be accelerated. For both the lift force and the added mass force, only few correlations are available that relate the coefficients to hold up and other dimensionless numbers.

Depending on the nature of the dispersed phase, some forces are important while others might be negligibly small. For example, for heavy particles, the drag force is the dominant force. For light particles, droplets, and bubbles, buoyancy forces will become important too. For bubbles, the added mass force will be much larger than the acceleration term in Newton's equation. Therefore, for accelerating bubbles, the acceleration is mostly determined by the added mass. In the topical sections (Sections 4–6), details of fluid–particle forces and their correlations will be discussed.

The forces exerted by a particle on the fluid obey Newton's third law $\mathbf{f}_{i\to f} = -\mathbf{f}_{f\to i}$ and by means of

$$\mathbf{S}_{p\to f} = \sum_{i} D(\mathbf{x} - \mathbf{x}_i)\mathbf{f}_{i\to f} \tag{13}$$

give rise to the source term in the fluid equation. Note that the buoyance term gives a contribution $\varepsilon_p \nabla p$ that can be combined with the pressure term already present in the fluid equation to give $-\nabla p + \varepsilon_p \nabla p = -\varepsilon_f \nabla p$. Very often, the pressure term of the fluid equation is already given in this form, and then the remaining source term is the sum of all forces excluding the buoyancy force.

In the closure relations for the fluid-to-particle interactions, it is usually assumed that the force on particle i only depends on particle i (but not on its neighbors) and on the macroscopic fluid velocity at the position of the particle. For some hydrodynamic forces, this is not a good assumption. For example, lubrication forces also arise from interactions between particles and the fluid phase, but these forces become dominant when particles are close. These are therefore better modeled as particle–particle interactions where the parameters are the distance between two particles and their relative velocities.

Closure relations can originate from three sources: experiments, theory, and computation. Usually, input from all three sources is used. For example, for the drag on a single sphere, very accurate experimental data are available for a very large range of Reynolds numbers. For the low Reynolds number limit, a rigorous theoretical result is available, i.e., the Stokes drag. Computations have confirmed these drag correlations. For the drag on a particle in the presence of many other particles, experiments are more difficult to perform. For example, for drag forces on solid particles, most commonly drag correlations are derived from pressure drop measurements for fixed arrays of particles. In this case, resolved DNSs become an attractive alternative. In theory, it enables the exact computation of the forces on individual particles. In practice, there are still computational bottlenecks that limit the accuracy of the results and have caused different authors to propose different correlations.

As discussed above, the fluid equations are derived using the assumption that macroscopic quantities like the solids volume fraction vary slowly in space. In that case, basing closure relations on local macroscopic variables seems reasonable. However, in reality, the solids volume fraction can change quite abruptly, e.g., at emulsion–bubble interfaces or when particle clusters are formed. It might be essential to incorporate these heterogeneities that arise on mesoscopic length scales into closure relations. We will return to this point in Section 4.1.

## 2.4 From the Eulerian Grid to Lagrangian Positions and Back

As we have seen in the preceding section, the coupling between the continuous and dispersed phase is treated through the interphase forces $\mathbf{f}_{f\to i}$ and corresponding force density $\mathbf{S}_{p\to f}$. To evaluate these forces, we need local properties, such as the local solids volume fraction and the local slip velocity, at the location of the particles. We also need a way to distribute the point force $\mathbf{f}_{f\to i}$ over a small but finite volume as a force density on the continuous phase. Theoretically, the smoothing function or filter kernel $D(\mathbf{x}-\mathbf{x}_i)$ introduced in Section 2.1 should be used to perform both these tasks. However, in numerical simulations, where the continuous-phase equations have been spatially discretized on a grid, it is rather tedious (though not impossible) to work with partial integrals of the smoothing function over the grid volumes. Therefore, in practice, it is preferred to use mapping functions which directly interpolate Eulerian grid point information to a specific Lagrangian point and, vice versa, distribute Lagrangian point information across the surrounding Eulerian grid points. We will refer to this as grid mapping. To avoid overly use of different symbols, we will still denote the mapping function with the symbol $D$, with the understanding that now a sum across the surrounding Eulerian grid points needs to be taken, instead of a volume integral.

In detail, an interpolated local property $\phi_f$ of the continuous fluid phase at the position $\mathbf{x}_i$ of a dispersed particle can be evaluated as:

$$\phi_f(\mathbf{x}_i) \approx \sum_f \widetilde{D}(\mathbf{x}_f - \mathbf{x}_i)\phi_f(\mathbf{x}_f) \tag{14}$$

$$\widetilde{D}(\mathbf{x}_\mathrm{f} - \mathbf{x}_\mathrm{i}) = \frac{D(\mathbf{x}_\mathrm{f} - \mathbf{x}_\mathrm{i})}{\sum_\mathrm{f} D(\mathbf{x}_\mathrm{f} - \mathbf{x}_\mathrm{i})} \tag{15}$$

where the f under the sum indicates that the sum must be performed over all fluid grid positions $\mathbf{x}_\mathrm{f}$ in the space in which the influence is accounted for. If the mapping function has a width of $n$, this leads to a mapping volume, $V_n = (2n)^3$ around the particle position $\mathbf{x}_\mathrm{i}$. Note that the normalization in the denominator is necessary to ensure exact conservation of properties during the mapping procedure because the sum of the mapping function values across the mapping volume is not necessarily equal to unity. In the often-used case of trilinear interpolation (*vide infra*), this sum is unity by construction, and only the numerator needs to be evaluated.

Vice versa, the (averaged) local property $\phi_\mathrm{p}$ of the dispersed phase at a particular grid position $\mathbf{x}_\mathrm{f}$ of the continuous fluid phase is expressed as:

$$\phi_\mathrm{p}(\mathbf{x}_\mathrm{f}) \approx \sum_\mathrm{i} \widetilde{D}(\mathbf{x}_\mathrm{f} - \mathbf{x}_\mathrm{i}) \phi_\mathrm{p}(\mathbf{x}_\mathrm{i}) \tag{16}$$

where the sum must be taken over all particles i in the mapping volume around the grid position $\mathbf{x}_\mathrm{f}$. This gives a partition of unity for the quantity defined at particle position i. Note that this is needed for conservation: $\sum_\mathrm{f} \phi_\mathrm{p}(\mathbf{x}_\mathrm{f}) = \sum_\mathrm{i} \phi_\mathrm{p}(\mathbf{x}_\mathrm{i})$. This way of dealing with the normalization has two advantageous properties: (1) for interpolation of fluid properties, a constant field gives the same constant value at the particle positions, and (2) for distribution of particle quantities to the fluid nodes, the quantities are partitioned such that the total sums do not change. When the *density* of a property is needed at the grid position (e.g., a force density in the case of interphase momentum transfer), the above expression should be divided by the grid cell volume $\Delta V$.

Usually, the mapping function $D$ is assumed to be identical and uncoupled in all three coordinate directions $\xi \in (x, y, z)$, so it can be written in a general form like:

$$D(\mathbf{x}_\mathrm{f} - \mathbf{x}_\mathrm{i}) = \begin{cases} \prod_\xi D_\xi(x_{\xi,\mathrm{f}} - x_{\xi,\mathrm{i}}) & \text{if } |x_{\xi,\mathrm{f}} - x_{\xi,\mathrm{i}}| < n \\ 0 & \text{else} \end{cases} \tag{17}$$

Often the choice of the mapping function is coupled to the computational grid. The most common method of this kind is trilinear interpolation:

$$D_\xi = 1 - \frac{\left|x_{\xi,\mathrm{f}} - x_{\xi,\mathrm{i}}\right|}{\Delta x_\xi}, \quad n = \Delta x_\xi \tag{18}$$

This method is very popular, particularly because it is cheap to compute and easy to implement and also because no further normalization is necessary. The biggest downside of this method is that its form is inherently grid dependent. Like with any other parts of Euler–Lagrange models, one would require the grid mapping to be grid independent. For this reason, several other mapping schemes have been suggested that are typically based on the following considerations (Deen et al, 2004):

1. The function should be smooth, i.e., the first derivative should be continuous.
2. It should have an absolute maximum around the particle position.
3. For practical reasons, it should have a small and finite domain. At the boundaries of the domain, the function should be zero.

Sometimes an additional requirement is given that the integral of the mapping function over the entire domain should equal unity. The important property in a discretized system is actually that when summing the mapping function over all grid points, the sum equals 1. This makes sure that conserved quantities remain conserved also in the discrete case and that interpolation of constant functions gives the same constant values.

We enforce this by means of the normalization. Clearly, if the integral of the function equals 1, the sum used for the normalization will be almost equal to 1 too and thus will not have a large influence.

Here, we discuss two examples of mapping functions that have been commonly used. The first one is a simple rectangular function:

$$D_\xi = \frac{1}{2n} \tag{19}$$

The second function is a cheap clipped fourth-order polynomial:

$$D_\xi = \frac{15}{16}\left[\frac{\left(x_{\xi,\mathrm{f}} - x_{\xi,\mathrm{i}}\right)^4}{n^5} - 2\frac{\left(x_{\xi,\mathrm{f}} - x_{\xi,\mathrm{i}}\right)^2}{n^3} + \frac{1}{n}\right] \tag{20}$$

Note that the rectangular function does not satisfy the first condition at $\left|x_{\xi,\mathrm{f}} - x_{\xi,\mathrm{i}}\right| = n$. We illustrate all functions in Fig. 2.

The influence of particularly the width of the mapping functions has been investigated by Link et al (2005) for gas–solid flows and by Darmana et al (2005) for gas–liquid bubbly flows. It was found that good

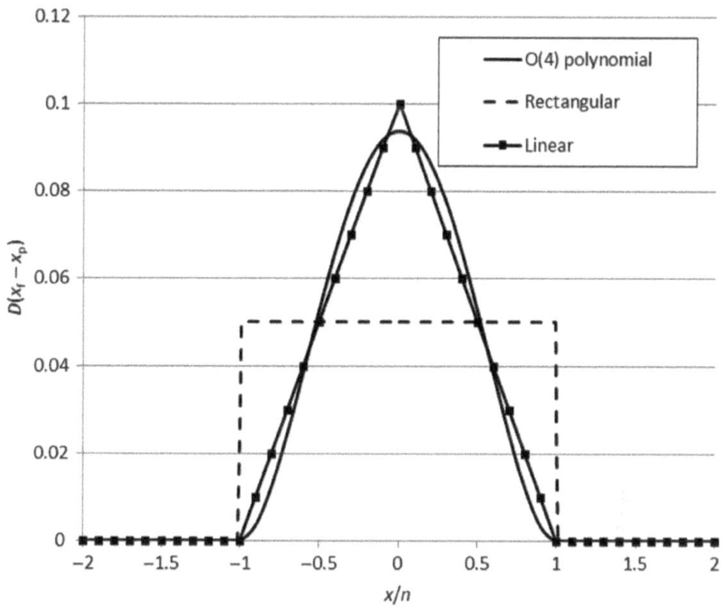


**Figure 2** Mapping functions $D_i$ for mapping Eulerian quantities to Lagrangian positions and vice versa.

results are obtained when a mapping window of $2n = 3–5d_p$ is used. This size of the mapping window is very similar to the grid resolution that is required for obtaining grid-independent results. Wang et al (2009) found that for grid sizes smaller than $\Delta x \approx 3d_p$, the main fluidization features do no longer change.

## 2.5 Numerical Techniques: Deterministic Approach

The governing momentum equations of the continuous phase are solved numerically after discretization in space and time. In almost all multiphase flows, the coupling between the continuous and dispersed phase dominates the transport equations of the continuous phase. When trying to solve these equations, it is therefore useful to locally linearize the interphase momentum force density as follows:

$$\mathbf{S}_{p\to f} = \frac{1}{\Delta V}\sum_{i} \widetilde{D}(\mathbf{x} - \mathbf{x}_i)\mathbf{f}_{i\to f} = \boldsymbol{\alpha} - \beta \mathbf{u}_f \tag{21}$$

The division by the grid cell volume $\Delta V$ turns the averaged force into a force density. The term $\beta$ represents the $\mathbf{u}_f$-dependent part of the interphase forces, while $\boldsymbol{\alpha}$ contains the $\mathbf{u}_f$-independent part of the interphase forces (both within the linearization approximation). For example, for the case of gas–solid flows, usually the only relevant interphase force is the drag force, in which case we find, with the help of Section 2.3:

$$\boldsymbol{\alpha} \equiv \frac{1}{\Delta V} \sum_i \widetilde{D}(\mathbf{x} - \mathbf{x}_i)\boldsymbol{\alpha}_i = \frac{1}{\Delta V} \sum_i \widetilde{D}(\mathbf{x} - \mathbf{x}_i)3\pi\mu_f F d_{p,i}\mathbf{v}_i \tag{22}$$

$$\beta \equiv \frac{1}{\Delta V} \sum_i \widetilde{D}(\mathbf{x} - \mathbf{x}_i)\beta_i = \frac{1}{\Delta V} \sum_i \widetilde{D}(\mathbf{x} - \mathbf{x}_i)3\pi\mu_f F d_{p,i} \tag{23}$$

where $\boldsymbol{\alpha}_i$ and $\beta_i$ represent the different parts of the interfacial force acting on particle i. Using this notation for the interphase momentum transfer, a first-order backward Euler scheme can be used to discretize the governing equations of the continuous phase in time:

$$\begin{aligned}(\varepsilon_f\rho_f\mathbf{u}_f)^{n+1} = &\ (\varepsilon_f\rho_f\mathbf{u}_f)^n \\ &+ \Delta t\left\{-\varepsilon_f\nabla p^{n+1} + \mathbf{C}^n + \mathbf{D}^n + \varepsilon_f\rho_f\mathbf{g} + \boldsymbol{\alpha}^n - \beta^n\mathbf{u}_f^{n+1}\right\}\end{aligned} \tag{24}$$

where the indices $n$ and $n+1$ denote the old and the new time level, respectively. **C** and **D** represent the convection and diffusion terms, which are usually discretized with a total variation diminishing (TVD) and central differencing scheme, respectively. Particularly, the use of a TVD scheme for the convection term is important, as it suppresses numerical diffusion in the fluid phase. Optionally, the diffusion terms can be treated implicitly (Patil et al, 2014).

Because the interphase momentum transfer is usually dominant, the interphase forces can best be treated in a semi-implicit manner. In particular, the last term on the RHS can be combined with the LHS to obtain:

$$\mathbf{u}_f^{n+1} = \frac{(\varepsilon_f\rho_f\mathbf{u}_f)^n + \Delta t\{-\varepsilon_f\nabla p^{n+1} + \mathbf{C}^n + \mathbf{D}^n + \varepsilon_f\rho_f\mathbf{g} + \boldsymbol{\alpha}^n\}}{(\varepsilon_f\rho_f)^{n+1} + \Delta t\beta^n} \tag{25}$$

Note that the denominator of this equation is positive, which has a stabilizing effect on the solution of the fluid velocity field. This is particularly important for cases where $\beta^n$ is very large, which is often the case in dense multiphase flows at large particle Reynolds numbers.

To solve the velocity field, usually a two-step approach is used where first a tentative velocity field is calculated using the pressure at the old time level. Subsequently, the complete expression using the tentative velocities is substituted in the continuity equation, leading to a Poisson equation. The latter is solved through an efficient solver methodology such as the incomplete Cholesky conjugate gradient method.

The momentum equations for the dispersed phase are treated in a somewhat similar way, with the main difference that due to the Lagrangian framework there are less terms in the equation:

$$m_\mathrm{i}\mathbf{v}_\mathrm{i}^{n+1} = m_\mathrm{i}\mathbf{v}_\mathrm{i}^{n} + \Delta t\left\{ -V_\mathrm{i}\nabla p^{n+1} + m_\mathrm{i}\mathbf{g} + \sum_\mathrm{p}\mathbf{F}_\mathrm{pi}^{n} + \boldsymbol{\alpha}_\mathrm{i}^{n+1} - \beta_\mathrm{i}^{n}\mathbf{v}_\mathrm{i}^{n+1}\right\} \tag{26}$$

Here, $\mathbf{F}_\mathrm{pi}$ is the contact force on particle i due to its interaction with particle p, and the terms $\boldsymbol{\alpha}_\mathrm{i}$ and $\beta_\mathrm{i}$ arise from the linearized versions of the interphase force acting on particle i, as introduced in Eqs. (21)–(23). The momentum equation for the dispersed phase can now be rewritten as:

$$\mathbf{v}_\mathrm{i}^{n+1} = \frac{m_\mathrm{i}\mathbf{v}_\mathrm{i}^{n} + \Delta t\left\{ -V_\mathrm{i}\nabla p^{n+1} + m_\mathrm{i}\mathbf{g} + \sum_\mathrm{p} F_\mathrm{pi}^{n} + \boldsymbol{\alpha}_\mathrm{i}^{\prime n+1}\right\}}{m_\mathrm{i} + \Delta t\beta_\mathrm{i}^{\prime n}} \tag{27}$$

Subsequently the positions of the particles are updated as:

$$\mathbf{x}_\mathrm{i}^{n+1} = \mathbf{x}_\mathrm{i}^{n} + \mathbf{v}_\mathrm{i}^{n+1}\Delta t \tag{28}$$

This is the scheme employed in the topical section (Section 5) on bubbly flows and Section 6 on droplet flows. Alternatively, one could use a second-order accurate time integration schemes to allow for possible rapid changes in the contact forces occurring between the particles. This is routinely done in flows with solid particles with stiff particle interactions (topical Section 4).

## 2.6 Numerical Techniques: Stochastic Approach

With an increasing number of particles (which can be solid particles, droplets, or bubbles), an increasing fraction of the computational time is spent on collision detection and calculating the contact forces $\mathbf{F}_\mathrm{pi}$. Even when optimal use is being made of spatial hashing techniques and neighbor lists, for systems containing more than a million particles usually the majority of the computational effort is spent on handling the particle–particle interactions. Fortunately, the very large number of particles can also be an advantage. It can be argued that at larger time and length scales, the details of individual particle collisions no longer matter, but rather the *statistics* of the particle interactions such as the particle spatial and relative velocity distribution functions. By making use of these statistics, the details of all individual particle interactions need no longer be taken into account explicitly. There are two approaches to this so-called coarse graining of the collision dynamics.

In the first coarse-grained approach, the discrete phase is treated as an Eulerian continuum, interpenetrating with the real continuous phase. The particle–particle interactions are then captured by an effective particle phase rheology obtained from kinetic theory of granular flows. These so-called two-fluid (Euler–Euler) models have been very successful at predicting the dynamic properties of, e.g., gas–solid fluidized beds (see Van der Hoef et al, 2008; Verma et al, 2013). Despite their success, two-fluid models also have their limitations: they are usually limited to idealized cases of monodisperse hard sphere particles, while extensions to polydisperse mixtures (e.g., in size or in contact properties) are difficult to make. Also, because no particles are explicitly tracked, it is difficult to include particle properties which may vary from particle to particle, such as particle temperature, surface moisture concentration, or chemical surface species concentrations.

These limitations can be alleviated in the second coarse-grained approach, where the particles are still explicitly tracked, but collisions with neighboring particles are determined stochastically through collision probability theory. This can be combined with a parcel approach where, to save computational costs, only a small subset of all real particles are simulated. For the calculations of the collision probabilities, the nonsimulated particles are then assumed to be in the neighborhood of the simulated particles and have the same statistical properties as the local ensemble of simulated particles. Such an approach is advantageous when dealing with extremely high numbers of particles. Two important methods that fall into this class are multiphase particle in cell (see Andrews and O'Rourke, 1996; Xie et al, 2013) and DSMC (see Bird, 1976; O'Rourke, 1981). Primarily because of our own experience with the method, here we will focus on DSMC.

DSMC was first introduced by Bird (1976) for the simulation of molecular gas flows, after which it was adapted to CFD simulations of granular particles or droplets by O'Rourke (1981). The method is highly efficient because the path of each simulated particle is updated independently from that of other particles in a sequence of smaller particle time steps, until a predetermined relatively large time step is reached. The duration of each particle time step is automatically adjusted to be less than the mean free time between collisions, to ensure that the particle will collide with at most one other neighboring particle. An actual collision with a neighboring particle within the same CFD cell may or may not be executed, depending on a probability determined by the particle number density, particle sizes, and particle relative velocity. If the collision is executed, momentum exchange is assumed to take place instantaneously. When the predetermined large time step is reached for the particle in question, the next particle is treated in the same way. And so forth, until all particles have progressed by one large time step. Because particle configurations do not change appreciably during one time step of the continuous-phase flow solver, the large DSMC time step can routinely be chosen equal to the flow solver time step $\Delta t$.

To avoid lattice artifacts caused by a lack of particle collisions across CFD cell boundaries, the approach of O'Rourke was later modified to have the neighbor searching region decoupled from the CFD grid and shaped like a sphere around the particle in question (Du et al, 2011; Pawar et al, 2014). The radius $R_s$ of this sphere, called the searching scope, is updated after each particle time step to ensure sufficiently that many neighboring particles are considered to accurately estimate the collision frequency. In particular, the collision frequency $f_i$ for particle i is estimated from:

$$f_i = \sum_{j \in R_{s,i}} f_{ij} \tag{29}$$

$$f_{ij} = |\mathbf{v}_{ij}| \frac{\pi}{4} \left(d_i + d_j\right)^2 \frac{n}{(4/3)\pi R_{s,i}^3} \tag{30}$$

Here, $|\mathbf{v}_{ij}|$ is the magnitude of the relative velocity between particle i and its neighboring particle j, $(\pi/4)\left(d_i + d_j\right)^2$ is the effective collision area for this pair, and $n$ is the parcel size. The sum runs over all neighboring particles j within the range $R_{s,i}$ of the particle in question. From the collision frequency $f_i$, the particle time step $\Delta t_{p,i}$ is calculated such that the displacement is less than a fraction of the mean free path; usually a fraction 1/3 is used to enforce a rigorous decoupling of free streaming and collision with just a single neighboring particle. Finally, a collision partner j is accepted with probability $P_{ij} = f_{ij} \Delta t_p$, and the searching scope radius is updated. For more technical details, the reader is referred to Pawar et al (2014).

The stochastic nature of the DSMC approach applies only to the particle–particle interactions. The equations of the continuous phase are treated exactly as in the deterministic approach explained in Section 2.5. Also, in between consecutive collisions, the particle positions and velocities are updated deterministically as explained in Section 2.5, with the only difference that the computationally expensive particle–particle forces $\mathbf{F}_{pi}$ are not included. Particle–particle interactions are handled instead by the DSMC algorithm, which probabilistically determines collision pairs, after which momentum exchange between two colliding particles is taking place.

The assumption of instantaneous momentum exchange brings forward an important limitation of the DSMC approach: the real duration of a particle–particle collision must be small relative to the mean free time between two consecutive collisions. Therefore, the method should not be applied to very dense systems in which prolonged particle–particle contacts take place such as in hopper flows or dead zones in fluidized systems. On the other hand, the method is extremely efficient for dilute and semi-dilute systems such as pneumatically conveyed powders, riser flows, and liquid sprays.

A requirement of the DSMC method is that expressions are available for the postcollisional velocities of two particles, given their precollisional positions and velocities. For solid particles, there are well-known expressions in terms of normal and tangential coefficients of restitution and friction coefficients (Van der Hoef et al, 2008). However, for droplets and bubbles (as well as wet particles), additional correlations are needed detailing the outcome of a binary collision, such as bouncing, coalescence, or breakup. In the topical sections, we will discuss these correlations for some representative systems.

## 3. DIMENSIONLESS NUMBERS FOR MULTIPHASE FLOWS

By analyzing dimensionless numbers, it is possible to estimate whether certain effects or forces are important or can safely be neglected in the model. Here and in the following, we denote the continuous phase by subscript f (a fluid, which can be either a gas or a liquid) and the discrete phase by subscript p (a generalized particle, which can be a solid particle, a liquid droplet, or a gas bubble). The main forces acting on the discrete phase are the following (Tomiyama, 2004):

1. Fluid inertial forces: $F_{if} = c_1 \rho_f |\mathbf{v}_p - \mathbf{u}_f|^2 d_p^2$
2. Particle inertial forces: $F_{ip} = c_2 \rho_p |\mathbf{v}_p - \mathbf{u}_f|^2 d_p^2$
3. Fluid viscous forces: $F_{\mu f} = c_3 \mu_f |\mathbf{v}_p - \mathbf{u}_f| d_p$
4. Particle viscous forces: $F_{\mu p} = c_4 \mu_p |\mathbf{v}_p - \mathbf{u}_f| d_p$
5. Surface tension forces: $F_\sigma = c_5 \sigma d_p$
6. Buoyancy forces: $F_b = c_6 |\rho_p - \rho_f| g d_p^3$

Here, $c_i$ are constants, $d_p$ is the particle diameter, $\mathbf{v}_p - \mathbf{u}_f$ is the relative velocity between particle and fluid, $\rho_f$ and $\mu_f$ are the mass density and viscosity of the fluid, $\rho_p$ and $\mu_p$ the mass density and viscosity of the particle, $\sigma$ the surface tension between fluid and particle, and $g$ is the gravitational acceleration. Depending on the case at hand, some forces are more dominant than others, and some can be neglected altogether. This is best appreciated by investigating dimensionless ratios from the above forces for different multiphase systems.

The relative importance of inertia between the particle and fluid phase is expressed by the density ratio $\rho_p/\rho_f$. For solid particles or droplets in a gas, this ratio is typically $O(10^3)$, for solid particles in a liquid $O(1)$, while for bubbles in a liquid it is $O(10^{-3})$. Inertial effects of bubbles can therefore be completely neglected relative to the inertial effects in the much denser surrounding liquid. For the converse case of solid particles or droplets moving through a gas, the inertial effects in the lower density gas cannot be neglected, however. This subtlety is caused by the fact that inertial forces are actually a measure of the amount of acceleration experienced by each phase. This is best appreciated when considering droplets or particles moving at a high but relatively slowly changing velocity through a stagnant gas. In that case, the most important inertial effect is the quick acceleration of the gas from zero velocity to a very high velocity as the droplet or particle surface passes by.

The relative importance of viscosity between the particle and fluid phase is expressed by the viscosity ratio $\mu_p/\mu_f$. For solid particles, this ratio is irrelevant, because they have (in an idealized sense) an infinite viscosity. For bubbles in a liquid, the viscosity ratio is typically $O(10^{-2})$, while for droplets in a gas, it is $O(10^2)$. This implies that for steadily falling droplets, in the absence of transient surface deformations, the internal shear rates inside the droplet are much smaller than the outside shear rates in the immediately surrounding gas phase.

Other characteristic numbers are usually named after famous scientists. Probably, the best known is the Reynolds number, which expresses the ratio of fluid inertial to fluid viscous forces:

$$Re_p = \frac{F_{if}}{F_{\mu f}} = \frac{\varepsilon_f \rho_f |\mathbf{v}_p - \mathbf{u}_f| d_p}{\mu_f} \quad \text{particle Reynolds number} \tag{31}$$

According to convention in the field of multiphase flow, we have included the fluid volume fraction $\varepsilon_f$ in the definition of the particle

Reynolds number. If we consider 0.1–1 mm glass particles falling through air, we have $Re_p = 10^1 - 10^3$. Larger solid particles in a gas therefore have relatively high particle Reynolds numbers. Conversely, if we consider the same glass particles falling through water, we have $Re_p = 10^0 - 10^2$. Solid particles in a liquid therefore have relatively lower particle Reynolds numbers. For sufficiently small particles (less than 0.1 mm for glass particles in water), we can neglect the fluid inertia entirely, in which case the flow is fully laminar in accordance with Stokesian dynamics.

The Weber number expresses the ratio of particle inertial to surface tension forces:

$$We = \frac{F_{ip}}{F_\sigma} = \frac{\rho_p |\mathbf{v}_p - \mathbf{u}_f|^2 d_p}{\sigma} \quad \text{Weber number} \tag{32}$$

Note that we choose the inertia of the discrete phase in the numerator here; in the literature sometimes the inertia of the continuous phase is used instead, in which case the Weber number is changed by a factor equal to the mass density ratio. Alternatively, when characterizing the outcome of binary collisions in droplet sprays, the relevant velocity is the relative velocity between two colliding droplets, i.e., $We = \rho_p (\Delta v)^2 d_p / \sigma$. In any of these cases, when a droplet hits a nonmoving surface or when two droplets hit each other, the droplet deformation will be relatively small if $We < 1$, while it will significantly deform for $We \gg 1$. If we consider a typical water droplet (0.1–1 mm) falling through air, we have $We = 10^0 - 10^3$, which shows that the amount of deformation in a collision depends very sensitively on the droplet diameter.

The Eötvös number expresses the ratio of buoyant to surface tension forces:

$$Eo = \frac{F_b}{F_\sigma} = \frac{\Delta\rho g d_p^2}{\sigma} \quad \text{Eötvös number} \tag{33}$$

This number is usually used in the context of bubbly flows but may equally be defined for droplets moving through a gas. If we consider bubbly flows of 1–10 mm air bubbles rising through water, this number is of the order of $Eo = 10^{-1} - 10^1$. For the smaller bubbles ($Eo < 10^0$), the shape of the bubble is almost spherical during its rise, and the drag forces and lift forces depend primarily on the Reynolds number. For larger bubbles ($Eo > 10^0$), the surface tension forces are not sufficient to maintain a spherical bubble shape, in which case the drag forces and lift forces are also determined by the Eötvös number.

From the six forces, only five independent dimensionless numbers can be formed. This could be the set of numbers discussed up to this point ($\rho_p/\rho_f$, $\mu_p/\mu_f$, *Re*, *We*, and *Eo*), but other sets could be chosen as well. For example, for bubbly flows, the second independent dimensionless number next to the Eötvös number may be the Reynolds number, but is often chosen to be the Morton number:

$$\mathrm{Mo} = \frac{F_\mu^4 F_b}{F_\sigma^3 F_{if}^2} = \frac{g\mu_f^4 \Delta\rho}{\rho_f^2 \sigma^3} \quad \text{Morton number} \tag{34}$$

The advantage of using Mo is that this number is a gas and fluid property independent of the actual bubble size and velocity. The Morton number is of the order of $\mathrm{Mo} = 10^{-11}$ for air bubbles in water.

For completeness, we mention the most common other dimensionless numbers which are frequently encountered in the literature on multiphase flow. These are the Archimedes number, Capillary number, and Ohnesorge number, defined as

$$\mathrm{Ar} = \frac{F_b F_{if}}{F_{\mu f}^2} = \frac{g d_p^3 \rho_f \Delta\rho}{\mu_f^2} \quad \text{Archimedes number} \tag{35}$$

$$\mathrm{Ca} = \frac{F_{\mu f}}{F_\sigma} = \frac{\mu_f |\mathbf{v}_p - \mathbf{u}_f|}{\sigma} \quad \text{Capillary number} \tag{36}$$

$$\mathrm{Oh} = \frac{F_{\mu f}}{\sqrt{F_{if} F_\sigma}} = \frac{\mu_f}{\sqrt{\rho_f \sigma d_p}} \quad \text{Ohnesorge number} \tag{37}$$

In summary, the use of the above dimensionless numbers has two advantages. First, it shows whether certain effects or forces can safely be neglected. For example, if the Eötvös number is sufficiently small, we can neglect the deformation of a bubble from its spherical equilibrium shape. Second, it allows us to generalize the correlations found for forces in a particular (experimentally convenient) system to apply to a much wider class of multiphase flow problems. Particular examples of such correlations will be given in the next topical sections.

# 4. FLOWS WITH PARTICLES: DISCRETE PARTICLE MODELS (CFD–DEM)

## 4.1 Particle–Fluid Interactions

For dispersed gas–solid systems, the dominant interaction between the fluid (i.e., gas) and the particles (solids) is the drag force. This force requires a correlation. Usually, the functional form of a correlation follows from theoretical reasoning combined with "playing" with data. The first point is to link the dimensional drag force (expressed as $C_D$ or $F$) to the appropriate dimensionless groups. Commonly, the dimensionless drag force is expressed as a function of the particle Reynolds number and the fluid volume fraction. For nonspherical particles also, dimensionless variables that characterize the particle shape and orientation have been considered.

The final functional form contains fitting parameters. Data for determining these parameters come from two sources, namely, experiments and fully resolved flow simulations. One of the most commonly used empirical correlations for drag based on experimental data was introduced by Gidaspow (1994) and combines the Wen and Yu (1966) expression for solids volume fractions less than 0.2 with the Ergun (1952) equation for larger solids volume fractions. Experimentally, the drag force is found by interpreting experimental data, e.g., pressure drops versus superficial velocities and average solids volume fractions. Up to a certain engineering accuracy, these models are very valuable, but experimental nonidealities, measurement accuracies, and the used interpretation model all give rise to uncertainties.

From about the year 2000 onward, a second class of drag correlations became available. These correlations were constructed using data obtained from fully resolved flow simulations (commonly referred to as DNS). Well-known examples are the correlations of Hill et al (2001) and Beetstra et al (2007). Compared to experiments, simulations can be much better controlled, but also these correlations have accuracy issues mainly due to relative low grid resolutions that were attainable at that time. Currently, much more accurate simulations are available that provide more accurate drag correlations such as presented by Tenneti et al (2011) and Tang et al (2015a).

In Fig. 3, four porosity profiles are shown obtained using Euler–Lagrange CFD–DEM simulations of a pseudo-2D fluidized bed at a superficial gas velocity of 2.5 $U_{mf}$ with different (DNS) drag correlations. For the four cases, the computed average bed heights are markedly different, i.e., lower for the newer/more accurate drag laws. Upon experimental validation, it appears that the drag laws obtained from higher resolution DNSs that are expected to be better actually give worse predictions. The likely underlying reason is not that the DNS results are too inaccurate, but rather that the systems and variables that are analyzed in the DNSs do not properly characterize the situation that is simulated by the CFD–DEM model.

The system that is usually considered in DNS is a random, but statistically homogeneous, configuration of static spheres. In CFD–DEM, the drag

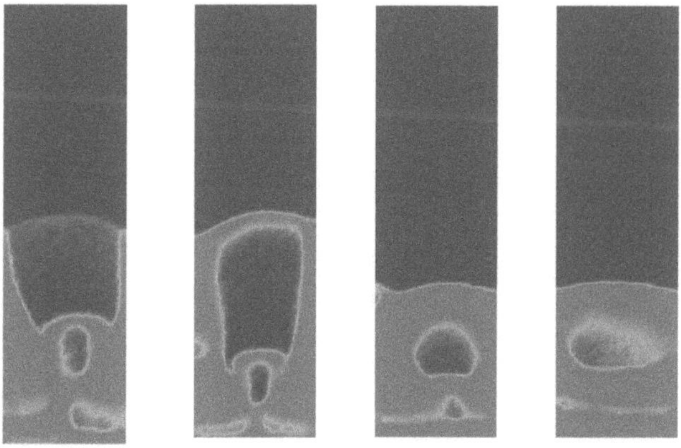


**Figure 3** Snapshots of a CFD–DEM simulation of pseudo-2D fluidized bed using all the same settings except for the used drag correlation. Correlations are obtained from Beetstra et al (2007), Hill et al (2001), Tenneti et al (2011), and Tang et al (2015a).

correlation is used for moving particles with possibly severe local heterogeneities such as emulsion–bubble interfaces or clustering. Besides in DNS, what is usually measured is a stationary, time-averaged pressure drop (or force averaged over all particles) as a function of the applied superficial gas velocity. In CFD–DEM, the relation is, however, used to compute forces on individual particles using their individual slip velocities.

Figure 4 shows a result from Kriebitzsch et al (2013a). Within a DNS of a pseudo-2D fluidized bed, the drag force on each particle is computed. Also, smoothed fluid velocity and porosity fields are computed from the resolved DNS data. These fields would correspond to the velocities and porosities on the macroscopic scale. Using these smoothed fields, drag correlations are applied for each particle. The difference of the magnitude of these forces is normalized by the buoyant weight (which balances the average drag force). Note that no CFD–DEM simulation is performed here. Two drag correlations, namely, that of Ergun and Beetstra et al (BHK) are considered. The legend entry $v_i$ indicates that the drag correlation used a slip force computed using the particle velocity, while $v$ indicates the use of a smoothed average velocity. All give about the same result that the drag force computed from the drag correlation is smaller on average and that the "real" drag forces computed from DNS have large fluctuations relative to the average value.

The results of Kriebitzsch et al illustrate that, even for drag correlations obtained from DNS, the drag forces found in a DNS for moving (fluidizing) particles are markedly different from the ones computed using the

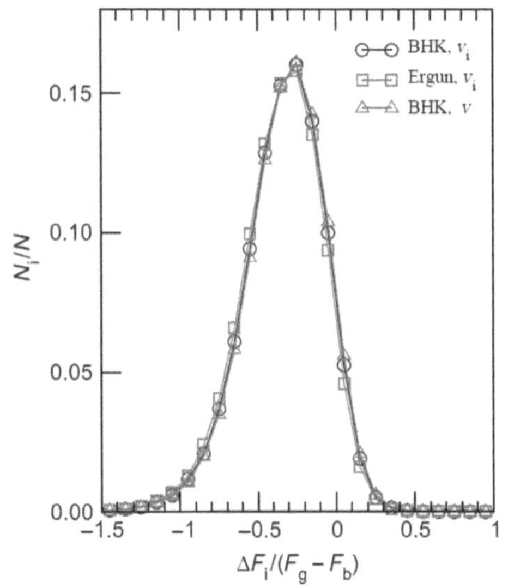


**Figure 4** Histogram of the magnitudes of the "real" drag force as computed in DNS and the force computed from drag correlations normalized by the buoyant weight. *Figure reproduced from Kriebitzsch et al (2013a) with permission.*

correlation. Not only do the "real" forces have more variation but also the average forces are different, as can be seen by the fact that the histogram in Fig. 4 is not centered on zero. Kriebitzsch et al (2013a) suggest that for moving particles, the drag law should be made dependent on the granular temperature of the particles. In the work of Wylie et al (2003) and Tenneti et al (2010), an increase in the drag force due to particle motion as characterized by the granular temperature has been reported. Recently, a drag correlation that includes the dependence on granular temperature has been proposed on the basis of DNSs by Tang et al (2015b). This addition indeed increases the average drag force computed by the correlation, but a CFD–DEM validation using this extension is not available yet.

Within CFD–DEM, the granular temperature is a smoothed macroscopic variable. So this addition will shift the average, but not influence the variability. Clearly, the difference between variability of the individual particle drag force experienced in DNS and CFD–DEM is due to the smoothed nature of the unresolved flow field in CFD–DEM.

At this moment, it is unclear if this difference of variability is important. It might be. As an analogy, consider Brownian motion. In a coarse-grained description, fluctuating forces of Brownian particles due to molecular interaction can be modeled as random fluctuating forces. In that case, the variability of the forces also in the coarse-grained situation is maintained although the precise motion of the Brownian particles will be different in a fully resolved molecular simulation or a coarse-grained Brownian dynamics description. When discarding the Brownian motion, essential physics such as diffusion is not accounted for. In analogy, modeling the variability of the fully resolved drag force on the CFD–DEM level by including random forces might provide a more faithful description of the typical motion of particles. At the moment, this idea is largely unexplored.

Another interesting set of results was obtained by Radl and Sundaresan (2014). They performed CFD–DEM simulations of the same periodic system at different resolutions. In a periodic domain with a set average solids volume fraction, the average gas velocity is such that average drag balances the buoyancy forces on the particles. As shown in Fig. 5, they found that depending on the mesh size (or "filter size"), the average slip velocity can change. For two smaller mesh sizes of 1.67 $d_p$ and 3.33 $d_p$, the results superimpose, while for a larger mesh size of 26.67 $d_p$, a lower slip velocity is found.

One important observation from the simulations of Radl and Sundaresan is that the CFD–DEM results are within a small range of filter widths independent of this width. It might suggest that this is the case for all small filter

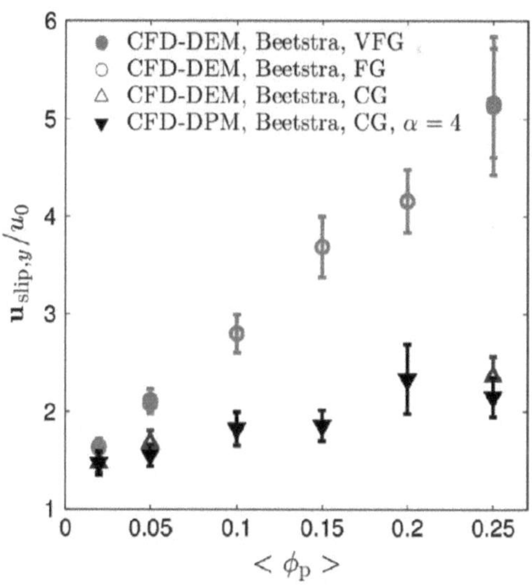


**Figure 5** Slip velocities found in CFD–DEM simulations using the same settings except for the mesh size used for the Eulerian grid. The three resolutions are labels very fine grained (VFG), fine grained (FG), and coarse grained (CG). *Figure reproduced from Radl and Sundaresan (2014) with permission.*

widths. However, Kriebitzsch et al (2013b) have shown that if the fluid fraction is computed in a very localized way, e.g., on the particle level by means of Voronoi tessellation, then using the drag correlation on that localized way gives worse predictions. So, the filter width also needs a minimum size. This is in accordance to the separation of length scales argument that was also addressed in Section 2.1.

What the results of Radl and Sundaresan (2014) show is that the range of small filter sizes for which "converged" results are obtained is rather limited. The underlying reason is that for somewhat larger filter sizes, the particle distribution already is heterogeneous: particles cluster due to the inelastic nature of their collisions. When the filter size is too large, the structure of these clusters is not resolved and therefore its influence should be incorporated into the drag correlation.

So, already on a relatively small length scale mesoscopic structures form. For their studied system, Radl and Sundaresan (2014) propose an extended drag correlation that includes the filter size. Using this drag correlation, they obtain simulation results that are invariant for a wider range of filter sizes. They study one system with a fixed set of material properties and two average solids volume fractions. The clustering dynamics will be dependent on the collisional properties of the particles, etc. Therefore, the mesoscopic structure will depend on these properties, and this structure is only partly resolved. So it is expected that the mesoscale structure needs to be effectively accounted for in the drag correlation at larger filter widths. General drag correlations for CFD–DEM that are applicable for larger filter widths are currently not available.

From a computational point of view for CFD–DEM, in most cases, considering larger filter sizes is not beneficial. The computational bottleneck is handling particle–particle interactions and therefore relatively small mesh sizes for the fluid phase are not the major issue. In Euler–Euler approaches such as the two-fluid model, mesh sizes are usually much larger than the particle size. Therefore, mesoscopic structures due to heterogeneous clustering are a subgrid phenomenon, and therefore incorporating it into the drag correlation is studied more extensively. A very important development in this direction is the energy minimization multiscale approach (see, e.g., Shah et al, 2015; Wang and Li, 2007). In the Euler–Lagrange setting, considering larger grid sizes becomes relevant if also the particle phase is coarse grained. This is the case when the Lagrangian discrete elements are parcels that model a set of particles. Within this setting, drag correlation that incorporate mesoscopic structures may become of importance. This is an emerging field of research. In the paper of Radl and Sundaresan (2014), a first example of a drag correlation for parcels is also discussed.

## 4.2 Particle–Particle Interaction

In most applications, two types of particle–particle interactions are of importance, namely collisions and long-lasting interactions when particles are packed. In reality, when particles interact they deform and internal stresses develop. Most of the deformation is elastic, but a part is dissipative due to plastic or viscous effects. These interactions are described by the field of contact mechanics.

However, collisions are very fast compared to other process in gas-fluidized flow. Therefore, only the net effect of a collision is of importance. These can be characterized by normal and tangential restitution coefficients. If (hard) particles are packed, also the detailed deformation of particles is not of importance. The force balances are determined by constraints and friction coefficients.

Collisions can be modeled as events at a specific point in time. Particle velocities and angular velocities subsequent to the collisions can be related to the values before collisions using masses, moments of inertia, and restitution coefficients. If, however, also long-lasting interactions occur, this approach breaks down, because this would correspond to an infinite amount of collisions.

For the situation that both collisions and long-lasting contacts can occur, the soft-sphere model is appropriate. In this model, particle forces are computed using particle overlap and relative velocities. This model was first introduced by Cundall and Strack (1979). The spheres are called soft because they are allowed to overlap much more (say 1%) than deformation as, e.g., computed by contact mechanics would allow for. This has the benefit that the spring stiffness can be chosen relatively small. Therefore, Newton's equations of motion for collisions and contacts can be numerically solved using a relatively large time step. So, this model is largely artificial such that numerical solution is facilitated. However, the effective results of collisional events are realistic. A recipe to relate spring constants and damping factors to material properties such as restitution coefficients and friction coefficients is given by Van der Hoef et al (2006).

Most CFD–DEM simulations are performed using spherical particles. Implementation and simulation of accurate soft-sphere models for non-spherical particles are more complicated and a topic of current research (see, e.g., Pournin et al, 2005; Zhou et al, 2011).

Another current development in CFD–DEM is the collision of wet particles. In many industrial processes, liquids are introduced in a fluidized system, e.g., to establish coating, agglomeration, or cooling. The presence of liquid covered particles heavily influences the collisional properties. Sutkar et al (2015) analyzed collisions of 2.5 mm glass particles on a flat plate covered with a thin layer of a water–glycerol mixture. These authors tried to establish a correlation for a wet restitution coefficient from these experiments. Such a restitution coefficient can next be used in a CFD–DEM simulation using the soft-sphere model.

In Fig. 6, a typical experiment by Sutkar et al (2015) is shown. They divided the collision process into different subsequent phases that are governed by different physical phenomena. Each phase has its own dissipative mechanism. By assuming that these phases do not influence each other, the total restitution coefficient can be written as the product of restitution coefficients of the separate phases, i.e., $e_{\text{wet}} = e_{\text{I}} e_{\text{II}} e_{\text{III}} e_{\text{IV}}$. As a simplified description for the restitution coefficient, they give a parameterization as proposed by Davis et al (2002),

$$e_{\text{wet}} = \begin{cases} 0, & \text{St} < \text{St}_c \\ e_\infty \left(1 - \dfrac{\text{St}_c}{\text{St}}\right), & \text{St} \geq \text{St}_c \end{cases} \tag{38}$$

where $\text{St} = \rho_p d_p v / 9\mu_l$ is the Stokes number that compares the particle inertia with the viscous dissipation of the liquid. Below a critical Stokes number particles stick, above that the collisions are more efficient for increasing Stokes number.

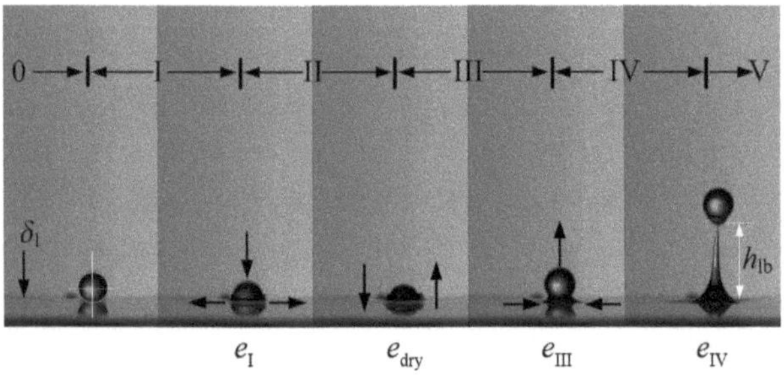


**Figure 6** Particle impact on a wetted plate analyzed in four phases: (I) penetration through liquid layer, (II) collision with solid plate, (III) rebound through liquid layer, and (IV) formation and breakage of liquid bridges. *Figure reproduced from Sutkar et al (2015) with permission.*

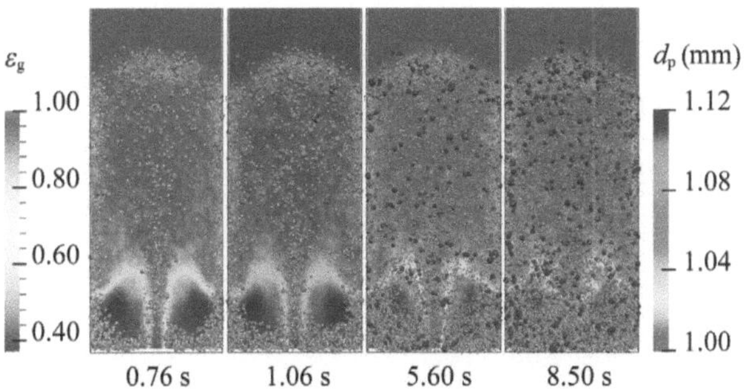


**Figure 7** CFD–DEM time series for spout fluidized bed simulations with droplet injections. Due to collisions between solid particles and droplets, the particles grow in size.

### 4.3 Application: Gas–Solid Fluidized Beds

As an example of the novel applications of CFD–DEM, we present results of a spout fluidized bed with liquid injection. In this simulation, both solid particles and liquid drops are modeled as "particles." The droplets are injected at a low Weber number such that droplet breakage can be neglected. When a droplet and solid particle collide, the liquid is assumed to form a uniform liquid layer around the particle. When wetted particles collide, the restitution coefficient is as given in the previous subsection.

In Fig. 7, a typical time series of simulation snapshots are used. The background coloring shows the gas hold up. The particles are color coded to indicate their size. Due to collisions between droplets and solid particles, the particles grow in size. Due to the presence of a wet layer, the collisional properties of the particles change. The collisions become much more dissipative. The formation of (semi) permanent liquid bridges between particles has not yet been accounted for in this simulation.

## 5. FLOWS WITH BUBBLES: DBMs

### 5.1 Bubble Interactions

At low gas volume fraction ($<0.01$), the forces acting on bubbles in a liquid are similar to those acting on a single bubble. Due to their low inertia as compared with the liquid phase, bubbles are subject to a large number of forces, i.e., forces due to drag, lift, and virtual mass. In the definition of the interfacial forces, it is customary to characterize the bubble size with the sphere equivalent diameter. That is, all effects due to nonsphericity are lumped in the closure. The closures that are most commonly applied were proposed by Tomiyama (2004) and are the drag closure for slightly contaminated liquids:

$$C_{D,\infty} = \max\left(\min\left(\frac{24}{Re}\left(1+0.15Re^{0.687}\right), \frac{72}{Re}\right), \frac{8}{3}\frac{Eo}{Eo+4}\right) \tag{39}$$

and the lift closure:

$$C_L = \begin{cases} \min\left(0.288\tanh\left(0.121Re\right), f(Eo_d)\right); & Eo_d < 4 \\ f(Eo_d); & 4 < Eo_d < 10 \\ -0.29; & Eo_d > 10 \end{cases} \tag{40}$$

where $Eo_d = Eo/E^{2/3}$, $E = \left(1+0.136Eo^{0.757}\right)^{-1}$, and $f(Eo_d) = 0.00105Eo_d^3 - 0.0159Eo_d^2 - 0.0204Eo_d + 0.474$. The lift coefficients obtained with this closure yield values between $\pm 0.29$ and change sign once bubbles start to deform ($Eo_d \approx 6$). Usually, a constant virtual mass constant of $C_{VM} = 0.5$ is used. However, Tomiyama (2004) also proposed virtual mass coefficients that account for the bubble shape. To our knowledge, only very few researchers have actually applied these closures for the virtual mass force though.

At higher gas volume fractions (>0.01), bubbles start interacting with each other and the liquid flow can become much more vigorous. This results in an enhancement of the drag force due to the reduced space for the liquid to flow around bubbles. Roghair et al (2011) performed numerical experiments with a detailed front tracking model to measure the increase of the drag force depending on the local gas volume fraction and the shape of the bubbles (characterized by *Eo*). He found that the increase in the drag force can be well described by:

$$\frac{C_D}{C_{D,\infty}\varepsilon_f} = 1 + \frac{18}{Eo}\left(1-\varepsilon_f\right) \tag{41}$$

Apart from an increased drag force, high gas volume fractions can also lead to occurrence of coalescence and breakup of bubbles. Although the closures derived for these kinds of phenomena are rather mature for droplet–droplet interactions, this is not the case for bubble–bubble interactions. The main reason is probably the role of surfactants, which can have a considerable effect on the rigidness of the bubble surface and hence on the processes occurring on that scale. Given the fact that many closures were derived for water–air systems makes things worse, as the water quality and in particular the amount of surfactants varies considerably from lab to lab or even from day to day.

Lau et al (2014) have extensively studied the modeling of coalescence and breakup in bubbly flows. Here, we only present the main closures. More details can be found in the work of Lau et al (2014). Bubble coalescence occurs if two bubbles get very close together and the thin liquid film separating them from each other is drowned. In the Euler–Lagrange framework, this is typically modeled as a competition between phenomena, each characterized by their own length scale. The drainage of the bubbles is usually modeled by a lubrication model (Prince and Blanch, 1990), yielding a drainage time, whereas the interaction time or contact time is characterized by the ratio of the bubble size and the relative velocity of the bubbles (Sommerfeld et al, 2003):

$$t_{\text{drainage}} = \sqrt{\frac{d_{\text{eq}}^3 \rho_l}{128\sigma}} \ln(\theta_0/\theta_{\text{f}}); \quad t_{\text{contact}} = \frac{C_{\text{co}} d_{\text{eq}}}{2|\mathbf{v}_{\text{rel}}|} \tag{42}$$

The ratio of these two timescales squared yields a Weber number, along with a criterion for coalescence to happen ($We_{\text{co}} \leq We_{\text{crit}}$):

$$We_{\text{co}} = \frac{\rho_{\text{f}} |\mathbf{v}_{\text{rel}}|^2 d_{\text{eq}}}{\sigma}; \quad We_{\text{crit}} = 2\left(\frac{4C_{\text{co}}}{\ln(\theta_0/\theta_{\text{f}})}\right)^2 \tag{43}$$

where $C_{\text{co}}$ represents the bubble deformation distance normalized by the effective bubble diameter and $\theta_0$ and $\theta_{\text{f}}$ represent the initial and final film thickness just during drainage. To see whether coalescence takes place, the Weber criterion is checked every time two bubbles collide.

The breakup of bubbles can have different causes, but in most bubbly flow applications, it is induced by the turbulent liquid flow field that exerts shear forces on the bubble surface. If these forces are larger than the surface tension, the bubble will break up:

$$We_{\text{BU}} = \frac{\rho_{\text{f}} |\mathbf{\Delta u}(d_{\text{eq}})|^2 d_{\text{eq}}}{\sigma}; \quad We_{\text{crit}} = 12 \tag{44}$$

where the stresses on the bubble surface are characterized by the velocity differences $\mathbf{\Delta u}$ occurring on the length scale equal to the equivalent bubble diameter $d_{\text{eq}}$. If the Weber number is larger than the critical Weber number, the bubble will break. This criterion is checked every time step.

In most cases, it is assumed that bubble breakup is binary, i.e., leading to two daughter bubbles, although in reality bubbles may also break in more than two daughter bubbles. The size of the daughter bubbles is described by a daughter size distribution (DSD). In literature, different shapes of the DSD have been proposed. Lau et al (2014) tested different closures to investigate the effect of the different shapes of the DSD on the bubble size distribution in a bubble column. It was found that the choice of the DSD only has a marginal effect on the prevailing bubble size distribution. For practical reasons, it is therefore fair to assume equal size breakup.

## 5.2 Application: Bubble Column Reactors

Euler–Lagrange models for bubbly flows are also known as DBMs and can be used to investigate various features of bubble column reactors. In this section, we will give two examples of studies that were made with the aid of a DBM. In the first example, we will highlight how the geometry affects the flow features inside the bubble column, which can be characterized by axial dispersion coefficients. In the second example, we will highlight the interplay between flow, mass transfer, and chemical reaction inside a bubble column.

Bai et al (2012) have carried out DBM simulations to computationally measure the key quantities of the bubbly flow in a square bubble column in terms of the turbulent liquid dispersion coefficient. This was done by releasing two sets of neutrally buoyant tracer particles in the liquid phase and recording their dispersion with time; one at the top of the column (tracer 0) and one set at the bottom. It was found that with increasing superficial gas velocity, the dynamics of the liquid in the column is enhanced, and hence the turbulent dispersion coefficient increases. The obtained dispersion coefficients are shown in Fig. 8 and are well within the spread of literature correlations. Note that this spread can be attributed to differences in geometry of the studied bubble columns. The power of the DBM is that it accounts for the details of the geometry and thereby provides a predictive capability that is hard to match when using empirical correlations.

As pointed out by Darmana et al (2005), reactive bubbly flows exhibit a strong interplay between fluid flow, mass transfer, and chemical reactions, which is illustrated in Fig. 9. That is, in many gas–liquid processes, mass transfer from the bubbles to the liquid takes place. The mass transfer rate ($k_l a$) depends on the bubble size, but in turn the bubble size also depends on the mass transfer rate, particularly if the mass transfer is enhanced due to a chemical reaction taking place in the liquid film surrounding the bubble ($E$). As the liquid flow is largely driven by the bubble movement, one can

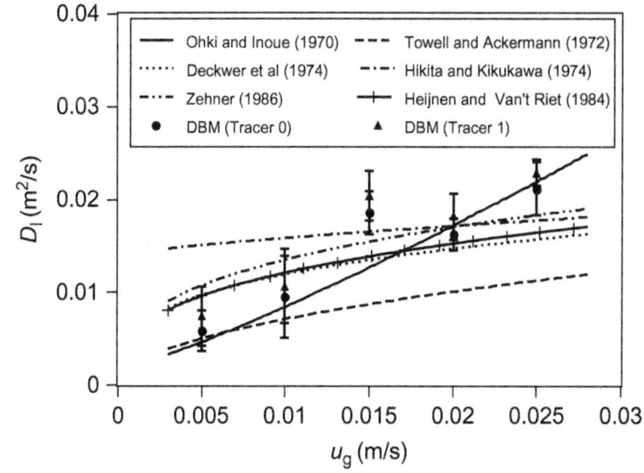


**Figure 8** Comparison of the simulated liquid phase dispersion coefficient with literature correlations. *Figure reproduced from Bai et al (2012) with permission.*

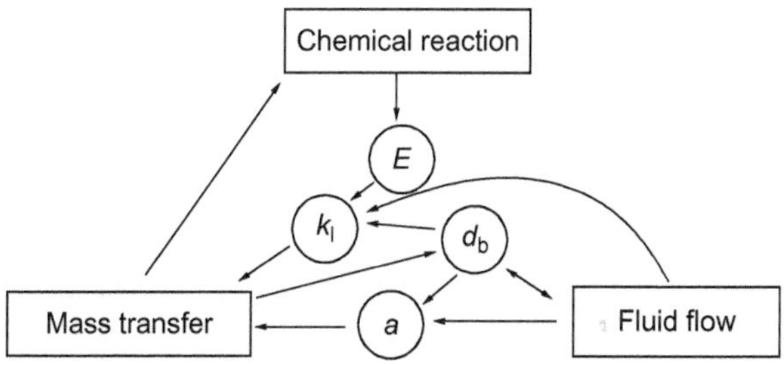


**Figure 9** Interdependency of fluid flow, mass transfer, and chemical reaction in a reactive bubbly flow. *Figure reproduced from Darmana et al (2007) with permission.*

recognize that it is difficult to decouple the description of fluid flow, mass transfer, and chemical reactions. In our view, the best way to capture this complicated interplay of phenomena is to solve all processes simultaneously. This means that apart from the momentum equations, one also needs to solve species balances for both phases. This has been done by Darmana et al (2005, 2007) for the reactive bubbly flow in a bubble column and by Jain et al (2015) for a similar system; however, this time equipped with wire meshes to cut large bubbles into smaller pieces. An example result of Jain et al (2015) is shown in Fig. 10. Due to the presence of the wire meshes, bubbles are cut into smaller pieces, which suppresses the effect of bubble coalescence. Moreover, the presence of wire meshes leads to a kind of staging of the liquid in the column, which can be utilized to promote plug flow-like behavior. Jain et al (2015) used the same system to study the

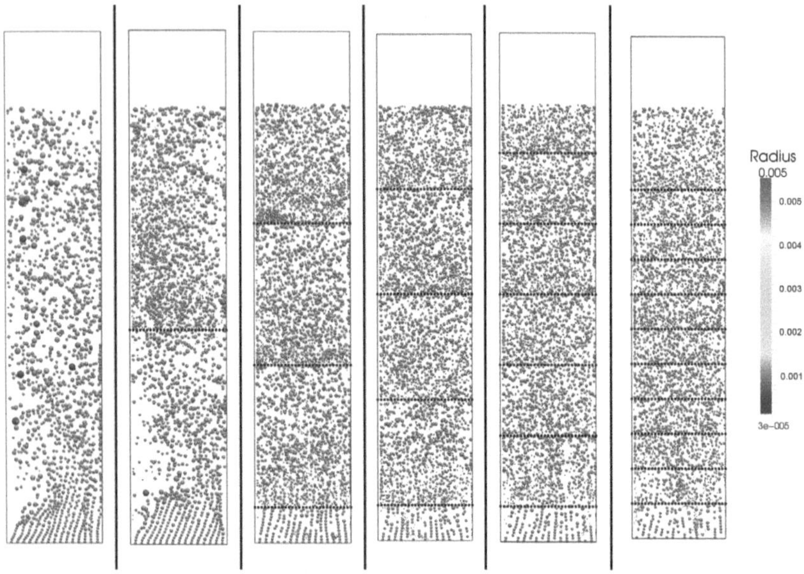


**Figure 10** Instantaneous snapshots of the bubble configuration in a bubble column equipped with a varying number of wire meshes to cut bubbles in to smaller pieces. From left to right, there are 0, 1, 3, 4, 6, 10 wire meshes. The positions of the wire meshes are indicated by the dotted lines. No mass transfer or chemical reaction is taking place in this case. *Figure reproduced from Jain et al (2015) with permission.*

chemisorption of $CO_2$ in an alkaline solution of sodium hydroxide. The bubble size changes along the column height due to the chemically enhanced mass transfer. The cutting of the bubbles enhances this effect, as can be seen in Fig. 11. Jain et al (2015) found that using multiple wire meshes in a bubble column the time to neutralize the alkaline solution can be reduced by some 20% as compared with a conventional bubble column without wire meshes.

## 6. FLOWS WITH DROPLETS: DSMC

### 6.1 Binary Droplet–Droplet Interactions

The last topical section concerns gas flows with droplets. Such flows are encountered in various spray processes in internal combustion engines, fluid catalytic cracking, gas phase polymerization, and spray drying. Before turning to large-scale simulations of the dynamic evolution of such sprays, it is essential to understand the elementary process of a binary droplet collision in a gas. It has been established that in such systems, the outcome of a collision

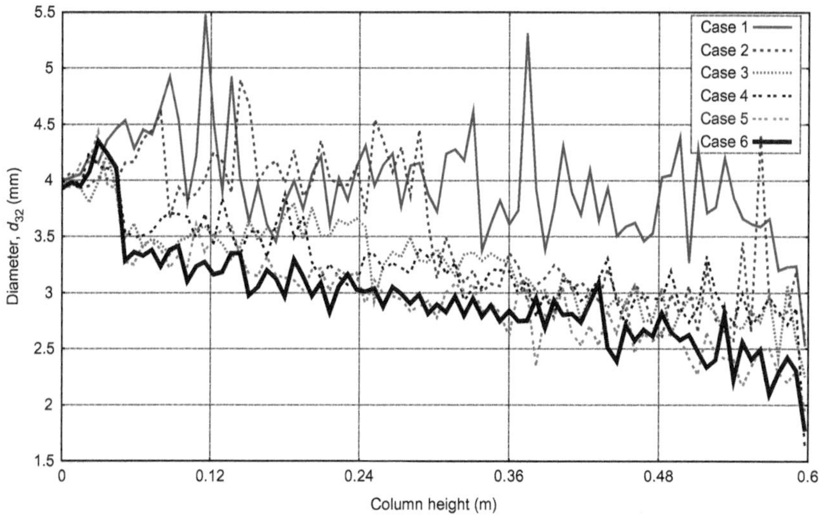


**Figure 11** Time-averaged vertical variation of the Sauter mean diameter for the chemisorption of $CO_2$ in sodium hydroxide in a bubble column equipped with a varying number of wire meshes to cut bubbles in to smaller pieces. Cases 1–6 correspond to 0, 1, 3, 4, 6, 10 wire meshes, as shown in Fig. 10. *Figure reproduced from Jain et al (2015) with permission.*

can be characterized by the Weber number (based on the droplet relative velocity) and the impact parameter $B$, where the latter is defined as the ratio of the projection of the center-to-center line onto the direction perpendicular to the relative droplet velocity to the average droplet diameter.

Due to the very complex topological changes of the free surface, the detailed study of droplet–droplet collision has been mostly limited to experimental studies. Ashgriz and Poo (1990) performed a detailed investigation on binary collision dynamics of water drops in air. Depending on the values of $We$ and $B$, they identified three different types of outcome or regimes for this process: coalescence, reflexive, and stretching separations. They also established the boundaries between the regimes. Qian and Law (1997) reported time-resolved photographic images of the droplet–droplet collision process, performed an extensive experimental investigation on binary collision dynamics of water drops in air and hydrocarbon drops in nitrogen, and distinguished five different regimes of collision outcomes, namely (i) coalescence after minor deformation, (ii) bouncing, (iii) coalescence after substantial deformation, (iv) coalescence followed by separation for near head-on collisions, and (v) coalescence followed by separation for off-center collisions.

Figure 12 shows a typical regime map and the observed topological changes of the free surfaces. In this regime map, no explicit distinction is made between coalescence after minor deformation and coalescence after substantial deformation. Let us first focus on the head-on collision of droplets. At very low *We* number bouncing prevails but with increasing value of *We*, the coalescence regime is entered. In this regime, the droplets merge and at sufficiently high *We* number substantial shape oscillations occur, but finally a single coalesced droplet results which attains a spherical shape. If the *We* number exceeds approximately 20, the reflexive separation regime is entered where the following phases can be discerned: coalescence, formation of a toroidal droplet (with a thin interior liquid disk (film) due to strong radial liquid flow) which after contraction forms an elongated cylindrical droplet along the main axis of the encounter process, stretching of the cylindrical droplet in the axial direction to form two outer droplets and an interior satellite droplet which increases in size with increasing *We* number. With further increase of *We*, an increasing number of satellite droplets are formed. For off-center collisions ($B > 0$), similar changes in encounter regime are

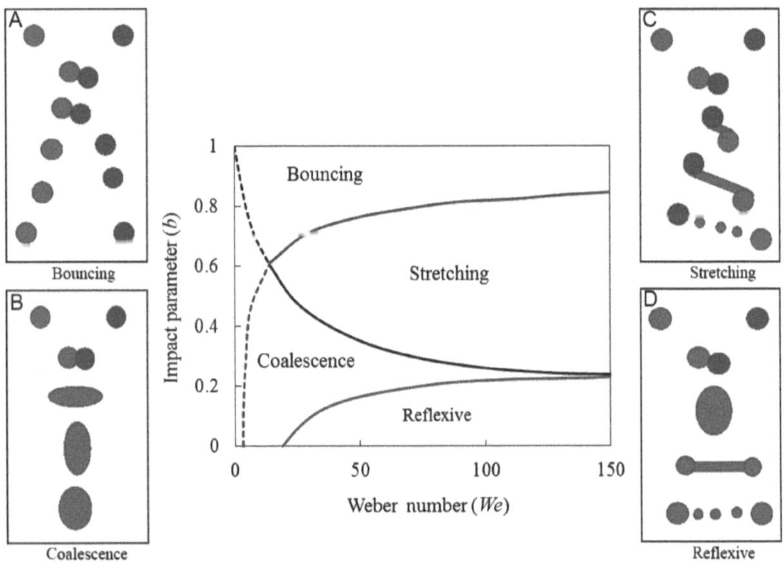


**Figure 12** Schematic illustration of the outcome of droplet collisions: (A) bouncing; (B) coalescence; (C) stretching separation; and (D) reflexive separation. Time is running from top to bottom in each schematic illustration. The center plot shows the lines demarcating the different regimes as a function of Weber number (based on the relative velocity between two droplets) and impact parameter for equal sized droplets. For unequal sized droplets, the lines are slightly shifted (see Ko and Ryou, 2005).

observed when *We* increases but in general the topological changes of the free surface are much more complex due to absence of a central axis of symmetry. Again at very low *We*, the bouncing regime prevails followed by the coalescence regime with increasing *We* number. With a further increase of *We*, the stretching separation regime is entered where the actual transition point depends on the value of the impact parameter.

We have conducted a series of DNSs of droplet–droplet collisions over a wide range of Weber numbers. Details of the method can be found in Appendix B. First, we compare our simulation results with data reported by Pan and Suga (2005) for near-head-on ($B=0.06$) colliding tetradecane droplets in nitrogen ($d_p=336\,\mu m$, $\rho_p=762\,kg/m^3$, $\mu_p=2.128\times 10^{-3}$ Pas, $\rho_f=1.225\,kg/m^3$, $\mu_f=1.827\times 10^{-5}$ Pas). Pan and Suga used a Level Set method and compared their simulation results with experimental data reported by Qian and Law (1997). According to the regime map (Fig. 12), a reflexive separation is expected with the formation of one satellite drop. As a first step, we investigated the effect of the grid size on the simulation results (not shown). We found that the differences between an intermediate grid $\left(d_p/\Delta x=45\right)$ and a fine grid $\left(d_p/\Delta x=56\right)$ are quite small. The biggest differences are observed with respect to the timing of pinch-off process of the satellite droplet. In Fig. 13, we show the predicted evolution of the shapes of the droplets during the encounter process for $d_p/\Delta x=45$, together with the experimental results of Qian and Law (1997) and the simulation results of Pan and Suga (2005). In accordance with the experimental observations, one satellite drop is formed. During the initial stage of the impact, the kinetic energy of the drops induces a strong radial movement of the liquid and formation of a disk.

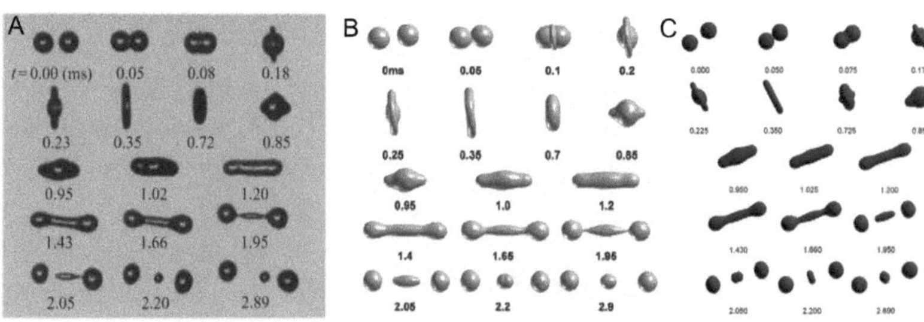


**Figure 13** Near head-on ($B=0.06$) collision of two 336 μm diameter tetradecane droplets in nitrogen at $We=61.35$: (A) experiment; (B) simulation; and (C) present simulation results ($d_p/\Delta x=45$). *Panel (A): Reproduced with permission from Qian and Law (1997). Panel (B): Reproduced with permission from Pan and Suga (2005).*

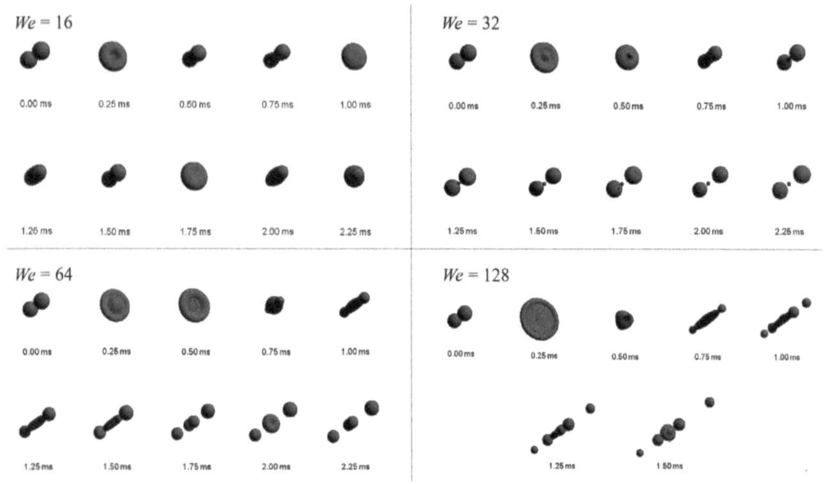


**Figure 14** VOF simulations of the effect of *We* number on head-on ($B=0$) collision of two 300 μm diameter water droplets.

Next, Fig. 14 shows the effect of *We* for the head-on ($B=0$) collision of two water droplets ($d_p = 300\,\mu\text{m}$, $\rho_p = 1000\,\text{kg/m}^3$, $\mu_p = 1.137 \times 10^{-3}\,\text{Pas}$, $\rho_f = 1.226\,\text{kg/m}^3$, $\mu_f = 1.780 \times 10^{-5}\,\text{Pas}$). The results are in good agreement with the regime map in Fig. 12. We do note that this regime map applies to surface tension-dominated droplets such as water droplets. For more highly viscous droplets, the regime boundaries will shift. This is a topic of current investigation.

## 6.2 Application: Droplet Sprays for Spray Drying

Having determined the correlations for the outcomes of droplet–droplet interactions, it is now possible to study the consequences of these interactions for the dynamics of liquid sprays containing very large amounts of droplets. Such liquid sprays can, for example, be generated by pushing a liquid through a swirling nozzle at high pressure, resulting in a conical liquid film which becomes unstable and breaks up into a very large number of small droplets (Fig. 15).

When modeling the droplet size, evolution, and other dynamical phenomena in a liquid spray, one quickly encounters the problem that the number of droplets in a realistic spray is very large, typically $10^9$ or an order of magnitude more. Deterministic modeling of each and every droplet is therefore out of the question. Fortunately, liquid sprays are usually also very dilute, making them perfect for modeling through stochastic techniques

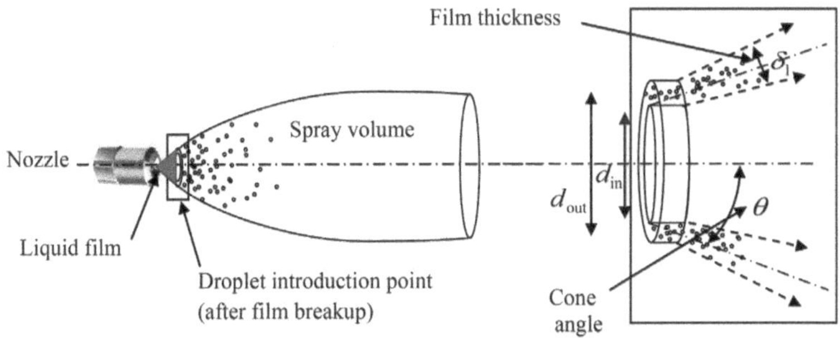


**Figure 15** Schematic diagram of the liquid spray system. Left: a pressure swirl nozzle atomizer produces a conical liquid film which breaks into a large number of droplets. Right: in the Euler–Lagrange model, new droplets are introduced in a cylindrical region with initial droplets sizes generated from a Rosin–Rammler distribution.

such as DSMC. In this section, we will show some results of DSMC simulations applied to an isothermal water spray in air, using a parcel size of $10^3$. For details, we refer to Pawar et al (2012, 2015).

A rectangular domain of size $0.6 \times 0.6 \times 0.7\ \mathrm{m}^3$ was simulated with a nozzle located close to the top wall. Water droplets with an average diameter of 51 μm were generated with a mass flow rate of 0.275 kg/s from a Rosin–Rammler distribution, which is known to be accurate for the droplet size distribution emerging at close distance from a high-pressure swirl nozzle (Han et al, 1997). We note that the region very close to the spray nozzle is not dilute. For this reason, collisions between droplets were disabled very close to the nozzle; based on the mass rate, average droplet size, initial droplet velocity, and the cone angle, the spray was estimated to be sufficiently dilute for the DSMC method to apply (less than $10^{-3}$; see Pawar et al, 2014) at 20 mm from the nozzle. Prescribed pressure (1 bar) boundary conditions were used at all boundaries, allowing for inflow and outflow of air. Droplets that cross the domain boundaries were automatically removed.

An important measure of the droplet size distribution in spray applications is the Sauter mean diameter $d_{32} = \langle d^3 \rangle / \langle d^2 \rangle$. This measure is so important because during evaporative drying the mass transfer happens at the interface of the droplets and the surrounding air. To enhance the evaporation of a population of droplets, one has to maximize the active surface areas and minimize the internal volumes. The DSMC simulations showed that the Sauter mean diameter is a very nontrivial function of the axial and radial position in the spray. Figure 16 shows that at a given axial position, with increasing distance from the central axis the mean droplet diameter first increases, then decreases, and finally increases again. Exactly, the same trends

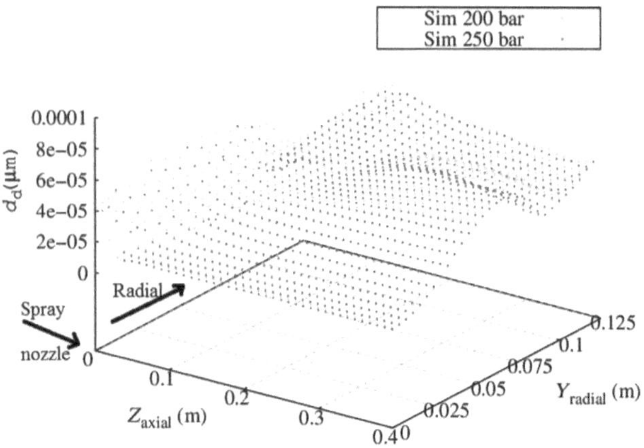


**Figure 16** Sauter mean diameter as a function of axial and radial position in an isothermal water spray, obtained from DSMC simulations at two different nozzle inlet pressures.

were observed both in PDI experiments (Pawar et al, 2015) and in SLIPI-LIF/Mie experiments (Mishra et al, 2014). Moreover, with increasing nozzle inlet pressure, smaller droplets are generated, as expected.

The DSMC model also provides the full details on the spatial and temporal evolution of the droplet size distribution. Figure 17 shows that the droplet size distribution predicted by the model is in near-quantitative agreement with the droplet size distribution obtained from PDI experiments (Pawar et al, 2015).

The advantage of a simulation model is that it can provide additional insight into processes which are very difficult to observe experimentally. An important example is the spatial distribution of the collision frequencies of the different collision events. Figure 18 shows these collision frequencies. Note that the color scale used in these plots is logarithmic to be able to show both high and low collision frequencies.

Bouncing of droplets (Figure 18A) is the most frequent event in this spray. Because of the high number density, the largest bouncing frequency occurs near the nozzle exit. The hollow shape of the spray density profile can easily be distinguished as well. The next most frequent event is coalescence (Figure 18B). There is a subtle difference with the bouncing event contours: coalescence seems to be relatively rarer near the central axis of the spray. Stretching and reflexive separations are again rarer and mostly limited to the region near the nozzle exit (Figure 18C and 18D). Here, the droplet velocities are very high. Droplets moving farther from the nozzle exit more much slower, and the Weber number is too low to induce a separation.

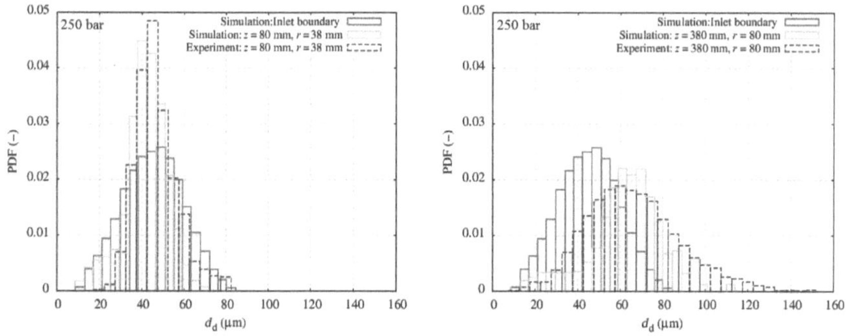


**Figure 17** Droplet probability distribution function for the total mass in each droplet size class, as measured with the PDI technique (blue; dashed gray in the print version) and in the DSMC simulation (green; light gray in the print version) for 250 bar inlet pressure. For reference, the red (solid dark gray in the print version) color histogram is the initial Rosin–Rammler droplet size distribution used in the inlet region. The left figure is from samples obtained at 80 mm from the nozzle exit and 38 mm from the central axis. The right figure is from samples obtained at 380 mm from the nozzle exit and 80 mm from the central axis.

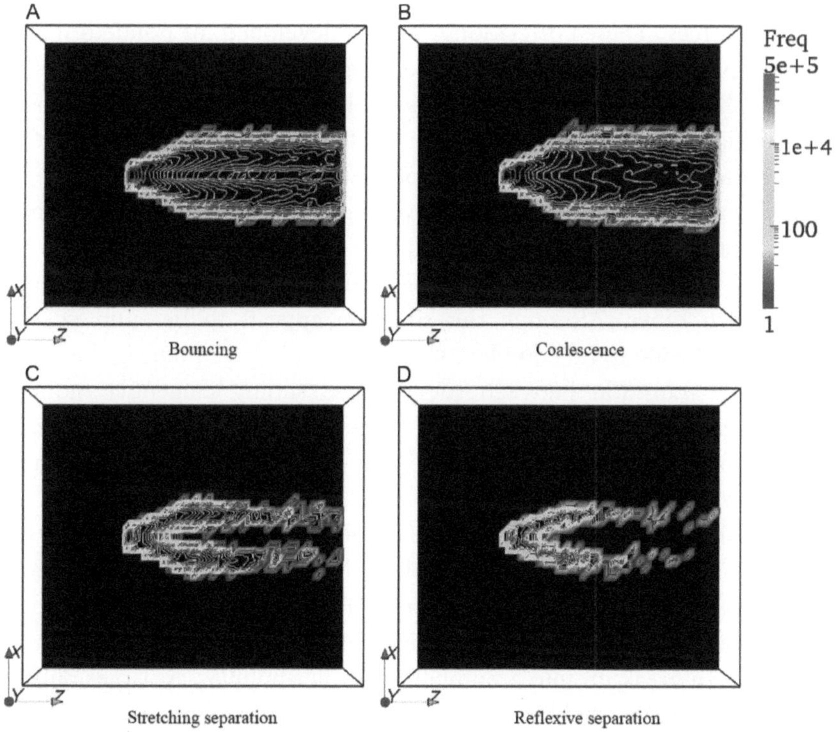


**Figure 18** Contour plots of the frequencies (number of collisions per second per cm$^3$) of different collision events on a cutting plane at the center of the domain in a spray for 250 bar inlet pressure. A: bouncing, B: coalescense, C: stretching separation, D: reflexive separation. Note the logarithmic color scale.

In summary, the DSMC simulations compare favorably to experimental results and show how frequencies of different collision events vary smoothly over length scales which are much larger than the droplet sizes, yet much smaller than the system size.

## 7. SUMMARY AND OUTLOOK

In this chapter, we have focused on relatively dense dispersed multiphase flows. Euler–Lagrange simulations make it possible to predict the flows and mesoscale structures appearing in such systems at industrial scales. Grid-averaged equations for the continuous-phase flow field are solved using CFD, where the grid size is larger than the discrete phase size, while the discrete phase is explicitly tracked and experiencing forces in a Lagrangian fashion. The advantage of the Euler–Lagrange approach is that much larger systems can be handled than is possible with fully resolved DNS. One key input for the method is an accurate quantitative description of the effective interphase momentum transfer, in terms of correlations for drag, lift, and virtual mass forces. These correlations are well developed for nondeformable spherical particles, such as encountered in some gas–solid flows, but less so for nonspherical and deformable particles, such as encountered in bubbly flows at high Eötvös number. Another key input is an accurate quantitative description of the binary interactions (or effects of the interactions) between members of the discrete phase. We have discussed a number of methods, connected to a number of particular multiphase systems.

In the DEMs (such as CFD–DEM), undeformable particles bounce against each other. Although this situation appears to be simplest, the response is already quite complex, with energy dissipated during the collision and a transition between sticking and sliding tangential motion, depending on the angle of impact. The details of these dissipative interactions ultimately determine the particle motion and inhomogeneities appearing at larger scale.

In the DBM, the centers of mass and volumes of individual bubbles are tracked. The outcome of a binary collision between two such bubbles is bounce or coalescence, depending on many factors including the surface tension and relative velocity. Under sufficiently strong deformations, bubbles can even break up into a number of smaller bubbles. At larger bubble volume fractions, the bubble motion leads to a strong mixing of the continuous fluid and therefore a large dispersion of dissolved chemicals.

In the DSMC method, collisions between the discrete phase particles are not detected by deterministic contact point calculations, but rather stochastically from the local number density, relative velocities, and sizes of neighboring particles. This method is applicable when the duration of a collision is much shorter than the mean free time between two consecutive collisions. We have shown how this method can be applied to predict the spatial and temporal evolution of a spray of liquid droplets in a gas, using correlations for binary droplet interactions obtained from experiment and DNSs.

Although much progress has been made in the field of Euler–Lagrange modeling, most models are still limited to near-spherical and nearly undeformable particles. We identify a number of problems which should be solved before further progress is possible. These are:

- Although highly accurate drag correlations can be obtained from DNSs, these correlations perform only fairly in CFD–DEM simulations. The probable cause is that the correlations have been obtained for non-moving particles. Particle mobility needs to be taken into account to obtain high accuracy drag closures for particles that move. One possible route is to include a dependency on the local granular temperature.
- Accurate correlations for the drag and lift forces in dense collections of nonspherical particles and polydisperse mixtures of particles are needed. For high aspect ratio particles, these correlations should include as a parameter the amount of mutual alignment between the particles.
- A systematic procedure to quantify the effects of subgrid scale particle clustering on the drag and lift forces is needed, not only for rigid spherical particles but also for deformable or highly nonspherical particles.
- The role of subgrid scale (pseudo) turbulent fluctuations on the collision probabilities and collision outcomes of deformable particles should be investigated. This is especially important for applications of stochastic models such as DSMC.
- Surfactants are known to play a very important role in bubbly flows. Yet, there is very little known on how to effectively characterize the presence of surfactants in such a manner that they can be used in the closure for the interfacial forces, as well as for coalescence and breakup.

One of the major difficulties will be to find the best suitable parameters that characterize the nonsphericity, the deformability, the polydispersity, the clustering, the alignment, and the surfactant fouling. These parameters should be sufficiently general to apply to a large range of dispersed particles, bubbles, or droplets, yet at the same time be sufficiently descriptive to distinguish between different systems and allow for accurate correlations to be formed. When such correlations become available, Euler–Lagrange models will have become truly powerful tools to predict the dynamics of nonideal, and therefore industrially relevant, multiphase systems.

## ACKNOWLEDGMENTS

The authors thank the European Research Council for its financial support, under its Advanced Investigator Grant scheme, Contract no. 247298 (MultiscaleFlows), its Starting Grant scheme, Contract no. 259521 (CuttingBubbles), and its Consolidator Grant scheme, Contract no. 615906 (NonSphereFlow).

## APPENDIX A. DERIVATION OF THE FILTERED FLUID EQUATION

In this appendix, we give a derivation of the macroscopic equations for the fluid phase. It is obtained from the microscopic equations using a filter or smoothing kernel. The approach is similar to the classical derivation of Anderson and Jackson (1967), but some of the results are new. In the derivation, there are some tricky parts. Important in the derivation is that explicit use is made of a separation of length scales. The scale on which macroscopic quantities vary is much larger than the filtered size, which in turn is much larger than the particle size. For all quantities (except the velocity), we will define the smoothed quantity of a microscopic property as,

$$a(\mathbf{x}) = \frac{\iiint\limits_{V_\mathrm{f}} D(\mathbf{x}-\mathbf{y})a^{\mathrm{micro}}(\mathbf{y})\,\mathrm{d}V_\mathrm{y}}{\iiint\limits_{V_\mathrm{f}} D(\mathbf{x}-\mathbf{y})\,\mathrm{d}V_\mathrm{y}} = \frac{1}{\varepsilon_\mathrm{f}(\mathbf{x})}\iiint\limits_{V_\mathrm{f}} D(\mathbf{x}-\mathbf{y})a^{\mathrm{micro}}(\mathbf{y})\,\mathrm{d}V_\mathrm{y} \tag{45}$$

For analyzing the filtering of the momentum balance, let us introduce the total stress or momentum flux as $\boldsymbol{\sigma}^{\mathrm{micro}} = p^{\mathrm{micro}}\mathbf{I} + \boldsymbol{\tau}_\mathrm{f}^{\mathrm{micro}}$. The source term of the smoothed momentum balance of the fluid phase is obtained as:

$$-\iiint\limits_{V_\mathrm{f}} D(\mathbf{x}-\mathbf{y})\nabla\cdot\boldsymbol{\sigma}^{\mathrm{micro}}(\mathbf{y})\,\mathrm{d}V_\mathrm{y}$$
$$= -\iint\limits_{A_\mathrm{fp}} D(\mathbf{x}-\mathbf{y})\mathbf{n}\cdot\boldsymbol{\sigma}^{\mathrm{micro}}(\mathbf{y})\mathrm{d}A_\mathrm{y} - \nabla\cdot\iiint\limits_{V_\mathrm{f}} D(\mathbf{x}-\mathbf{y})\boldsymbol{\sigma}^{\mathrm{micro}}(\mathbf{y})\,\mathrm{d}V_\mathrm{y}$$

$$
\begin{aligned}
&= \sum_i \oiint_{A_i} D(\mathbf{x}-\mathbf{y})\mathbf{n}\cdot\boldsymbol{\sigma}^{\text{micro}}(\mathbf{y})\,\mathrm{d}A_\mathrm{y} - \nabla\cdot\iiint_{V_\mathrm{f}} D(\mathbf{x}-\mathbf{y})\boldsymbol{\sigma}^{\text{micro}}(\mathbf{y})\,\mathrm{d}V_\mathrm{y} \\
&= \sum_i \oiint_{A_i} D(\mathbf{x}-\mathbf{y})\mathbf{n}\cdot\boldsymbol{\sigma}(\mathbf{y})\,\mathrm{d}A_\mathrm{y} + \oiint_{A_i} D(\mathbf{x}-\mathbf{y})\mathbf{n}\cdot\left(\boldsymbol{\sigma}^{\text{micro}}(\mathbf{y})-\boldsymbol{\sigma}(\mathbf{y})\right)\mathrm{d}A_\mathrm{y} \\
&\quad - \nabla\cdot\iiint_{V_\mathrm{f}} D(\mathbf{x}-\mathbf{y})\boldsymbol{\sigma}^{\text{micro}}(\mathbf{y})\,\mathrm{d}V_\mathrm{y} \\
&= \sum_i \oiint_{A_i} D(\mathbf{x}-\mathbf{y})\mathbf{n}\cdot\boldsymbol{\sigma}(\mathbf{y})\,\mathrm{d}A_\mathrm{y} + \oiint_{A_i} D(\mathbf{x}-\mathbf{y})\mathbf{n}\cdot\left(\boldsymbol{\sigma}^{\text{micro}}(\mathbf{y})-\boldsymbol{\sigma}(\mathbf{y})\right)\mathrm{d}A_\mathrm{y} \\
&\quad - \nabla\cdot(\varepsilon_\mathrm{f}\boldsymbol{\sigma})
\end{aligned} \tag{46}
$$

In the second line, the surface integral over the particle–fluid interface is split up into integrals over the individual particle surfaces and the normal is inverted as to become an outward pointing for each particle. For the determination of a good approximation of the surface integrals, the separation of length scales argument will be explicitly invoked. For the first of the two surface integrals over particle surface, the macroscopic, smoothed, field $\boldsymbol{\sigma}(\mathbf{y})$ varies slowly over the length scale associated with the filter width. Therefore, this macroscopic stress can be assumed constant for each particle integral, giving:

$$
\begin{aligned}
\sum_\mathrm{i} \oiint_{A_\mathrm{i}} D(\mathbf{x}-\mathbf{y})\mathbf{n}\cdot\boldsymbol{\sigma}(\mathbf{y})\mathrm{d}A_\mathrm{y} &= \sum_\mathrm{i}\iiint_{V_\mathrm{i}} \nabla\cdot(D(\mathbf{x}-\mathbf{y})\boldsymbol{\sigma}(\mathbf{y}))\,\mathrm{d}V_\mathrm{y} \\
&= -\nabla\cdot\sum_\mathrm{i}\iiint_{V_\mathrm{i}} D(\mathbf{x}-\mathbf{y})\boldsymbol{\sigma}(\mathbf{y})\,\mathrm{d}V_\mathrm{y} \\
&\quad + \sum_\mathrm{i}\iiint_{V_\mathrm{i}} D(\mathbf{x}-\mathbf{y})\nabla\cdot\boldsymbol{\sigma}(\mathbf{y})\,\mathrm{d}V_y \\
&\approx -\nabla\cdot((1-\varepsilon_\mathrm{f})\boldsymbol{\sigma}) + (1-\varepsilon_\mathrm{f})\nabla\cdot\boldsymbol{\sigma} \\
&= -\nabla(1-\varepsilon_\mathrm{f})\cdot\boldsymbol{\sigma}
\end{aligned} \tag{47}
$$

On the other hand, the variation of the filter over a particle diameter is negligible compare to the variation of microscopic fluctuations (i.e., the microscopic field minus the smoothed field). Therefore, for each particle, the smoothing filter can be Taylor expanded with respect to the center position of the particle indicated by $\mathbf{x}_\mathrm{i}$,

$$\begin{aligned}
&\sum_{\mathrm{i}} \oiint_{A_{\mathrm{i}}} D(\mathbf{x}-\mathbf{y})\mathbf{n}\cdot\left(\boldsymbol{\sigma}^{\mathrm{micro}}(\mathbf{y})-\boldsymbol{\sigma}(\mathbf{y})\right)\mathrm{d}A_{\mathrm{y}} \\
&\approx \sum_{\mathrm{i}} D(\mathbf{x}-\mathbf{x}_{\mathrm{i}}) \oiint_{A_{\mathrm{i}}} \mathbf{n}\cdot\left(\boldsymbol{\sigma}^{\mathrm{micro}}(\mathbf{y})-\boldsymbol{\sigma}(\mathbf{y})\right)\mathrm{d}A_{\mathrm{y}} \\
&= \sum_{\mathrm{i}} D(\mathbf{x}-\mathbf{x}_{\mathrm{i}})(\mathbf{f}_{\mathrm{i}\to\mathrm{f}} - V_{\mathrm{i}}\nabla\cdot\boldsymbol{\sigma}) \\
&= \mathbf{S}_{\mathrm{p}\to\mathrm{f}} - (1-\varepsilon_{\mathrm{f}})\nabla\cdot\boldsymbol{\sigma}
\end{aligned} \tag{48}$$

Here, the total force that a particle applies to the fluid is computed by integrating the total microscopic stress, $\boldsymbol{\sigma}^{\mathrm{micro}}$ over the particle surface. When distributing this term over the fluid by means of the filter kernel, it provides the source term $\mathbf{S}_{\mathrm{p}\to\mathrm{f}}$ to the macroscopic momentum balance of the fluid. An improved expansion would be a second-order Taylor expansion of the filter function where the extra contribution is

$$\begin{aligned}
&-\sum_{\mathrm{i}} \left( \oiint_{A_{\mathrm{i}}} \mathbf{n}\cdot\left(\boldsymbol{\sigma}^{\mathrm{micro}}(\mathbf{y})-\boldsymbol{\sigma}(\mathbf{y})\right)(\mathbf{y}-\mathbf{x}_{\mathrm{i}})\mathrm{d}A_{\mathrm{y}} \right)\cdot\nabla D(\mathbf{x}-\mathbf{x}_{\mathrm{i}}) \\
&= -\nabla\cdot\sum_{\mathrm{i}} D(\mathbf{x}-\mathbf{x}_{\mathrm{i}}) \oiint_{A_{\mathrm{i}}} (\mathbf{y}-\mathbf{x}_{\mathrm{i}})\left(\boldsymbol{\sigma}^{\mathrm{micro}}(\mathbf{y})-\boldsymbol{\sigma}(\mathbf{y})\right)^{\mathrm{T}}\cdot\mathbf{n}\mathrm{d}A_{y} \\
&= -\nabla\cdot\left(\sum_{\mathrm{i}} D(\mathbf{x}-\mathbf{x}_{\mathrm{i}})\mathbf{s}_{\mathrm{i}}\right)
\end{aligned} \tag{49}$$

The symmetric part of $\mathbf{s}_{\mathrm{i}}$ is the so-called stresslet of particle i, while the components of the antisymmetric part is the torque supplied to the fluid by the particle. Usually, this second-order correction of the particle–fluid interaction is not taken into account.

Taken all terms (up to first order) together, we get to good approximation

$$\begin{aligned}
&-\iiint_{V_{\mathrm{f}}} D(\mathbf{x}-\mathbf{y})\nabla\cdot\boldsymbol{\sigma}^{\mathrm{micro}}(\mathbf{y})\mathrm{d}V_{\mathrm{y}} \approx -\nabla(1-\varepsilon_{\mathrm{f}})\cdot\boldsymbol{\sigma} \\
&\quad + \left(\mathbf{S}_{\mathrm{p}\to\mathrm{f}} - (1-\varepsilon_{\mathrm{f}})\nabla\cdot\boldsymbol{\sigma}\right) - \nabla\cdot(\varepsilon_{\mathrm{f}}\boldsymbol{\sigma}) \\
&= \mathbf{S}_{\mathrm{p}\to\mathrm{f}} - \nabla\cdot\boldsymbol{\sigma} = \mathbf{S}_{\mathrm{p}\to\mathrm{f}} - \nabla p - \nabla\cdot\boldsymbol{\tau}_{\mathrm{f}}^{\mathrm{Newt}}
\end{aligned} \tag{50}$$

This final result is deceptively simple, but proper accounting of the $\varepsilon_{\mathrm{f}}$ is subtle as was illustrated by the preceding derivation. For example, one might expect a factor $\varepsilon_{\mathrm{f}}$ in front of the pressure gradient and stress terms. Some authors in fact have proposed that such a factor should be present. This is usually referred to as model A, while our form is more close to model B. To increase the confusion, there is also a possible difference in the interpretation of $\mathbf{S}_{p\rightarrow f}$. This is more extensively discussed in Section 2.3. The fact that there is no factor $\varepsilon_f$ in front of the pressure gradient (and stress term) in our formula denotes that all pressure is assigned to the fluid phase and none to the discrete particle phase. The physical reason is that the discrete phase cannot bear a load because of its disconnected nature.

Most authors do have an extra $\varepsilon_f$ term inside the stress divergence and give a Newtonian stress expression in terms of the macroscopic velocities. If we rigorously smooth the microscopic Newtonian stress for the case of constant density and constant viscosity, we get

$$\begin{aligned}
\boldsymbol{\tau}_f^{\text{Newt}} &= \frac{1}{\varepsilon_f}\iiint_{V_f} D(\mathbf{x}-\mathbf{y})\boldsymbol{\tau}^{\text{micro}}(\mathbf{y})\,\mathrm{d}V_y \\
&= \frac{\mu_f}{\varepsilon_f}\iiint_{V_f} D(\mathbf{x}-\mathbf{y})\left(\nabla\mathbf{u}^{\text{micro}}(\mathbf{y}) + \left[\nabla\mathbf{u}^{\text{micro}}(\mathbf{y})\right]^{\mathrm{T}}\right)\mathrm{d}V_y \\
&= \frac{\mu_f}{\varepsilon_f}\iiint_{V_f+V_p} D(\mathbf{x}-\mathbf{y})\left(\nabla\mathbf{u}^{\text{micro}}(\mathbf{y}) + \left[\nabla\mathbf{u}^{\text{micro}}(\mathbf{y})\right]^{\mathrm{T}}\right)\mathrm{d}V_y \\
&= \frac{\mu_f}{\varepsilon_f}\left(\left[\nabla\iiint_{V_f+V_p} D(\mathbf{x}-\mathbf{y})\mathbf{u}^{\text{micro}}(\mathbf{y})\,\mathrm{d}V_y\right] + [\cdots]^{\mathrm{T}}\right) \\
&= \frac{\mu_f}{\varepsilon_f}\left(\left[\nabla\left(\varepsilon_f\mathbf{u}+\varepsilon_p\mathbf{v}\right)\right] + [\cdots]^{\mathrm{T}}\right)
\end{aligned} \tag{51}$$

In the second line, the integral over the fluid domain is extended to also include the particles. This is allowed for solid particles because for rigid bodies the deformation rate equals zero. For a deformable discrete phase (i.e., bubbles or droplets), this is a possibly severe approximation. Note that the assumption that the fluid density is constant was used because the proper definition of $\mathbf{u}$ is a mass average, while here we performed a volume average. In the final result, we introduced the notation $\mathbf{v}$ for the smoothed particle velocity field, defined as

$$\mathbf{v}(\mathbf{x}) = \frac{1}{\varepsilon_p(\mathbf{x})}\iiint_{V_p} D(\mathbf{x}-\mathbf{y})\mathbf{v}^{\text{micro}}(\mathbf{y})\,\mathrm{d}V_y \approx \frac{1}{\varepsilon_p(\mathbf{x})}\sum_i D(\mathbf{x}-\mathbf{x}_i)\,V_i\mathbf{v}_i \tag{52}$$

The expression for the microscopic stress used was the one for incompressible fluids. For a compressible fluid, a contribution related to the divergence of the compressibility enters and the bulk viscosity comes into play. In the case of an incompressible liquid, we have $\nabla \cdot \mathbf{u}^{\mathrm{micro}} = 0$. At a macroscopic level, incompressibility gives, as a consequence of the continuity equation: $\nabla \cdot \left(\varepsilon_{\mathrm{f}}\mathbf{u} + \varepsilon_{\mathrm{p}}\mathbf{v}\right) = 0$. So, both the microscopic and the smoothed Newtonian stress have a trace zero for the incompressible case (as should be). Note that this property is not obeyed if the ordinary Newtonian form in terms of macroscopic fluid velocities is assumed for the viscous stress.

## APPENDIX B. DIRECT NUMERICAL SIMULATION OF DROPLET COLLISION OUTCOMES

DNSs can be used fruitfully to derive closures and correlations which serve as input for Euler–Lagrange multiphase flow models. In this appendix, we provide some technical details of the DNS used to determine droplet collision outcomes, as referred to in Section 6.

In the literature, relatively few papers have been published on DNS of droplet–droplet collisions. Pan and Suga (2005) used a Level Set (LS) method to study in detail the complete process of the collision of two equally sized liquid droplets. They could reproduce the main features of the droplet–droplet collision process from their DNS but volume conservation of the drops turned out to be problematic, especially at low *We* due to the time step constraint imposed by the explicit treatment of the surface tension. For the treatment of surface tension, they used the CSF model of Brackbill et al (1992). Nikolopoulos and Bergeles (2011) used the Volume-of-Fluid (VOF) method to simulate the head-on collision of two unequal-size drops in the reflexive regime assuming axisymmetrical flow. For the solution of the color function equation, they used the high-resolution CICSAM scheme proposed by Ubbink and Issa (1999). Kwakkel et al (2012) used an efficient multiple marker front-capturing method based on the combined LS–VOF method to simulate a large number of droplets dispersed in a turbulent carrier fluid. The reported advantage of the combined LS–VOF method is the more accurate representation of the surface tension with the LS method and the better intrinsic volume conservation offered by the VOF method.

We have applied our own VOF method to the problem of droplet–droplet collisions, which we will now briefly discuss. The method is based on the one-field formulation where a single set of conservation equations is used to describe the flow in two incompressible immiscible phases sharing an interface. The governing momentum equation for unsteady incompressible Newtonian flow is given by the following expression:

$$\rho\left[\frac{\partial}{\partial t}(\mathbf{u})+(\nabla\cdot\mathbf{uu})\right]=-\nabla p+\rho\mathbf{g}+\left(\nabla\cdot\mu\left[(\nabla\mathbf{u})+(\nabla\mathbf{u})^{\mathrm{T}}\right]\right)+\mathbf{f}_{\sigma} \quad (53)$$

where the local density $\rho$ and local viscosity $\mu$ are evaluated from the local distribution of the color or phase indicator function $F$ which is governed by:

$$\frac{DF}{Dt}=\frac{\partial F}{\partial t}+(\mathbf{u}\cdot\nabla F)=0 \quad (54)$$

expressing that any interface property is transported with the local fluid velocity. For the local average density $\rho$, linear weighing of the densities of the continuous fluid (f) and dispersed phase (p) is used:

$$\rho=F\rho_{\mathrm{f}}+(1-F)\rho_{\mathrm{p}} \quad (55)$$

In a similar way, the kinematic viscosities of the involved phases are harmonically averaged:

$$\frac{\rho}{\mu}=F\frac{\rho_{\mathrm{f}}}{\mu_{\mathrm{f}}}+(1-F)\frac{\rho_{\mathrm{p}}}{\mu_{\mathrm{p}}} \quad (56)$$

The volumetric surface tension force or force density $\mathbf{f}_{\sigma}$ appearing in the momentum equation acts only in the vicinity of the interface. The numerical solution of the governing equations mainly follows the approach of Van Sint Annaland et al (2005), but the treatment of the viscous interaction terms and surface tension is different from our previous work. By treating the viscous interaction terms in an implicit fashion, the time step constraint imposed by the viscosity stability condition, which is prohibitive for the simulation of collisions of highly viscous droplets, can be circumvented. Moreover, for the surface tension force, in the current implementation we do not use the CSF model (Brackbill et al, 1992) which requires the computation of the interface curvature in conjunction with (necessary) smoothing of the color function. Instead, the net surface force acting on each discretized surface segment is calculated directly from the cross-product of tangent vectors at the segment edges and normal vectors of the reconstructed interface segments. For more details, we refer to Dijkhuizen et al (2010) and Baltussen et al (2014).

We note that the ratio of droplet diameter $d_p$ to grid size ranged from 40 to 60, whereas the ratio of the droplet diameter and the domain size ranged from 6.0 up to 7.5 for the highest *We* values used. Since at high *We* numbers the elongation of the separating droplets is very substantial, the body diagonal of the simulation domain was taken as the main axis of collision for the impacting droplets. In all cases, the droplets were given an equal but opposite initial velocity.

## REFERENCES


Anderson TB, Jackson R: A fluid mechanical description of fluidized beds, *Ind Eng Chem Fundam* 6:527–539, 1967.

Andrews MJ, O'Rourke PJ: The multiphase particle-in-cell (MP-PIC) method for dense particulate flows, *Int J Multiphase Flow* 22:379–402, 1996.

Ashgriz N, Poo JY: Coalescence and separation in binary collisions of liquid drops, *J Fluid Mech* 221:183–204, 1990.

Bai W, Deen NG, Kuipers JAM: Numerical investigation of gas holdup and phase mixing in bubble column reactors, *Ind Eng Chem Res* 51:1949–1961, 2012.

Balachandar S, Eaton JK: Turbulent dispersed multiphase flow, *Annu Rev Fluid Mech* 42:111–133, 2010.

Baltussen MW, Kuipers JAM, Deen NG: A critical comparison of surface tension models for the volume of fluid method, *Chem Eng Sci* 109:65–74, 2014.

Beetstra R, van der Hoef MA, Kuipers JAM: Drag force of intermediate Reynolds number flow past mono- and bidisperse arrays of spheres, *AIChE J* 53:489–501, 2007.

Bird A: *Molecular gas dynamics and direct simulation of gas flow,* Oxford, 1976, Clarendon.

Bouillard JX, Lyczkowski RW, Gidaspow D: Porosity distributions in a fluidized-bed with an immersed obstacle, *AIChE J* 35:908–922, 1989.

Brackbill JU, Kothe DB, Zemach C: A continuum method for modeling surface tension, *J Comput Phys* 100:335, 1992.

Cundall PA, Strack ODL: A discrete numerical model for granular assemblies, *Géotechnique* 29:47–65, 1979.

Darmana D, Deen NG, Kuipers JAM: Detailed modeling of hydrodynamics, mass transfer and chemical reactions in a bubble column using a discrete bubble model, *Chem Eng Sci* 60:3383–3404, 2005.

Darmana D, Henket RLB, Deen NG, Kuipers JAM: Detailed modelling of hydrodynamics, mass transfer and chemical reactions in a bubble column using a discrete bubble model: chemisorption of $CO_2$ into NaOH solution, numerical and experimental study, *Chem Eng Sci* 62:2556–2575, 2007.

Davis RH, Ragerd DA, Good BT: Elastohydrodynamic rebound of spheres from coated surface, *J Fluid Mech* 468:107–119, 2002.

Deckwer WD, Burckhart R, Zoll G: Mixing and mass transfer in tall bubble columns, *Chem Eng Sci* 29:2177–2188, 1974.

Deen NG, van Sint Annaland M, Kuipers JAM: Multi-scale modeling of dispersed gas–liquid two-phase flow, *Chem Eng Sci* 59:1853–1861, 2004.

Deen NG, van Sint Annaland M, van der Hoef MA, Kuipers JAM: Review of discrete particle modeling of fluidized beds, *Chem Eng Sci* 62(1–2):28–44, 2007.

Dijkhuizen W, Roghair I, van Sint Annaland M, Kuipers JAM: DNS of gas bubbles behaviour using an improved 3D front tracking model—model development, *Chem Eng Sci* 65:1427–1437, 2010.

Du M, Zhao C, Zhao B, Guo H, Hao Y: A modified DSMC method for simulating gas-particle two-phase impinging streams, *Chem Eng Sci* 66:4922–4931, 2011.

Ergun S: Fluid flow through packed columns, *Chem Eng Prog* 48:89–94, 1952.

Gidaspow D: *Multiphase flow and fluidization: continuum and kinetic theory descriptions,* Boston, MA, 1994, Academic Press.

Gorokhovski M, Herrmann M: Modeling primary atomization, *Annu Rev Fluid Mech* 40:343–366, 2008.

Hadjiconstantinou NG: The limits of Navier–Stokes theory and kinetic extensions for describing small-scale gaseous hydrodynamics, *Phys Fluids* 18:111301, 2006.

Han Z, Parrish S, Farrell PV, Reitz RD: Modeling atomization processes of pressure-swirl hollow-cone fuel sprays, *Atomization Sprays* 7:663–684, 1997.

Heijnen JJ: Van't Riet K: Mass transfer, mixing and heat transfer phenomena in low viscosity bubble column reactors, *Chem Eng J* 28:B21–B42, 1984.

Hikita H, Kikukawa H: Liquid-phase mixing in bubble columns: effect of liquid properties, *Chem Eng J* 9:171–174, 1974.

Hill RJ, Koch DL, Ladd AJC: Moderate-Reynolds-number flows in ordered and random arrays of spheres, *J Fluid Mech* 448:243–278, 2001.

Jain D, Kuipers JAM, Deen NG: Numerical modeling of carbon dioxide chemisorption in sodium hydroxide solution in a micro-structured bubble column, *Chem Eng Sci* 137:685–696, 2015.

Jakobsen HA, Sannaes BH, Grevskott S, Svendsen HF: Modeling of vertical bubble-driven flows, *Ind Eng Chem Res* 36:4052–4074, 1997.

Jakobsen HA, Lindborg H, Dorao CA: Modeling of bubble column reactors: progress and limitations, *Ind Eng Chem Res* 44:5107–5151, 2005.

Jamaleddine TJ, Ray MB: Application of computational fluid dynamics for simulation of drying processes: a review, *Drying Technol* 28:120–154, 2010.

Jiang X, Siamas GA, Jagus K, Karayiannis TG: Physical modelling and advanced simulations of gas–liquid two-phase jet flows in atomization and sprays, *Prog Energy Combust Sci* 36:131–167, 2010.

Ko GH, Ryou HS: Modeling of droplet collision-induced breakup process, *Int J Multiphase Flow* 31:723–738, 2005.

Kolakaluri R, Subramaniam S, Panshagnula MV: Trends in multiphase modeling and simulation of sprays, *Int J Combust Dyn* 6:317–356, 2014.

Kriebitzsch SHL, van der Hoef MA, Kuipers JAM: Fully resolved simulation of a gas-fluidized bed: a critical test of DEM models, *Chem Eng Sci* 91:1–4, 2013a.

Kriebitzsch SHL, van der Hoef MA, Kuipers JAM: Drag force in discrete particle models—continuum scale or single particle scale? *AIChE J* 59:316–324, 2013b.

Kwakkel M, Breugem WP, Boersma BJ: An efficient multiple marker front-capturing method for two-phase flow, *Comput Fluids* 63:47–56, 2012.

Lau YM, Bai W, Deen NG, Kuipers JAM: Numerical study of bubble break-up in bubbly flow using a deterministic Euler-Lagrange framework, *Chem Eng Sci* 108:9–22, 2014.

Li J, Kuipers JAM: On the origin of heterogeneous structure in dense gas-solid flows, *Chem Eng Sci* 60:1251–1265, 2005.

Link JM, Cuypers LA, Deen NG, Kuipers JAM: Flow regimes in a spout-fluid bed: a combined experimental and simulation study, *Chem Eng Sci* 60:3425–3442, 2005.

Mehrabadi M, Tenneti S, Garg R, Subramaniam S: Pseudo-turbulent gas-phase velocity fluctuations in homogeneous gas-solid flow: fixed particle assemblies and freely evolving suspensions, *J Fluid Mech* 770:210–246, 2015.

Mishra YN, Kristensson E, Berrocal E: Reliable LIF/Mie droplet sizing in sprays using structured laser illumination planar imaging, *Opt Express* 22:4480–4492, 2014.

Nikolopoulos N, Bergeles G: The effect of gas and liquid properties and droplet size ratio on the central collision between two unequal-size droplets in the reflexive regime, *Int J Mass Heat Transf* 54:678–691, 2011.

Ohki Y, Inoue H: Longitudinal mixing of the liquid phase in bubble columns, *Chem Eng Sci* 25:1–16, 1970.

O'Rourke: *Collective drop effect on vaporizing liquid sprays,* Ph.D. thesis, 1981. Mechanical and Aerospace Engineering, Princeton University, Princeton, NJ.

Padding JT, Louis AA: Hydrodynamic interactions and Brownian forces in colloidal suspensions: coarse-graining over time and length scales, *Phys Rev E* 74:031402, 2006.

Pan Y, Suga K: Numerical simulation of binary liquid droplet collision, *Phys Fluids* 17:082105, 2005.

Patil AV, Peters EAJF, Kolkman T, Kuipers JAM: Modeling bubble heat transfer in gas-solid fluidized beds using DEM, *Chem Eng Sci* 105:121–131, 2014.

Pawar SK, Padding JT, Deen NG, Jongsma A, Innings F, Kuipers JAM: Eulerian–Lagrangian modelling with stochastic approach for droplet-droplet collisions. In Proceedings of the ninth international conference on CFD in the minerals and process industries CSIRO, Melbourne, Australia, *10–12*, 2012.

Pawar SK, Padding JT, Deen NG, Jongsma A, Innings F, Kuipers JAM: Lagrangian modelling of dilute granular flow-modified stochastic DSMC versus deterministic DPM, *Chem Eng Sci* 105:132–142, 2014.

Pawar S, Padding J, Deen N, Jongsma A, Innings F, Kuipers JAM: Numerical and experimental investigation of induced flow and droplet–droplet interactions in a liquid spray, *Chem Eng Sci* 138:17–30, 2015.

Portela LM, Oliemans RVA: Possibilities and limitations of computer simulations of industrial turbulent dispersed multiphase flows, *Flow Turbul Combust* 77:381–403, 2006.

Pournin L, Weber M, Tsukahara M, Ferrez J-A, Ramaioli M, Liebling ThM: Three-dimensional distinct element simulation of spherocylinder crystallization, *Granul Matter* 7:119, 2005.

Prince MJ, Blanch HW: Bubble coalescence and break-up in air-sparged bubble columns, *AIChE J* 36:1485–1499, 1990.

Qian J, Law CK: Regimes of coalescence and separation in droplet collision, *J Fluid Mech* 331:59–80, 1997.

Radl S, Sundaresan S: A drag model for filtered Euler-Lagrange simulations of clustered gas-particle suspensions, *Chem Eng Sci* 117:416–425, 2014.

Roghair I, Lau YM, Deen NG, et al: On the drag force of bubbles in bubble swarms at intermediate and high Reynolds numbers, *Chem Eng Sci* 66:3204–3211, 2011.

Shah MT, Utikar RP, Pareek VK, Tade MO, Evans GM: Effect of closure models on Eulerian-Eulerian gas-solid flow prediction in riser, *Powder Technol* 269:247–258, 2015.

Silva R, Garcia FAP, Faia PMGM, Rasteiro MG: Settling suspensions flow modelling: a review, *Kona Powder Part J* 32:41–56, 2015.

Sommerfeld M, Bourloutski E, Bröder D: Euler/Lagrange calculations of bubbly flows with consideration of bubble coalescence, *Can J Chem Eng* 81:508–518, 2003.

Subramaniam S: Lagrangian–Eulerian methods for multiphase flows, *Prog Energy Combust Sci* 39:215–245, 2013.

Sundaresan S: Modeling the hydrodynamics of multiphase flow reactors: current status and challenges, *AIChE J* 46:1102–1105, 2000.

Sutkar VS, Deen NG, Padding JT, et al: A novel approach to determine wet restitution coefficients through a unified correlation and energy analysis, *AIChE J* 61:769–779, 2015.

Tang Y, Peters EAJF, Kuipers JAM, Kriebitzsch SHL, van der Hoef MA: A new drag correlation from fully resolved simulations of flow past monodisperse static arrays of spheres, *AIChE J* 61:688–698, 2015a.

Tang Y, Peters EAJF, Kuipers JAM: Direct numerical simulations of dynamic gas-solid suspensions, *AIChE J* , 2015b. submitted for publication.

Tenneti S, Subramaniam S: Particle-resolved direct numerical simulation for gas-solid flow model development, *Annu Rev Fluid Mech* 46:199–230, 2014.

Tenneti S, Garg R, Hrenya CM, Fox RO, Subramaniam S: Direct numerical simulation of gas-solid suspensions at moderate Reynolds number: quantifying the coupling between hydrodynamic forces and particle velocity fluctuations, *Powder Technol* 203:57–69, 2010.

Tenneti S, Garg R, Subramaniam S: Drag law for monodisperse gas-solid systems using particle-resolved direct numerical simulation of flow past fixed assemblies of spheres, *Int J Multiphase Flow* 37:1072–1092, 2011.

Tomiyama A: Drag, lift and virtual mass forces acting on a single bubble. In Proceedings 3rd international symposium on two-phase flow modelling and experimentation, Pisa, 22–24 September, 2004.

Towell GD, Ackermann GH: Axial mixing of liquid and gas in large bubble column reactors. In *2nd international symposium on chemical reaction engineering, Amsterdam*, 1972, p. B 3-1.

Ubbink O, Issa RI: A method for capturing sharp fluid interfaces on arbitrary meshes, *J Comput Phys* 153:26–50, 1999.

Van der Hoef MA, Ye M, van Sint Annaland M, Andrews AT, Sundaresan S, Kuipers JAM: Multi-scale modeling of gas-fluidized beds, *Adv Chem Eng* 31:65–149, 2006.

Van der Hoef MA, Van Sint Annaland M, Deen NG, Kuipers JAM: Numerical simulation of dense gas-solid fluidized beds: a multiscale modeling strategy, *Annu Rev Fluid Mech* 40:47–70, 2008.

Van Sint Annaland M, Deen NG, Kuipers JAM: Numerical simulation of gas bubbles behaviour using a three-dimensional volume of fluid method, *Chem Eng Sci* 60:2999–3011, 2005.

Verma V, Deen NG, Padding JT, Kuipers JAM: Two-fluid modeling of three-dimensional cylindrical gas-solid fluidized beds using the kinetic theory of granular flow, *Chem Eng Sci* 102:227–245, 2013.

Wang W, Li JH: Simulation of gas-solid two-phase flow by a multi-scale CFD approach of the EMMS model to the sub-grid level, *Chem Eng Sci* 62:208–231, 2007.

Wang J, van der Hoef MA, Kuipers JAM: Why the two-fluid model fails to predict the bed expansion characteristics of Geldart A particles in gas-fluidized beds: a tentative answer, *Chem Eng Sci* 64:622–625, 2009.

Wen CY, Yu YH: Mechanics of fluidization, *Chem Eng Prog Symp Ser* 62:100–111, 1966.

Wylie JJ, Koch DL, Ladd AJC: Rheology of suspensions with high particle inertia and moderate fluid inertia, *J Fluid Mech* 480:95–118, 2003.

Xie J, Zhong WQ, Jin BS, Shao YJ, Huang YJ: Eulerian–Lagrangian method for three-dimensional simulation of fluidized bed coal gasification, *Adv Powder Technol* 24:382–392, 2013.

Xu BH, Yu AB: Comments on the paper "Numerical simulation of the gas-solid flow in a fluidized bed by combining discrete particle method with computational fluid dynamics"—reply, *Chem Eng Sci* 53:2646–2647, 1998.

Zehner P: Momentum, mass and heat transfer in bubble columns. part 2. axial blending and heat transfer, *Int Chem Eng* 26:29–35, 1986.

Zhang J: Lattice Boltzmann method for microfluidics: models and applications, *Microfluid Nanofluid* 10:1–28, 2011.

Zhou ZY, Kuang SB, Chu KW, Yu AB: Discrete particle simulation of particle–fluid flow: model formulations and their applicability, *J Fluid Mech* 661:482–510, 2010.

Zhou ZY, Pinson D, Zou RP, Yu AB: Discrete particle simulation of gas fluidization of ellipsoidal particles, *Chem Eng Sci* 66:6128, 2011.

Zhu HP, Zhou ZY, Yang RY, Yu AB: Discrete particle simulation of particulate systems: theoretical developments, *Chem Eng Sci* 62:3378–3396, 2007.

CHAPTER FOUR

# Particle Scale Study of Heat Transfer in Packed and Fluidized Beds

**Qinfu Hou*,†, Jieqing Gan*, Zongyan Zhou*, Aibing Yu*,[1]**

*Laboratory for Simulation and Modelling of Particulate Systems, Department of Chemical Engineering, Monash University, Clayton, Victoria, Australia

†Laboratory for Simulation and Modelling of Particulate Systems, School of Materials Science and Engineering, University of New South Wales, Sydney, New South Wales, Australia

[1]Corresponding author: e-mail address: aibing.yu@monash.edu

## Contents

1. Introduction 196
2. Model Description 198
   2.1 Governing Equations 198
   2.2 Heat Transfer Models 202
   2.3 Coupling Schemes 211
3. Model Application 211
   3.1 Packed Beds 211
   3.2 Fluidized Beds 221
4. Conclusions and Future Work 235
Acknowledgment 238
References 238

## Abstract

Understanding and modeling coupled flow and heat transfer in a particulate system at a particle scale is a rapidly developing research area, in connection with the development of discrete particle simulation techniques and computer technology. The approach based on the discrete element method plays an important role in this area. This approach can provide detailed dynamic information of particulate systems, such as realistic packing structure, transient/static deformation, and relative velocities between fluid and individual particles. The information is directly related to the prediction of thermal behavior. In this chapter, the development of this approach in our laboratory is briefly reviewed, together with the discussion of different case studies. It is concluded that this particle scale approach is effective for studying the coupled fluid flow and heat transfer in particulate systems, although further developments are necessary to be generally applied to industrial processes.

*Advances in Chemical Engineering*, Volume 46
ISSN 0065-2377
http://dx.doi.org/10.1016/bs.ache.2015.10.006

## NOMENCLATURE

$a$ parameter used to calculate the Nusselt number in Eq. (8), dimensionless
$A_i$ particle surface area, m$^2$
$A^*$ dimensionless contact area, m$^2$
$A_c$ maximum contact area, m$^2$
$A_f$ contact area, m$^2$
$A_i$ surface area of particles $i$, m$^2$
$A_{ii}$, $A_{jj}$ the face area of AA′ and BB′ in Fig. 4, m$^2$
$A_{ij}$ the face area of Voronoi polyhedron between particles $i$ and $j$, m$^2$
$b$ parameter used to calculate the Nusselt number in Eq. (8), dimensionless
$c$ correction coefficient for heat flux in Eq. (17), dimensionless
$c_{p,i}$ thermal capacity of particle $i$, J/(Kg K)
$C'$ correction coefficient for heat flux in Eq. (12), dimensionless
$C_1$, $C_2$, $C_3$ parameters to calculate the correction coefficient $C'$, dimensionless
$d_{ij}$ defined by Fig. 3, m
$d_{pi}$ diameter of particle $i$, m
$D$ hydraulic diameter, m
$E_i$ Young's modulus of particles $i$ used in the DEM, Pa
$E_{ij}$ mean Young's modulus of particles $i$ and $j$, Pa
$E_{ij,0}$ real value of Young's modulus of the materials, Pa
$f_{c,ij}$ elastic force between particles $i$ and $j$, N
$f_{d,ij}$ viscous damping force between particles $i$ and $j$, N
$f_{d,i}$ fluid drag force of particle $i$, N
$f_{\nabla p,i}$ the pressure gradient force of particle $i$, N
$f_{f,i}$ particle–fluid interaction force of particle $i$, N
$F_{fp}$ volumetric fluid–particle interaction force, N
$F$ radiation exchange factor, dimensionless
$F_o$ Fourier number, dimensionless
$h$ separation of surfaces along the line of the centers of particles $i$ and $j$, m
$h_{i,conv}$ convective heat transfer coefficient, W/(m$^2$ K)
$h_{f,wall}$ heat transfer coefficient of the wall, W/(m$^2$ K)
$H$ defined by Eq. (19), m
$H_a$ Hamaker constant, J
$H_b$ the height between the two layers with two constant temperatures at the top and the bottom in the packed bed, m
$I_i$ moment of inertia of particle $i$, kg m$^2$
$k_V$ the number of particles in a computational cell of volume $\Delta V$, dimensionless
$k_e$ effective thermal conductivity, equals to $k_s/k_r$, dimensionless
$k_f$ the fluid thermal conductivity, W/(m K)
$k_s$ the particle thermal conductivity, W/(m K)
$k_{e,c}$ effective conductivity without radiation contributions, dimensionless
$k_r$ the radiative conductivity without conduction contributions, W/(m K)
$m_i$ mass of particle $i$, kg
$m_{ij}$ mean mass of particles $i$ and $j$, kg
$M_{t,ij}$ the tangential torque acting on particle $i$ by particle $j$, N M
$M_{r,ij}$ the rolling friction torque, N M

**Nu** the Nusselt number, dimensionless
$\boldsymbol{P}$ fluid pressure, Pa
**Pr** the gas Prandtl number, dimensionless
$q$ heat exchange rate per unit of area, $W/m^2$
$\mathbf{Q}_{\mathbf{csfs}}$ heat conduction between two contacting particles through the stagnant fluid, W
$\mathbf{Q}_{\mathbf{css}}$ heat conduction between two contacting particles through the contact area, W
$\mathbf{Q}_{\mathbf{f,wall}}$ fluid–wall heat exchange rate, W
$\mathbf{Q}_{ij}$ total heat transferred during each impact, J
$\mathbf{Q}_{i,j}$ the heat exchange rate between particles $i$ and $j$ due to conduction, W
$\mathbf{Q}_{ij,0}$ asymptotic one-dimensional heat transferred during each impact, W
$\mathbf{Q}_{i,f}$ the heat exchange rate between particle $i$ and its local surrounding fluid, W
$\mathbf{Q}_{ij,\mathbf{rad}}$ radiant exchange between particles $i$ and $j$, W
$\mathbf{Q}_{i,\mathbf{rad}}$ the heat exchange rate between particle $i$ and its surrounding environment by radiation, W
$\mathbf{Q}_{i,\mathbf{wall}}$ particle–wall heat exchange rate, W
$\mathbf{Q}_{\mathbf{nsfs}}$ heat conduction between two noncontacting particles through the stagnant fluid, W
$r_c$ maximum contact radius, m
$r_f$ contact radius at time $t$, s
$r_{ij}$ the radius of the lens of fluid between two contacting or near contacting spheres, m
$R_c$ the radius of the isothermal core used in Model B, m
$\mathbf{Re}_i$ local relative Reynolds number for particle $i$, dimensionless
$R_i$ radius of particle $i$, m
$R_j$ radius of particle $j$, m
$R_{ij}$ mean radius of particles $i$ and $j$, m
$t_c$ collision contact duration between particles $i$ and $j$, s
$t_d$ static contact duration between particles $i$ and $j$, s
$T_e$ the environmental temperature, K
$T_b$ bed temperature, K
$T_{f,i}$ fluid temperature in a computational cell where particle $i$ is located, K
$T_i$ temperature of particle $i$, K
$T$ fluid temperature, K
$T_{\mathbf{local},i}$ averaged temperature of particles and fluid by volume fraction in an enclosed domain $\Omega$, K
$T_m$ mean temperature of particles in the considered packed bed, K
$T_0$ initial bed temperature, K
$T_{in}$ air temperature at the inlet, K
$T_s$ tube temperature, K
u fluid velocity, m/s
$u_{exc}$ excess gas velocity, m/s
$U_g$ superficial gas velocity, m/s
$U_{mf}$ minimum fluidization velocity, m/s
$v_i$ translational velocity of particle $i$, m/s
$V_{n,ij}$ normal relative velocity between particles $i$ and $j$, m/s
$V_{ij}$ the volume of Voronoi polyhedra between particles $i$ and $j$, $m^3$
$V_{p,i}$ The volume of particle $i$ (or part of the volume if the particle is not fully in the cell), $m^3$
$\Delta V$ the volume of a computational cell, $m^3$

## GREEK LETTERS

$\alpha$ particle thermal diffusivity, $m^2/s$
$\varepsilon_f$ fluid porosity, dimensionless
$\varepsilon_i$ local porosity corresponding to particle $i$, dimensionless
$\varepsilon_{pi}$ sphere emissivity, dimensionless
$\varepsilon_{r,i}$ emissivity of particles $i$, dimensionless
$\nu$ Possion's ratio, dimensionless
$\rho_f$ fluid density, $kg/m^3$
$\sigma$ Stefan–Boltzmann constant, equal to $5.67 \times 10^{-8}$ W/($m^2$ $K^4$)
$\sigma_T$ turbulence Prandtl number, dimensionless
$\mu_e$ fluid effective viscosity, kg/(m s)
$\tau$ dimensionless time, dimensionless
$\tau$ fluid viscous stress tensor, Pa
$\Gamma$ fluid thermal diffusivity, $m^2/s$
$\omega_i$ angular velocity of particle $i$, 1/s

# 1. INTRODUCTION

Fluid bed reactors are widely used in industries mainly due to their high heat and mass transfer capability (Kunii and Levenspiel, 1991). To achieve optimal design and control of such reactors, it is important to understand the fundamentals governing flow, heat, and mass transfer. In the past, many macroscopic or reactor scale correlations have been formulated to determine the heat transfer coefficient (HTC), as reviewed by various investigators (Botterill, 1975; Kunii and Levenspiel, 1991; Molerus and Wirth, 1997; Wakao and Kaguei, 1982). The heat transfer mechanisms have been identified for years (Yagi and Kunii, 1957). However, it is difficult to quantify them. To overcome this difficulty, different techniques have been proposed at a microscopic or particle scale, mainly by tracing the motion and temperature of a single particle in packed/fluidized beds (Agarwal, 1991; Baskakov et al., 1987; Collier et al., 2004; Parker et al., 1993; Parmar and Hayhurst, 2002; Patil et al., 2015; Prins et al., 1985; Scott et al., 2004). Such particle scale studies are useful but have different limitations in exploring the fundamental details. For example, the heat transfer of a particle is strongly affected by the local gas–solid flow structure which varies spatially and temporally; this effect is not considered because of the difficulty in experimentally quantifying such structural information.

Alternatively, mathematical modeling has been increasingly accepted as an effective method to study the heat transfer phenomena in particle–fluid systems. Generally speaking, the existing approaches to modeling particle flow and thermal behavior can be classified into two categories: the continuum approach at a macroscopic level and the discrete approach at a microscopic/particle level. In the continuum approach, the macroscopic behavior is described by balance equations, e.g., mass, momentum, and energy, closed with constitutive relations together with initial and boundary conditions (see, for example, Anderson and Jackson, 1967; Enwald et al., 1996; Gidaspow, 1994; Ishii, 1975). However, its effective use heavily depends on the constitutive or closure relations for the solid phase and the momentum exchange between phases which are often difficult to obtain within its framework. This modeling consideration also applies not only to the convective heat transfer between particles and fluid but also to the conductive and radiative heat transfer between particles.

Various discrete approaches have been used to study heat transfer in the past. For example, the mechanistic approach based on the packet model originally proposed by Mickley and Fairbanks (1955) is a typical one to study the heat transfer between a bubbling fluidized bed and an immersed object. One of the problems associated with such early studies is the lack of reliable estimation of pertinent parameters (Chen, 2003; Chen et al., 2005). This problem can be overcome by studying the heat transfer at a subparticle scale. For example, Zhou et al. (2008) investigated the heat transfer between colliding particles by means of finite element method (FEM) and based on the results, formulated an equation to describe the effects of some variables related to collision conditions and materials properties. Feng et al. (2008, 2009) proposed a method to evaluate the element thermal conductivity matrix based on the characteristics of the contact zones including contact positions and contact angles. To consider the effect of interstitial stagnant fluid, Cheng et al. (1999) proposed a method to quantify the effective thermal conductivity (ETC) of a packed bed based on the packing structure of particles.

Recently, DEM has emerged to be a popular approach to investigate the flow of particles. It can be coupled with computational fluid dynamics (CFD) to model particle–fluid flows under different conditions (Zhu et al., 2007, 2008). In this CFD–DEM approach, the motion of discrete particles is obtained by solving Newton's second law of motion, and the flow of continuum fluid by solving the Navier–Stokes equations based on the concept of local average, with the coupling of CFD and DEM made through particle–fluid interaction forces (Feng and Yu, 2004; Tsuji et al., 1993; Xu and Yu, 1997, 1998; Zhou et al., 2010a). This approach can produce information such as particle–particle and particle–wall contacts, local voidage, and gas–solid flow structure. Such information is essential in determining the heat transfer behavior of individual particles. Not surprisingly, it has been attempted by various investigators to study heat transfer in packed or fluidized beds, e.g., coal combustion (Peters, 2002; Rong and Horio, 1999; Zhou et al., 2003, 2004a), air drying (Li and Mason, 2000, 2002), olefin polymerization (Kaneko et al., 1999), and inserts in a fluidized bed (Di Maio et al., 2009; Zhao et al., 2009). Deficiency can be found in some of these studies. For example, the conduction heat transfer between particles was only partially considered. Different models were used for the radiative heat transfer, including those of Zhou et al. (2009), Feng and Han (2012), and Cheng and Yu (2013).

However, although not perfect at this stage of development, a comprehensive model does exist, taking into account most of the known heat transfer mechanisms (Hou et al., 2012b, 2013, 2015a; Yang et al., 2015a,b,c; Zhou et al., 2009, 2010b, 2011a), including particle–fluid convection, particle–particle conduction, and radiative heat transfer between fluid and its surrounding environment and between a solid particle and its surrounding environment. This model offers a useful numerical technique to generally study the heat transfer in packed/fluidized beds at a particle scale, which can provide useful information to formulate correlations for large-scale continuum modeling through proper averaging methods (Zhu and Yu, 2002; Zhu et al., 2009, 2011). This chapter aims to introduce this approach and demonstrates its applicability through some case studies.

# 2. MODEL DESCRIPTION

## 2.1 Governing Equations

The CFD–DEM approach has been well developed and documented (for example, see Feng and Yu, 2004; Tsuji et al., 1992, 1993; Xu and Yu, 1997, 1998; Zhou et al., 2010a). It has been widely used as reviewed by Zhu et al. (2007, 2008). It should however be noted that different formulations can be implemented in numerical simulations. Corresponding to those used in the two-fluid model, Zhou et al. (2010a) demonstrated that there are three sets of formulations: an original format (set I) and subsequent derivations of set II and set III. Sets I and II are essentially the same, with small differences resulting from different mathematical or numerical treatments of a few terms in the original equation. Set III is however a simplified version of set I, which should be used with caution. The present work is based on the formulation in Set I, as described below.

Generally, a particle in a particle–fluid flow system can have two types of motion: translational and rotational, which are determined by Newton's second law of motion. The corresponding governing equations for particle $i$ with radius $R_i$, mass $m_i$, moment of inertia $I_i$, and specific heat capacity $c_{p,i}$ can be written as:

$$m_i \frac{d\mathbf{v}_i}{dt} = \mathbf{f}_{f,i} + \sum_{j=1}^{k_c} \left( \mathbf{f}_{c,ij} + \mathbf{f}_{d,ij} \right) + m_i \mathbf{g} \tag{1}$$

$$I_i \frac{d\boldsymbol{\omega}_i}{dt} = \sum_{j=1}^{k_c} \left( \mathbf{M}_{t,ij} + \mathbf{M}_{r,ij} \right) \tag{2}$$

$$m_i c_{p,i} \frac{dT_i}{dt} = \sum_{j=1}^{k_c} Q_{i,j} + Q_{i,f} + Q_{i,rad} + Q_{i,wall} \tag{3}$$

where $\mathbf{v}_i$ and $\boldsymbol{\omega}_i$ are the translational and angular velocities of the particle, respectively, and $T_i$ is the particle temperature. The forces involved are: particle–fluid interaction force $\mathbf{f}_{f,i}$, the gravitational force $m_i\mathbf{g}$, and interparticle forces between particles which include elastic force $\mathbf{f}_{c,ij}$ and viscous damping force $\mathbf{f}_{d,ij}$. These interparticle forces can be resolved into the normal and tangential components at a contact point. The torque acting on particle $i$ by particle $j$ includes two components: $\mathbf{M}_{t,ij}$ which is generated by tangential force and causes particle $i$ to rotate, and $\mathbf{M}_{r,ij}$ commonly known as the rolling friction torque is generated by asymmetric normal forces and slows down the relative rotation between particles. A particle may undergo multiple interactions, so the individual interaction forces and torques are summed over the $k_c$ particles interacting with particle $i$. $Q_{i,j}$ is the heat exchange rate between particles $i$ and $j$ due to conduction, $Q_{i,f}$ is the heat exchange rate between particle $i$ and its local surrounding fluid, $Q_{i,rad}$ is the heat exchange rate between particle $i$ and its surrounding environment by radiation, and $Q_{i,wall}$ is particle–wall heat exchange rate. Different heat transfer models are adopted to determine the different heat exchange rates shown in Eq. (3) and described in Section 2.2.

Equations used to calculate the interaction forces and torques between two spheres have been well established in the literature (Zhu et al., 2007). In our work, the determination of particle–particle interaction is based on the nonlinear models, as listed in Table 1. This approach was also

**Table 1** Components of Forces and Torques on Particle $i$

| | |
|---|---|
| Normal elastic force, $\mathbf{f}_{\mathrm{cn},ij}$ | $-\frac{4}{3}E^*\sqrt{R^*}\delta_n^{3/2}\mathbf{n}$ |
| Normal damping force, $\mathbf{f}_{\mathrm{dn},ij}$ | $-c_n\left(8m_{ij}E^*\sqrt{R^*\delta_n}\right)^{1/2}\mathbf{v}_{n,ij}$ |
| Tangential elastic force, $\mathbf{f}_{\mathrm{ct},ij}$ | $-\mu_{\mathrm{s}}\lvert\mathbf{f}_{\mathrm{cn},ij}\rvert\left(1-\left(1-\delta_{\mathrm{t}}/\delta_{\mathrm{t,max}}\right)^{3/2}\right)\hat{\boldsymbol{\delta}}_{\mathrm{t}}$ |
| Tangential damping force, $\mathbf{f}_{\mathrm{dt},ij}$ | $-c_{\mathrm{t}}\left(6\mu_{\mathrm{s}}m_{ij}\lvert\mathbf{f}_{\mathrm{cn},ij}\rvert\sqrt{1-\delta_{\mathrm{t}}/\delta_{\mathrm{t,max}}}/\delta_{\mathrm{t,max}}\right)^{1/2}\mathbf{v}_{\mathrm{t},ij}$ |
| Coulomb friction force, $\mathbf{f}_{\mathrm{t},ij}$ | $-\mu_{\mathrm{s}}\lvert\mathbf{f}_{\mathrm{cn},ij}\rvert\hat{\boldsymbol{\delta}}_{\mathrm{t}}$ |
| Torque by tangential forces, $\mathbf{T}_{\mathrm{t},ij}$ | $R_{ij}\times\left(\mathbf{f}_{\mathrm{ct},ij}+\mathbf{f}_{\mathrm{dt},ij}\right)$ |
| Rolling friction torque, $\mathbf{T}_{\mathrm{r},ij}$ | $\mu_{\mathrm{r},ij}\lvert\mathbf{f}_{\mathrm{n},ij}\rvert\widehat{\boldsymbol{\omega}}_{ij}^{\mathrm{n}}$ |
| Particle–fluid drag force, $\mathbf{f}_{\mathrm{d},i}$ | $0.125C_{\mathrm{d0},i}\rho_{\mathrm{f}}\pi d_{\mathrm{p}i}^2\varepsilon_i^2\lvert\mathbf{u}_i-\mathbf{v}_i\rvert(\mathbf{u}_i-\mathbf{v}_i)\varepsilon_i^{-\chi}$ |
| Pressure gradient force, $\mathbf{f}_{\nabla\mathrm{p},i}$ | $-V_{\mathrm{p},i}\nabla\mathbf{P}_i$ |

where $1/m_{ij}=1/m_i+1/m_j$, $1/R^*=1/\lvert\mathbf{R}_i\rvert+1/\lvert\mathbf{R}_j\rvert$, $E^*=E/2\left(1-\nu^2\right)$, $\widehat{\boldsymbol{\omega}}_{ij}^{n}=\boldsymbol{\omega}_{ij}^{n}/\lvert\boldsymbol{\omega}_{ij}^{n}\rvert$, $\hat{\boldsymbol{\delta}}_{\mathrm{t}}=\boldsymbol{\delta}_{\mathrm{t}}/\lvert\boldsymbol{\delta}_{\mathrm{t}}\rvert$, $\delta_{\mathrm{t,max}}=\mu_{\mathrm{s}}\delta_n(2-\nu)/(2(1-\nu))$, $\mathbf{v}_{ij}=\mathbf{v}_j-\mathbf{v}_i+{}^{i}\boldsymbol{\omega}_j\times\mathbf{R}_j-\boldsymbol{\omega}_i\times\mathbf{R}_i$, $\mathbf{v}_{n,ij}=\left(\mathbf{v}_{ij}\cdot\mathbf{n}\right)\cdot\mathbf{n}$, $\mathbf{v}_{t,ij}=\left(\mathbf{v}_{ij}\times\mathbf{n}\right)\times\mathbf{n}$, $\chi=3.7-0.65\exp\left[-(1.5-\log_{10}\mathrm{Re}_i)^2/2\right]$, $C_{\mathrm{d0},i}=\left(0.63+4.8/\mathrm{Re}_i^{0.5}\right)^2$, $\mathrm{Re}_i=\rho_{\mathrm{f}}d_{\mathrm{p}i}\varepsilon_i\lvert u_i-v_i\rvert/\mu_{\mathrm{f}}$. Note that tangential forces ($\mathbf{f}_{\mathrm{ct},ij}+\mathbf{f}_{\mathrm{dt},ij}$) should be replaced by $\mathbf{f}_{\mathrm{t},ij}$ when $\delta_{\mathrm{t}}\geq\delta_{\mathrm{t,max}}$.

used by other investigators (see, for example, Langston et al., 1994, 1995; Yang et al., 2000; Zhou et al., 1999). Particle–fluid interaction force $\mathbf{f}_{\mathrm{f},i}$ is the sum of fluid drag $\mathbf{f}_{\mathrm{d},i}$ and pressure gradient force $\mathbf{f}_{\nabla\mathrm{p},i}$. Many correlations are available in the literature to calculate the fluid drag acting on the individual particles including, for example, Ergun equation (1952), Wen and Yu (1966), and Di Felice (1994). Particularly, Di Felice correlation (1994) has been widely used in the literature and also in our work (see, for example, Feng and Yu, 2004, 2007; Hou et al., 2012c; Xu and Yu, 1997; Xu et al., 2000; Zhou et al., 2009, 2010a,b). Rong et al. (2013) recently studied the particle–fluid interactions in packed beds of uniform spheres using the Lattice Boltzmann method and proposed an equation to estimate the mean drag force on particles and to study the effect of porosity on the internal fluid flow. This model is then extended to study the effects of particle size distribution, sphericity, and porosity on the fluid flow and quantify the drag force on particles in packed beds recently (Rong et al., 2014, 2015). Their equations are more accurate and should be used in future studies.

For fine particles, the van der Waals force as a typical cohesive force is usually considered by using the Hamaker theory (Hamaker, 1937). For spherical particles, the van der Waals force is given as:

$$\mathbf{f}_{\mathrm{vdw}} = -\frac{H_{\mathrm{a}}}{6}\left[\frac{64R_i^3R_j^3(h+R_i+R_j)}{(h^2+2R_ih+2R_jh)^2(h^2+2R_ih+2R_jh+4R_iR_j)^2}\right]\mathbf{n}_{ij} \quad (4)$$

where $H_{\mathrm{a}}$ is the Hamaker constant and $h$ is the separation of surfaces along the line of the centers of particles $i$ and $j$. The equation shows that the van der Waals force becomes infinite as two particles get into contact ($h=0$), which induces a singularity problem in DEM simulation. To solve this problem, a "cut-off" distance is assumed in the calculation of this force, and this distance in the range of 0.165–1 nm has been widely adopted (Zhu et al., 2007).

The continuum fluid field is calculated from the continuity and Navier–Stokes equations based on the local mean variables over a computational cell, which can be written as (Zhou et al., 2010a):

$$\frac{\partial \varepsilon_{\mathrm{f}}}{\partial t} + \nabla\cdot(\varepsilon_{\mathrm{f}}\mathbf{u}) = 0 \quad (5)$$

$$\frac{\partial(\rho_{\mathrm{f}}\varepsilon_{\mathrm{f}}\mathbf{u})}{\partial t} + \nabla\cdot(\rho_{\mathrm{f}}\varepsilon_{\mathrm{f}}\mathbf{uu}) = -\nabla p - \mathbf{F}_{\mathrm{fp}} + \nabla\cdot\varepsilon_{\mathrm{f}}\boldsymbol{\tau} + \rho_{\mathrm{f}}\varepsilon_{\mathrm{f}}\mathbf{g} \quad (6)$$

and by definition, the corresponding equation for heat transfer can be written as

$$\frac{\partial(\rho_{\mathrm{f}}\varepsilon_{\mathrm{f}}c_{\mathrm{p}}T)}{\partial t} + \nabla\cdot(\rho_{\mathrm{f}}\varepsilon_{\mathrm{f}}\mathbf{u}c_{\mathrm{p}}T) = \nabla\cdot(c_{\mathrm{p}}\Gamma\nabla T) + \sum_{i=1}^{k_V} Q_{\mathrm{f},i} + Q_{\mathrm{f,wall}} \quad (7)$$

where $\mathbf{u}$, $\rho_{\mathrm{f}}$, $p$, $T$, and $\mathbf{F}_{\mathrm{fp}}\left(=\left(\sum_{i=1}^{k_V}\mathbf{f}_{\mathrm{f},i}\right)/\Delta V\right)$ are the fluid velocity, density, pressure, temperature, and volumetric fluid–particle interaction force, respectively, and $k_V$ is the number of particles in a computational cell of volume $\Delta V$. $\Gamma$ is the fluid thermal diffusivity, defined by $\mu_{\mathrm{e}}/\sigma_{\mathrm{T}}$, and $\sigma_{\mathrm{T}}$ the turbulence Prandtl number. $Q_{\mathrm{f},i}$ is the heat exchange rate between fluid and particle $i$ which locates in the computational cell, and $Q_{\mathrm{f,wall}}$ is the fluid–wall heat exchange rate. $\boldsymbol{\tau}\left(=\mu_{\mathrm{e}}\left[(\nabla\mathbf{u})+(\nabla\mathbf{u})^{-1}\right]\right)$ and $\varepsilon_{\mathrm{f}}\left(=\left(1-\left(\sum_{i=1}^{k_V}V_{\mathrm{p},i}\right)/\Delta \mathrm{V}\right)\right)$ are the fluid viscous stress tensor and porosity, respectively. $V_{\mathrm{p},i}$ is the volume of particle $i$ (or part of the volume if the particle is not fully in the cell), $\mu_{\mathrm{e}}$ is the fluid effective viscosity determined by the standard $k-\varepsilon$ turbulent model (Launder and Spalding, 1974).

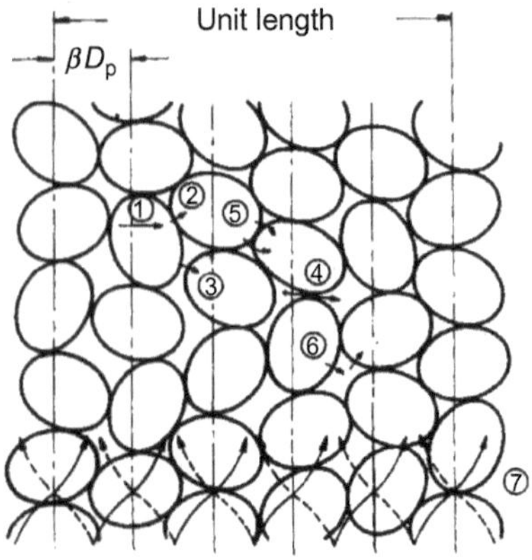


**Figure 1** Heat transfer mechanisms in a packed bed: Conduction (1: heat transfer through solid; 2: heat transfer through the contact surface of solid; 5: heat transfer through the fluid film near the contact surface; and 6: heat transfer through solid–fluid–solid between noncontacting solid), convection (7: heat transfer by lateral mixing of fluid), and radiation (3: radiant heat transfer between surfaces of solid and 4: radiant heat transfer between adjacent voids). *Reprinted from Yagi and Kunii (1957) with permission from John Wiley and Sons.*

## 2.2 Heat Transfer Models

The heat transfer in fluidized beds of monodisperse particle has been extensively investigated in the past. Heat transfer in a packed/fluidized bed with an interstitial fluid may involve many mechanisms as shown in Fig. 1 (Yagi and Kunii, 1957). These mechanisms can be classified into three heat transfer modes in fluidized beds: fluid–particle or fluid–wall convection; particle–particle or particle–wall conduction; and radiation. Different heat transfer models are developed for these mechanisms, as described in the following.

### *2.2.1 Convective Heat Transfer*

Convective heat transfer between particles and fluid has been extensively investigated since 1950s, and different equations have been proposed (Botterill, 1975; Kunii and Levenspiel, 1991; Molerus and Wirth, 1997; Wakao and Kaguei, 1982). Often, the convective heat transfer rate between particle $i$ and fluid is calculated according to $Q_{i,f}=h_{i,\mathrm{conv}}\cdot A_i\cdot(T_{f,i}-T_i)$, where $A_i$ is the particle surface area, $T_{f,i}$ is the fluid temperature in a computational cell where particle $i$ is located, and $h_{i,\mathrm{conv}}$ is the convective HTC. $h_{i,\mathrm{conv}}$ is associated with the Nusselt number (Nu), which is usually a function of particle Reynolds number and gas Prandtl number, given by

$$\mathrm{Nu}_i = h_{i,\mathrm{conv}} d_{\mathrm{p}i}/k_{\mathrm{f}} = 2.0 + a\mathrm{Re}_i^b \mathrm{Pr}^{1/3} \tag{8}$$

where $k_{\mathrm{f}}$ and $d_{\mathrm{p}i}$ are the fluid thermal conductivity and particle diameter, respectively. $\mathrm{Re}_i$ is the local relative Reynolds number for particle $i$ (see Table 1 for the definition). The gas Prandtl number, Pr, is a material property. The constant, 2.0, represents the contribution by particle–fluid natural convection. $a$ and $b$ are two parameters that need to be evaluated. As suggested by Kunii and Levenspiel (1991), $b=0.5$, and $a$ could range from 0.6 to 1.8, depending on bed conditions. Correlations have also been established to consider bed porosity (Gunn, 1978) and particle shape (Kishore and Gu, 2011; Sparrow et al., 2004; Wadewitz and Specht, 2001).

For fluid–wall heat transfer between turbulent flow and a smooth tube, the HTC $h_{\mathrm{f,wall}}$ can be determined by

$$\mathrm{Nu}_D = h_{\mathrm{f,wall}} D/k_{\mathrm{f}} = 0.023\mathrm{Re}^{0.8}\mathrm{Pr}^n \tag{9}$$

where $D$ is the hydraulic diameter, and the exponent $n$ is 0.4 for heating, and 0.3 for cooling (Holman, 1981). When the insert tube is treated as a flat plate a different Nusselt number relation, $\mathrm{Nu} = 0.037\mathrm{Re}^{0.8}\mathrm{Pr}^{1/3}$ should be used for turbulent flow (Incropera and Dewitt, 2002).

### 2.2.2 Conductive Heat Transfer

Conductive heat transfer has different mechanisms mainly through two paths: (1) particle–fluid–particle path and (2) particle–particle path, described as follows.

Two approaches can be used for calculating interparticle and particle surface collision heat transfer (Amritkar et al., 2014). The first approach is based on the quasi-steady state solution of the collisional heat transfer between two spheres (Vargas and McCarthy, 2002). The other approach is based on the analytical solution of the one-dimensional unsteady heat conduction between two semi-infinite objects. This approach was proposed by Sun and Chen (1988) based on the analysis of the elastic deformation of the spheres in contact.

According to Hertz's theory of elastic collision, the change rate of the contact area $A_{\mathrm{f}}$ during the collision is given by

$$\frac{\mathrm{d}A_{\mathrm{f}}}{\mathrm{d}t} = \left[\left(\pi v_{n,ij} R_{ij}\right)^2 - \frac{4}{5\sqrt{\pi}}\frac{E_{ij}}{m_{ij}} A_{\mathrm{f}}^{5/2}\right]^{1/2} \tag{10}$$

where $R_{ij}=R_iR_j/(R_i+R_j)$, $E_{ij}=4/\left[3(1-\nu_i^2)/E_i+3\left(1-\nu_j^2\right)/E_j\right]$, and $m_{ij}=m_im_j/(m_i+m_j)$. Integrating Eq. (10) yields $\tau=\int_0^{A*}\left[1/\left(1-x^{5/2}\right)^{1/2}\right]dx$, where $\tau$ is the dimensionless time, defined by $\tau=\left[4E_{ij}/(5m_{ij})\right]^{2/5}\left(R_{ij}\nu_{n,ij}\right)^{1/5}t$, and $A^*$ is the dimensionless contact area, defined by $A^*=A_f/A_c=r_f^2/r_c^2$, where $r_f$ is the contact radius at time $t$, $A_c$, and $r_c$ are the maximum contact area and the maximum contact radius, respectively. They are related, given by $A_c=\pi r_c{}^2=\pi\left[5m_{ij}R_{ij}{}^2/(4E_{ij})\right]^{2/5}\left(\nu_{n,ij}\right)^{4/5}$.

Sun and Chen (1988) derived an analytical equation to calculate the heat exchange ($Q_{ij}$) between colliding spheres based on the well-established "semi-infinite-media assumption." The equation is given by $Q_{ij}=CQ_{ij,0}$, where

$$Q_{ij,0}=\frac{0.87\left(T_j-T_i\right)\pi r_c^2 t_c^{1/2}}{\left(\rho_ic_ik_i\right)^{-1/2}+\left(\rho_jc_jk_j\right)^{-1/2}} \tag{11}$$

where parameter $C$ can be determined graphically as given by Sun and Chen (1988).

Zhou et al. (2008) compared the calculated heat exchange obtained by the FEM simulation and by a analytical model (Sun and Chen, 1988). Figure 2 shows the symmetric coordinate system used, where the mesh is generated automatically according to the contact area at each time step. The results obtained by the FEM simulation agree well with those by the analytical model for the cases of small Fourier number $F_o$. However, when the Fourier number is high, the analytical model will overestimate the heat exchange attributing to the semi-infinite-media assumption. Therefore, Zhou et al. (2008) provided a modified equation by introducing a modified coefficient $C'$ by fitting the FEM results,

$$Q_{ij}=\frac{C'\left(T_j-T_i\right)\pi r_c^2 t_c^{1/2}}{\left(\rho_ic_ik_i\right)^{-1/2}+\left(\rho_jc_jk_j\right)^{-1/2}} \tag{12}$$

where,

$$C'=0.435\cdot\left(\sqrt{C_2^2-4C_1(C_3-F_o)}-C_2\right)/C_1 \tag{13}$$

$$C_1=-2300\cdot\left(\rho_{pi}c_{pi}/\rho_{pj}c_{pj}\right)^2+8.909\cdot\left(\rho_{pj}c_{pj}/\rho_{pi}c_{pi}\right)-4.235 \tag{14}$$

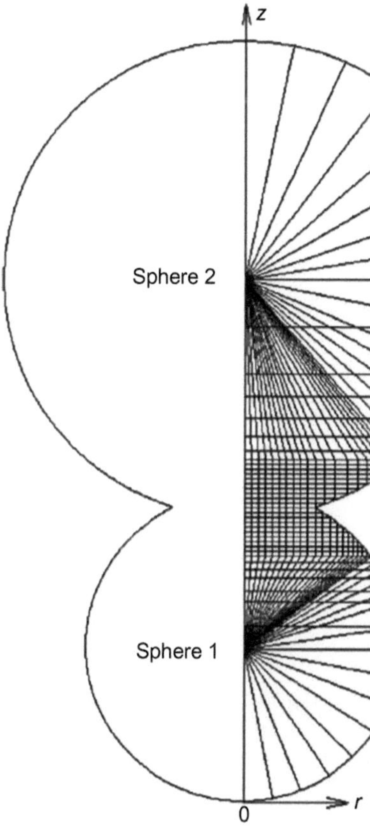


**Figure 2** The coordinate system and space discretization. *Reprinted from Zhou et al. (2008) with permission from Elsevier.*

$$C_2 = 8.169 \cdot \left(\rho_{pi} c_{pi} / \rho_{pj} c_{pj}\right)^2 - 33.770 \cdot \left(\rho_{pj} c_{pj} / \rho_{pi} c_{pi}\right) + 24.885 \quad (15)$$

$$C_3 = -5.758 \cdot \left(\rho_{pi} c_{pi} / \rho_{pj} c_{pj}\right)^2 + 24.464 \cdot \left(\rho_{pj} c_{pj} / \rho_{pi} c_{pi}\right) - 20.511 \quad (16)$$

where $F_o$ is the Fourier number, is defined as $\alpha \cdot t_c / \left(r_c^2\right)$, and $a$ is particle thermal diffusivity. $t_c$ is the maximum collision contact duration, given by $t_c = 2.94\left[5m_{ij}/\left(4E_{ij}\right)\right]^{2/5}\left(R_{ij}v_{n,ij}\right)^{-1/5}$. $v_{nij}$ is the normal relative velocity between particles $i$ and $j$.

For particle–wall static or collision contact, a wall can be treated as a particle with an infinite diameter and mass, as commonly used in the DEM. For two colliding particles, if $t_c > t_d$, only collisional heat transfer applies. If $t_c < t_d$, two particles will keep in touch after collision. In such a case, collision heat transfer applies first during the time of $t_c$, and then static heat transfer during the time of $(t_d - t_c)$ (Zhou et al., 2009, 2010b).

It should be noted that the contact radius $r_c$ mentioned above is obtained from the DEM simulation conditions, based on the Hertz elastic contact theory. However, in DEM simulation, the Young's modulus is usually set to $1 \sim 100$ MPa to reduce the computing effort, while the Young's modulus of real hard materials like glass beads would be much larger than this value range, e.g., around 50 GPa. Therefore, an additional correction coefficient $c$ is introduced by Zhou et al. (2010b):

$$c = \frac{r_{c,0}}{r_c} = \left(\frac{E_{ij}}{E_{ij,0}}\right)^{1/5} \tag{17}$$

where $c$ varies between 0 and 1, depending on the magnitude of Young's modulus used in the DEM. Note that here $E_{ij} = 4/3\left[\left(1-\nu_i^2\right)/E_i + \left(1-\nu_j^2\right)/E_j\right]$, $E_{ij,0} = 4/3\left[\left(1-\nu_i^2\right)/E_{i,0} + \left(1-\nu_j^2\right)/E_{j,0}\right]$, $\nu$ is the Possion's ratio, and $E_i$ is the Young's modulus used in the DEM. It can be observed that, to determine the introduced correction coefficient $c$, two parameters are required: $E_{ij}$, the value of Young's modulus used in the DEM simulation and $E_{ij,0}$, the real value of Young's modulus of the materials considered.

Particle–fluid–particle heat transfer has been examined by various investigators. Kunii and Smith (1960) assumed that the heat conduction in the cell is through two parallel paths: the conduction through the fluid-filled voids and the conduction through the solid and fluid phases. Argento and Bouvard (1996) assumed the heat conduction through the contact area between particles to be the only heat transfer mechanism, with the effect of fluid phase completely ignored. Cheng et al. (1999) proposed a model based on the packing structure quantified through the Voronoi tessellation. Therefore, the heat transfer can be quantified within a Voronoi polyhedron and between neighboring Voronoi polyhedra.

Figure 3 shows the contact conditions between two neighboring particles $i$ and $j$, with their representative temperatures expressed as $T_i$ and $T_j$, respectively. Depending on the contact conditions, two cases can be identified, which give different heat transfer paths corresponding to the above heat transfer mechanisms:

Case I: The two particles are not or just in contact (Fig. 3A and B). In this case, the heat transfer path is the heat conduction through the solid

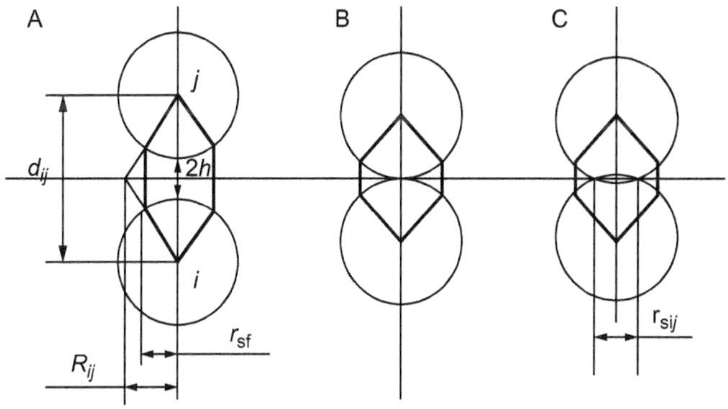


**Figure 3** Different contact conditions between two particles: (A) Noncontact, (B) Point-contact and (C) Area-contact. *Reprinted from Cheng et al. (1999) with permission from Elsevier.*

particles and stagnant fluid in-between, i.e., the particle–fluid–particle conduction between noncontacted or point-contacted particles.

Case II: There is a contact area between two particles, due to their deformation (Fig. 3C). In this case, the heat transfer is through two parallel paths: the conduction through the solid particles and stagnant fluid in-between, and the conduction through the contact area, i.e., the particle–fluid–particle conduction and the particle–particle conduction between contacted particles.

Boundary conditions have to be specified in order to quantify the heat transfer process, i.e., $Q_{ij}$, the heat flux between particles $i$ and $j$, for these cases. To solve this problem, Cheng et al. (1999) proposed two simplified and physically acceptable models—Models A and B. Model A assumes that the surface of a taper cone is isothermal with the value equal to its corresponding representative temperature, and the conduction through the stagnant fluid in the void region (the region outside the fluid lens in Fig. 3A) can be ignored. Model B assumes that each particle has an isothermal core of radius $R_c$ and representative temperature. According to Model A, the heat transfer flux between spheres $i$ and $j$ is written as

$$Q_{i,j} = \left(T_j - T_i\right) \times \int_{r_{sij}}^{r_{sf}} \frac{2\pi \cdot r\mathrm{d}r}{\sqrt{R^2 - r^2} - r(R+H)/r_{ij} \cdot \left(1/k_{pi} + 1/k_{pj}\right) + 2\left[(R+H) - \sqrt{R^2 - r^2}\right]/k_f} \tag{18}$$

where

$$H = (d_{ij} - 2R)/2 \tag{19}$$

$$r_{sf} = \frac{R \cdot r_{ij}}{\sqrt{r_{ij}^2 + (R + H)^2}} \tag{20}$$

$$r_{ij} = \sqrt{3V_{ij}/(\pi \cdot d_{ij})} \tag{21}$$

where $k_{pi}$ and $k_{pj}$ are the thermal conductivities of particles $i$ and $j$, respectively. $R$ is the particle radius of uniform spheres. $H$ is half of the distance between two uniform sphere surfaces ($2h$ in Fig. 3A). Parameter $r_{sij} = 0$ when $H \geq 0$ and $r_{sij} = r_c$ when $H < 0$, where $r_c$ is the radius of the contact area of two contacting particles. $r_{ij}$ is the radius of the lens of fluid between two contacting or near contacting spheres, determined by Eq. (21). $V_{ij}$ is the volume of Voronoi polyhedra between particles $i$ and $j$. The structure of fluidized beds varies, and the determination of the transient Voronoi polyhedra for a system composed of a large number of particles is very time-consuming.

Actually, there is only one parameter that requires in this heat transfer modeling, i.e., the face area $A_{ij}$ of Voronoi polyhedron between particles $i$ and $j$. Based on the results of Yang et al. (2002), $A_{ij}$ can be written as

$$A_{ij} = \pi r_{ij}^2 = 0.985 R_i^2 (1 - \varepsilon_i)^{-2/3} \tag{22}$$

or

$$r_{ij} = 0.560 R_i (1 - \varepsilon_i)^{-1/3} \tag{23}$$

where $\varepsilon_i$ is the local porosity corresponding to particle $i$, as used in the CFD–DEM calculation. Parameter $2h$, as indicated in Fig. 3, is determined by Eq. (19), where $d_{ij}$ can be obtained from the DEM simulation. Generally, the larger the distance between two noncontacting spheres, the less the heat flux.

Conduction heat transfer always occurs through the contacted area between particles or between particles and wall. Generally, such conduction heat transfer due to elastic deformation includes two mechanisms: conduction due to particle–particle static contact (particularly common in a packed bed), and conduction due to particle–particle collision (often occurs in a moving or fluidized bed).

In fact, when the contact radius $r_{sij}$ is small compared to the particle radius $R$, the distribution of temperature inside the two particles is approximately the same as that of the velocity potential in irrotational flow of incompressible fluid through a circular hole in a plane wall. Based on this argument,

Batchelor and O'Brien (1977) obtained an equation to calculate the heat flux between two contacted particles. Note that the thermal properties of particles are not necessarily same. The heat flux through the contact area of two particles is here calculated by a slightly modified version of Batchelor's equation (Batchelor and O'Brien, 1977):

$$Q_{i,j} = 4r_c(T_j - T_i)/(1/k_{pi} + 1/k_{pj}) \tag{24}$$

### 2.2.3 Radiative Heat Transfer

As done in the case for heat transfer by conduction (Cheng et al., 1999), a double taper cone model (Fig. 4d) is proposed by Cheng and Yu (2013) to

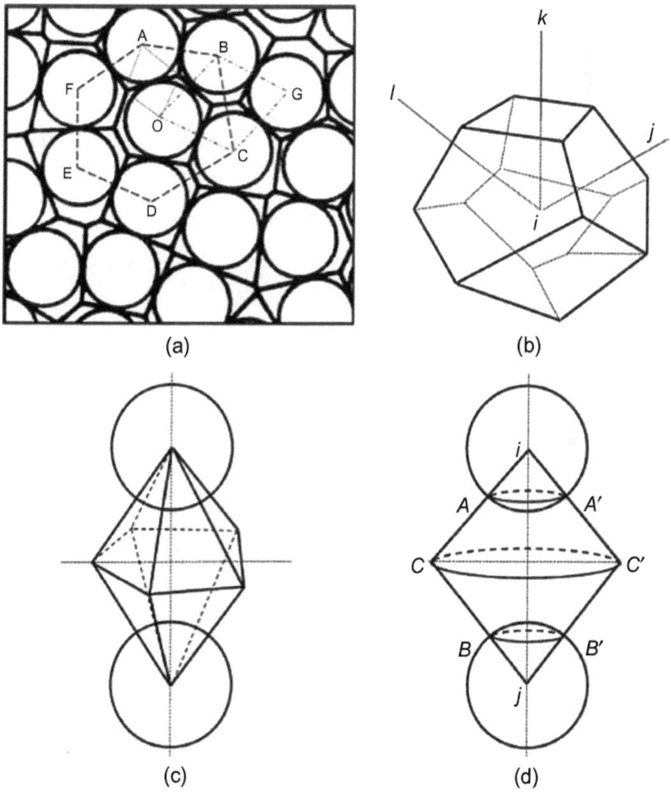


**Figure 4** Schematic illustrations: (a) a two-dimensional packing and its Voronoi elements, with dotted lines highlighting particle O and its surrounding particles A–F, and different connections between particles O and A, and between particles O and G; (b) a single Voronoi element *i* together with its neighbors *j*, *k*, *l*, and so on; (c) the connection between two neighboring Voronoi polyhedra as a double pyramid model; and (d) the simplified connection as a double taper cone model. *Reprinted from Cheng and Yu (2013) with permission from ACS.*

calculate the radiation heat transfer between particles $i$ and $j$. The packed bed represented by a Voronoi element network to establish its particle-to-particle connections is shown in Fig. 4a. Under the three-dimensional condition, a Voronoi element may have a relatively complicated geometry (Fig. 4b and c).

Apart from the assumptions made in the work of Cheng et al. (1999), a few more assumptions are made for the calculation of the radiant heat transfer between the two neighboring Voronoi elements, including larger sphere diameter than radiative wavelength, gray surface, opaque particles, much smaller $\Delta T/T$ across a sphere layer, and the perfectly insulated and diffusively reflective surface $R$ (ACBB′C′A′). Under these assumptions, the radiant exchange $Q_{ij,\mathrm{rad}}$ between particles $i$ and $j$ can be calculated by the following equation:

$$Q_{ij,\mathrm{rad}} = \frac{\sigma\left(T_i^4 - T_j^4\right)}{\dfrac{1-\varepsilon_{\mathrm{r},i}}{\varepsilon_{\mathrm{r},i}A_{ii}} + \dfrac{1}{A_{ii}F_{ij} + \left[\left(1/A_{ii}F_{i\mathrm{R}}\right) + \left(1/A_{jj}F_{j\mathrm{R}}\right)\right]^{-1}} + \dfrac{1-\varepsilon_{\mathrm{r},j}}{\varepsilon_{\mathrm{r},j}A_{jj}}} \tag{25}$$

where $A_{ii}$ and $A_{jj}$ are the areas of surfaces AA′ and BB′ on the spheres. $\varepsilon_{\mathrm{r},i}$ and $\varepsilon_{\mathrm{r},j}$ are the emissivities of particles $i$ and $j$. $F_{ij}$, $F_{i\mathrm{R}}$, and $F_{j\mathrm{R}}$ are the view factors between the surfaces AA′ and BB′, AA′ and ACBB′C′A′, and BB′ and ACBB′C′A′, respectively. For uniform spheres (i.e., spheres of the same geometrical and physical properties), we have $A_{ii}=A_{jj}$, $\varepsilon_{\mathrm{r},i}=\varepsilon_{\mathrm{r},j}$, $F_{ij}+F_{i\mathrm{R}}=1$, $F_{ji}+F_{j\mathrm{R}}=1$, and $F_{ij}=F_{ji}$. Equation (25) can hence be simplified to

$$Q_{ij,\mathrm{rad}} = \frac{\sigma\left(T_i^4 - T_j^4\right)}{2\dfrac{1-\varepsilon_{\mathrm{r},i}}{\varepsilon_{\mathrm{r},i}A_{ii}} + \dfrac{A_{ii}\left(1-F_{ij}\right)}{2}} \tag{26}$$

In a fixed or fluidized bed, a particle is surrounded by other particles and fluid. In a specified enclosed cell, an environmental temperature is assumed to represent the enclosed surface temperature around such a particle. Thus, the equation used by Zhou et al. (2004b) is slightly modified to calculate the heat flux due to radiation using a local environmental temperature to replace the bed temperature and is written as (Zhou et al., 2009):

$$Q_{i,\mathrm{rad}} = \sigma\varepsilon_{\mathrm{p}i}A_i\left(T_{\mathrm{local},i}^4 - T_i^4\right) \tag{27}$$

where $\sigma$ is the Stefan–Boltzmann constant, equal to $5.67 \times 10^{-8}$ W/($m^2$ $K^4$), $\varepsilon_{pi}$ is the sphere emissivity, and $A_i$ is the particle surface area. The parameter $T_{local,i}$ is the average temperature of particles and fluid by volume fraction in an enclosed spherical domain $\Omega$ given by Zhou et al. (2009):

$$T_{local,i} = \varepsilon_f T_{f,\Omega} + (1 - \varepsilon_f) \frac{1}{k_\Omega} \sum_{j=1}^{k_\Omega} T_j (j \neq i) \quad (28)$$

where $T_{f,\Omega}$ and $k_\Omega$ are the fluid temperature and the number of particles located in the domain $\Omega$, respectively, with its radius of 1.5 $d_p$. To be fully enclosed, a larger radius can be used. Gas radiation is considered similarly, usually ignored due to low gas emissivity (Hou et al., 2012b).

### 2.3 Coupling Schemes

The methods for numerical solution of CFD and DEM have been well established in the literature. For the DEM model, an explicit time integration method is used to solve the translational and rotational motions of discrete particles (Cundall and Strack, 1979). For the CFD model, the conventional SIMPLE method is used to solve the governing equations for the fluid phase (Patankar, 1980). The modeling of the solid flow by DEM is at the individual particle level, while the fluid flow by CFD is at the computational cell level. The coupling methodology of the two models at different length scales has been well documented (Feng and Yu, 2004; Xu and Yu, 1997; Zhou et al., 2010a; Zhu et al., 2007). The present model simply extends that approach to include heat transfer, and more details can be seen in the reference of Zhou et al. (2009).

## 3. MODEL APPLICATION

### 3.1 Packed Beds

#### *3.1.1 Predictions Versus Measurements*

ETC is an important parameter describing the thermal behavior of packed beds with a stagnant or dynamic fluid and has been extensively investigated experimentally and theoretically in the past. Various mathematical models, including continuum models and microscopic models, have been proposed to help solve this problem, but they are often limited by the homogeneity assumption in a continuum model (Wakao and Kaguei, 1982; Zehner and Schlünder, 1970) or the simple assumptions in a microscopic model

(Argento and Bouvard, 1996; Kobayashi et al., 1991). Cheng et al. (1999) and Cheng (2003) proposed a structure-based approach to evaluate the ETC of the packing structure of a packed bed of monosized spheres in the presence of a stagnant fluid. The evaluation is based on two aspects: (i) the heat transfer between particles, which is obtained under some simplified conditions; and (ii) the connectivity of particles in a packing, which is determined from the packing structure measured by Finney (1970). Three heat transfer mechanisms or paths are considered in this approach, including the conduction through the solid particles and stagnant fluid between non-contacted particles, the conduction through the solid particles and stagnant fluid, and the conduction through the contact area between contacted particles. The validity of this approach is verified by the good agreement between measured and calculated results for a packed bed over a wide range of solid-to-fluid thermal conductivity ratio and for packed beds with particles of different thermal conductivities, as shown in Fig. 5. The reason for the selection of variables in the figure is discussed here. Note that the dimensionless variables are given in term of $k_f$ because $k_f$ can be obtained readily, and more importantly, this treatment can generalize the results, e.g., independent of fluid properties. Note here $c=0.5$ for Model B, the calculated effective thermal conductivities are in good agreement with those measured and the predictions by Model A.

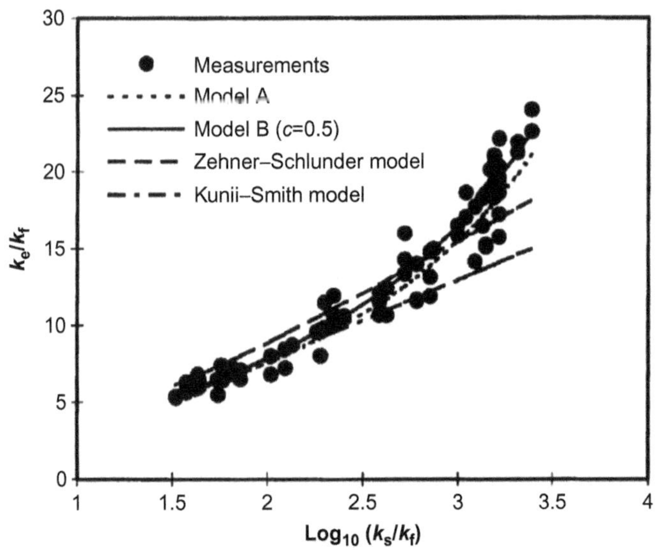


**Figure 5** Comparison of ETC between model predictions and experimental measurements. *Reprinted from Cheng et al. (1999) with permission from Elsevier.*

Cheng et al. (1999) considered three heat transfer mechanisms or paths. Their relative contributions to the overall heat transfer as a function of $k_s/k_f$ can be easily identified. Such an analysis presents a striking difference from the models proposed by other investigators (for example, see Kunii and Smith, 1960; Zehner and Schlunde, 1970). As demonstrated in Fig. 6, the solid–solid heat conduction increases, while the other two decrease with the increase of $k_s/k_f$ ratio. Therefore, when $k_s/k_f$ is low, the dominant heat transfer mechanism is the solid–fluid–solid conduction between contacted particles; but as $k_s/k_f$ increases, the solid–solid conduction between contacted particles becomes important.

The combined conduction and radiation heat transfer through a packed bed of spheres is more often studied because of its practical importance. The quasi-homogeneous theories predict that ETC ($k_e$) is the sum of the effective conductivity $k_{e,c}$ without radiation contributions and the radiative conductivity $k_r$ without conduction contributions (Vortmeyer, 1978). That is

$$k_e = k_{e,c} + k_r \tag{29}$$

where the radiative conductivity $k_r$ can be calculated according $k_r = 4Fd\sigma T_m^3$, where $F$ is called the radiation exchange factor and $T_m = \sum_{i=1}^{n} T_i/n$ is the

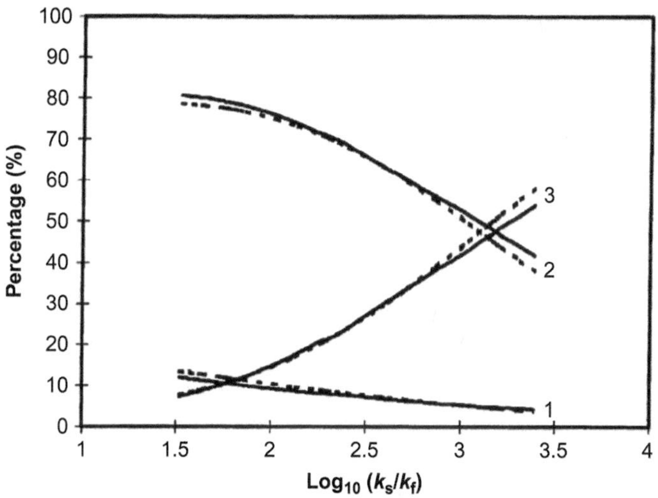


**Figure 6** Relative contribution of the heat transfer mechanism to the overall heat transfer. line 1, the conduction through the solid particles and stagnant fluid between non-contacted particles; line 2, the conduction through the solid particles and stagnant fluid between contacted particles; and line 3, the conduction through the contact area between contacted particles: solid line, Model A; dash line, Model B ($c=0.5$). *Reprinted from Cheng et al. (1999) with permission from ACS.*

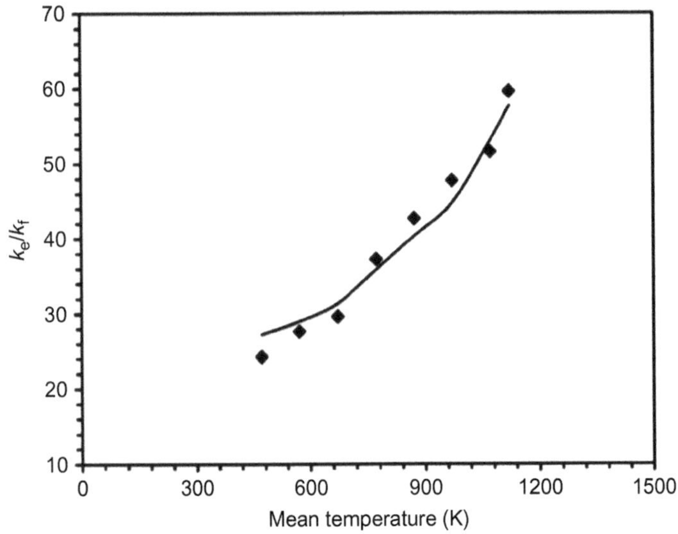


**Figure 7** Comparison between the calculated and measured ETC (using a packed bed of iron spheres of $d=11$ mm and porosity $\varepsilon=0.4$): ◆, measurements of Yagi and Kunii (1957); line, calculation of Cheng and Yu (2013). *Reprinted from Cheng and Yu (2013) with permission from ACS.*

mean temperature of particles in the considered packed bed. Cheng and Yu (2013) compared the calculated and measured ETC by Yagi and Kunii (1957), and good agreement was achieved, as shown in Fig. 7.

Cheng and Yu (2013) compared the relative contributions of the heat transfer mechanisms to the overall heat transfer as a function of $k_s/k_f$ at a high temperature where the radiative heat transfer cannot be ignored, as shown in Fig. 8. Four mechanisms or modes of heat transfer are considered: heat conduction between two noncontacting particles through the stagnant fluid $Q_{nsfs}$, heat conduction between two contacting particles through the stagnant fluid $Q_{csfs}$, heat conduction between two contacting particles through the contact area $Q_{css}$, and heat radiation between two particle surfaces. The relative contribution of heat conduction increases with an increase of the solid conductivity $k_s$ (note that $k_f$ here is fixed), while the relative contribution of radiation decreases. The reason is that the absolute contribution of conduction increases more than the absolute contribution of radiation as the solid conductivity $k_s$ increases. Also, it is clear that under the given conditions ($T_m=1173$ K), radiation heat transfer is the dominant heat transfer mechanism (>75%).

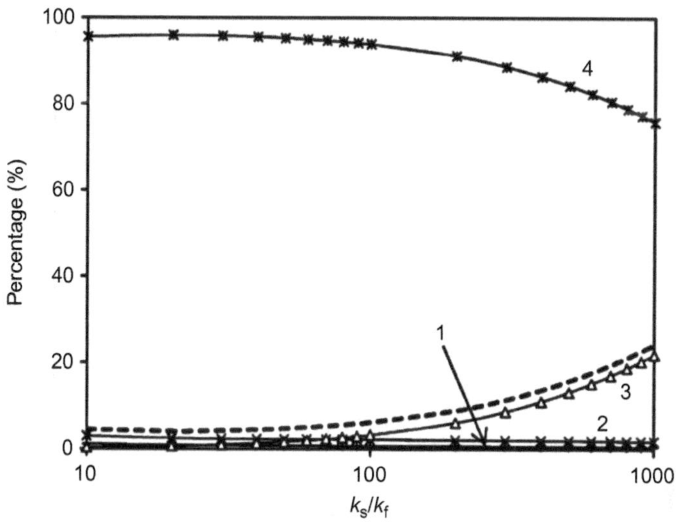


**Figure 8** Relative contributions of the heat transfer modes considered to the overall heat transfer as a function of $k_s/k_f$: line 1, heat conduction $Q_{nsfs}$; line 2, heat conduction $Q_{csfs}$; line 3, heat conduction $Q_{css}$; line 4, the solid–solid radiation between particle surfaces; dashed-line, the percentage of total conduction. *Reprinted from Cheng and Yu (2013) with permission from ACS.*

The structure-based approach of Cheng et al. (1999) has limited application because of its complexity in the determination of the packing structure and the ignorance of fluid flow in a packed bed. To overcome this problem, Zhou et al. (2009, 2010b) extended the CFD–DEM approach to model the heat transfer, as described in Section 2. In this approach, a packed bed can be generated quite readily by DEM. The ETC is determined by the following method: the temperatures at the bed bottom and top are set constants. A uniform heat flux, $q$ (W/m$^2$), is generated and passes from the bottom to the top. Thus, the bed ETC is calculated by $k_e = q \cdot H_b/(T_b - T_t)$, where $H_b$ is the height between the two layers with two constant temperatures at the top and the bottom, respectively.

It is noted that Young's modulus is an important parameter affecting the particle–particle overlap, hence the particle–particle heat transfer (Zhou et al., 2010b). Figure 9 shows the predicted ETC for different Young's modulus varying from 1 MPa to 50 GPa. When the Young's modulus $E$ is around 50 GPa, which is in the range of hard materials like glass beads, the predicted ETC are comparable with experiments. The high ETC for low Young's modulus is caused by the overestimated particle–particle

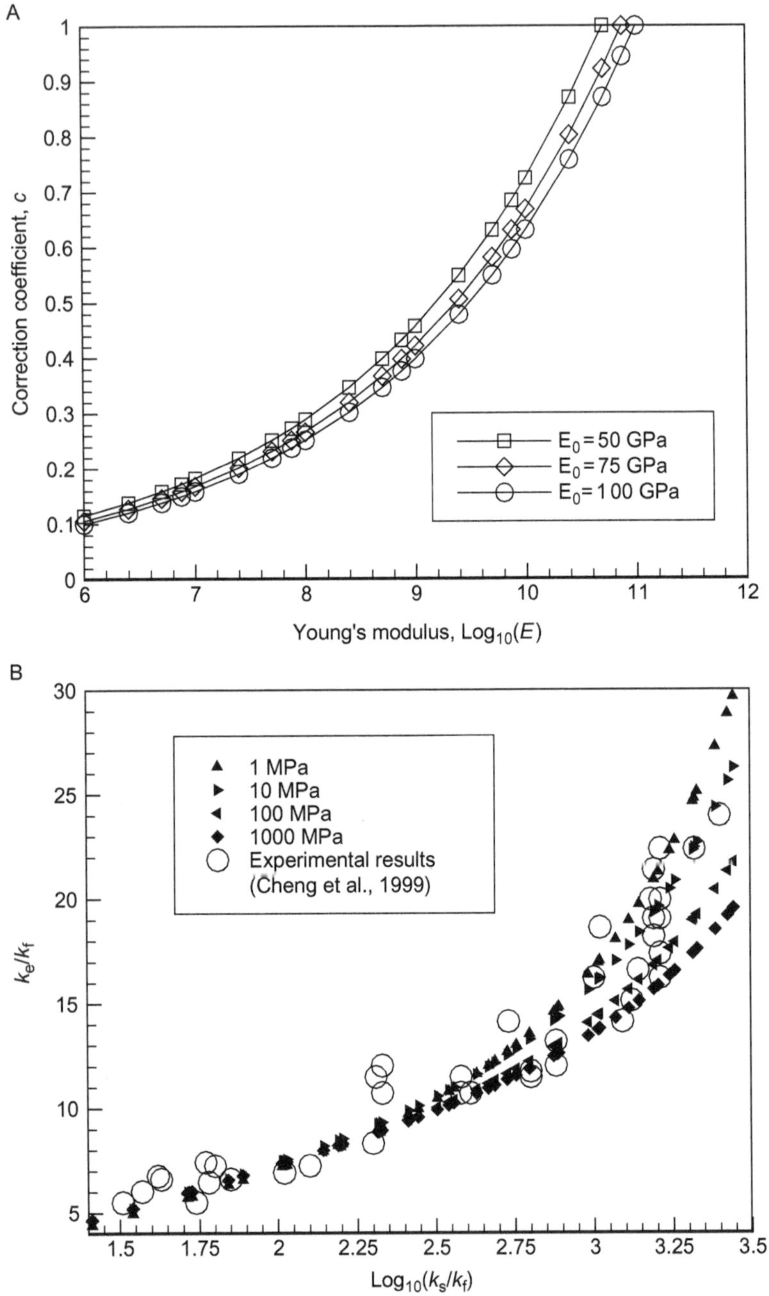


**Figure 9** (A) Relationship between correction coefficient and Young's modulus $E$ used in the DEM, and (B) the predicted ETCs as a function of $k_s/k_f$ ratios for different $E$ using the obtained correction coefficients according to Eq. (14) where $E_0 = 50$ GPa (Zhou et al., 2010b).

overlap in the DEM based on the soft-sphere approach. A large overlap significantly increases the heat flux $Q_{ij}$. However, in the DEM, it is computationally demanding to carry out the simulation using a high Young's modulus (often at an order of $10^3 \sim 10^5$ MPa), particularly when involving a large number of particles. This is because a high Young's modulus requires extremely small time steps to obtain accurate results, resulting in a high computational cost which may not be tolerated under the current computational capacity. Zhou et al. (2010b) proposed a correction coefficient to the contact radius $r_c$, the obtained correction coefficients are shown in Fig. 9A. Figure 9B further shows the applications of the obtained correction coefficients in other cases, where the particle thermal conductivity varied from 1.0 to 80 W/(m K); gas thermal conductivities varied from 0.18 to 0.38 W/(m K); Young's modulus used in the DEM varies from 1 MPa to 1 GPa, and the real value of Young's modulus is set to 50 GPa. The results show that the predicted ETCs are comparable with experiments.

The approach of introducing a correction coefficient has also been applied to gas fluidization to test its applicability. An example of flow patterns is shown in Fig. 13, which illustrates a heating process of the fluidized bed by hot gas (Zhou et al., 2009). It can be seen that the approach can reproduce those general features of solid flow patterns and temperature evolution with time using low Young's modulus, and the obtained results are comparable to those reported by Zhou et al. (2009) using a high Young's modulus. Moreover, Zhou et al. (2010b) compared the obtained average convective and conductive HTCs by three treatments: (1) $E=E_0=50$ GPa, and $c=1.0$; (2) $E=10$ MPa, and $c=1.0$; and (3) $E=10$ MPa, and $c=0.182$. Treatment 1 corresponds to the real materials, and its implementation requires a small time step. Treatments 2 and 3 reduce the Young's modulus so that a large time step is applicable. The difference between them is one with reduced contact radius ($c=0.182$ in treatment 3), and another not ($c=1$ in treatment 2). The results are shown in Fig. 10. The convective HTC is not affected by those treatments (Fig. 10A). Particle–particle contact only affects the conduction heat transfer (Fig. 10B). The results are very comparable and consistent between the models using treatments 1 and 3, but they are quite different from the model using treatment 2. If the particle thermal conductivity is high, such difference becomes even more significant. The comparison in Fig. 10B indicates that the modified model by treatment 3 can be used in the study of heat transfer not only in packed beds but also in fluidization beds. It must be pointed out that the significance of proposed modified model (treatment 3) is to save computational cost. For the current

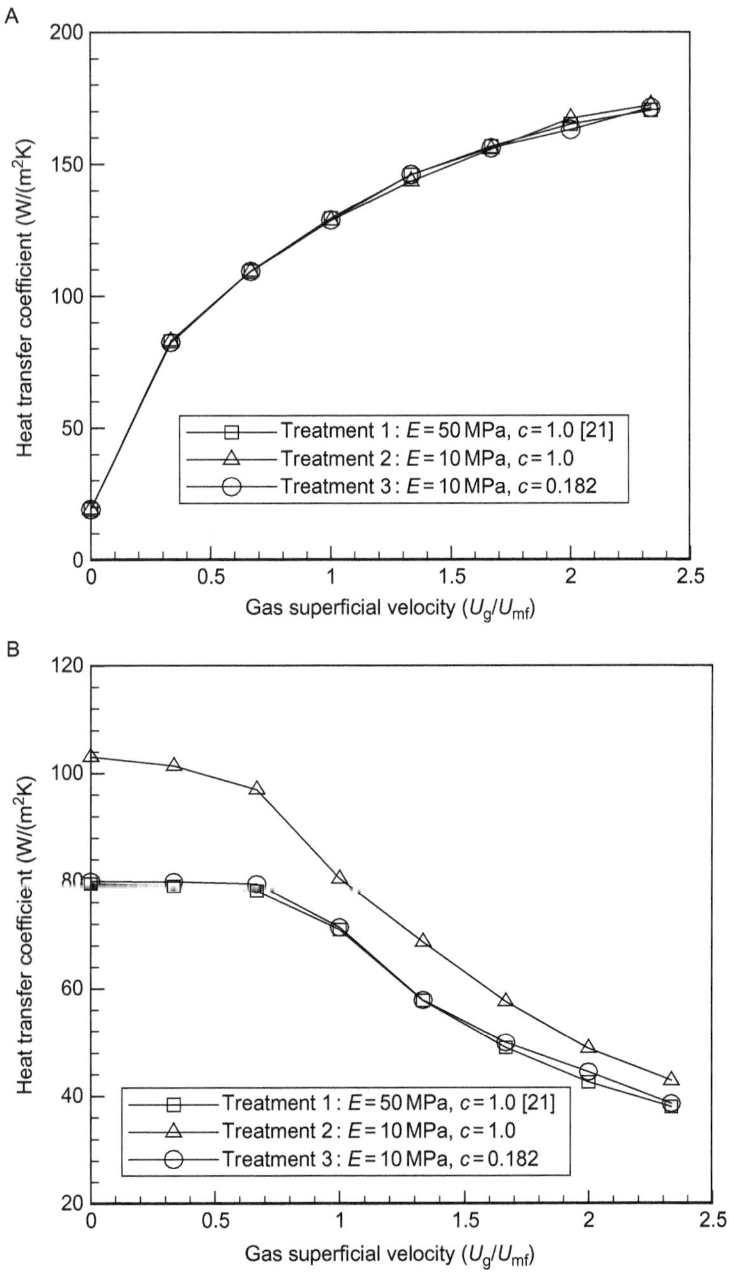


**Figure 10** Average convective heat transfer coefficient (A) and conductive heat transfer coefficient (B) of bed particles with different gas superficial velocities ($k_s = 0.84$ W/(m K)) (Zhou et al., 2010b).

case shown in Fig. 10, the use of a low Young's modulus significantly reduces the computational time, i.e., 4 ~ 5 times faster with 16,000 particles. Such a reduction becomes more significant for a larger system involving a large number of particles.

### 3.1.2 Effects of Some Variables

There are many factors influencing the ETC of a packed bed. The main factors are the thermal conductivities of the solid and fluid phases. Other factors include particle size, particle shape, packing method that gives different packing structures, bed temperature, fluid flow, and other properties. Zhou et al. (2010b) examined the effects of some parameters on ETC, and revealed the ETC is not sensitive to particle–particle sliding friction coefficient which varies from 0.1 to 0.8. ETC increases with the increase of bed average temperature, which is consistent with the observation in the literature (Wakao and Kaguei, 1982). The predicted ETC at 1475 °C can be about five times larger than that at 75 °C. The effect of particle size on ETC is more complicated (Fig. 11). At low thermal conductivity ratios of $k_s/k_f$, the ETC varies little with particle size from 250 μm to 10 mm. But it is not the case for particles with high thermal conductivity ratios, where the ETC increases with

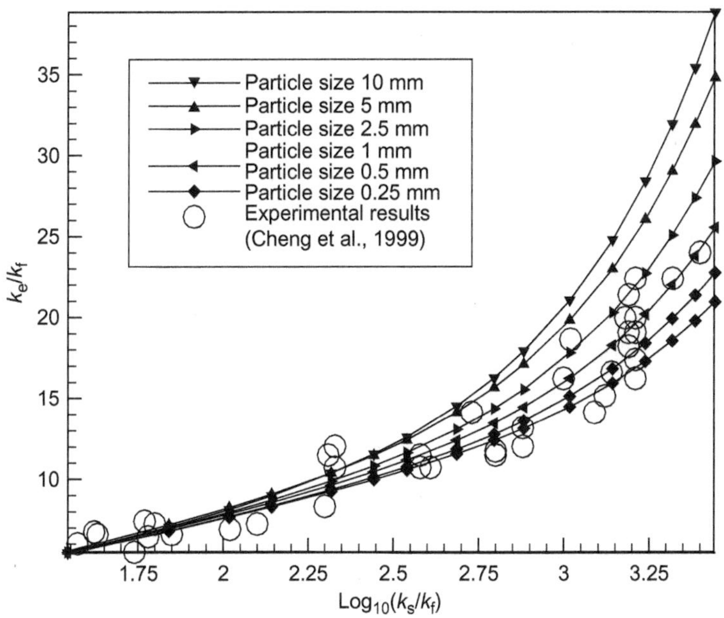


**Figure 11** Effect of particle size on the bed ETC (Zhou et al., 2010b).

particle size. The main reason could be that the particle–particle contact area is relatively large for large particles, and consequently, the increase of $k_s/k_f$ enhances the conductive heat transfer between particles. However, that ETC is affected by particle size offers an explanation as to why the literature data are so scattered. This is because different sized particles were used in experiments. For particles smaller than 500 μm, the predicted ETC is lower than that measured for high $k_s/k_f$ ratios. This is because large particles were used in the reported experiments. Further studies are required to quantify the effect of particle size on the bed ETC under more complex conditions with moving fluid, size distributions, or high bed temperature, as done experimentally (Fjellerup et al., 2003; Khraisha, 2002; Moreira et al., 2005).

Particle shape is another important parameter affecting heat transfer. Recently, the effect of particle shape on ETC has been examined (Gan, 2015). Here, ellipsoids are used as they can represent a wide range of particle shapes from platy to elongate. Interestingly, the shape can be described by one parameter called as aspect ratio. Aspect ratio is less than 1.0 for oblate, equal to 1.0 for spherical, and larger than 1.0 for prolate particles.

Figure 12 shows the variation of bed ETC with aspect ratio. It can be observed that with the deviation of aspect ratio from 1.0, the bed ETC increases significantly, especially for a high particle thermal conductivity. Spheres have the lowest ETC. Such a feature can be explained by the difference of packing structures. For example, with particles being more

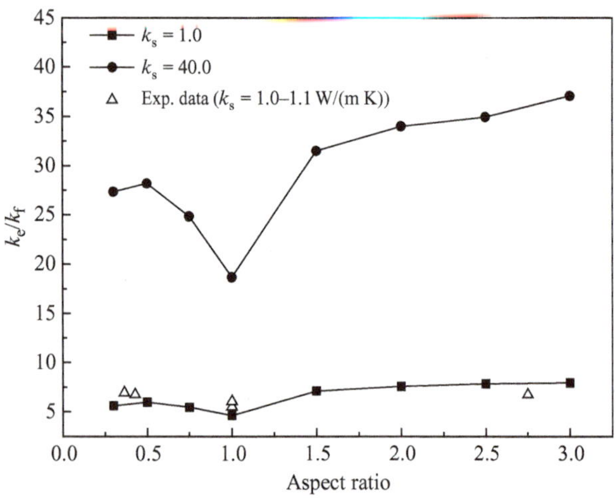


**Figure 12** Variation of ETC with aspect ratio (Gan, 2015), compared with experiment data from Verma et al. (1991).

nonspherical, packing density generally increases, indicating denser packing structure. The contact area of particles with different aspect ratios is also different under the same conditions. Spheres have the smallest contact area, while oblate and prolate particles have larger contact area, with the values more than doubled than spheres when aspect ratios are 0.25 or 3.0. This is because ellipsoids tend to lay flat in the horizontal direction in packed beds (Zhou et al., 2011b). This directly contributes to the high ETC according to the particle–fluid–particle and particle–particle conductive models.

## 3.2 Fluidized Beds

### 3.2.1 Coarse Particles

Gas fluidization is observed when solid particles are transformed into a fluid-like state at a proper gas velocity (Kunii and Levenspiel, 1991). By varying gas velocity, different flow patterns can be generated from a fixed bed ($U_g < U_{mf}$) to a fluidized bed. The solid flow patterns and particle temperatures in a fluidized bed are transient and vary spatiotemporally, as shown in Fig. 13. Particles located at the bottom are heated first, and flow upward mainly dragged by gas. Particles at the top with low velocities descend. Due to the strong mixing and high gas–particle contact surface area, the whole bed is heated quickly and reaches the gas inlet temperature at around 70 s. The general features observed are qualitatively in good agreement with those reported in the literature, confirming the predictability of the proposed CFD–DEM model in dealing with the gas–solid flow and heat transfer in gas fluidization.

The cooling of copper spheres at different initial locations in a gas-fluidized bed was examined (Zhou et al., 2009). In physical experiments, the temperature of hot spheres is measured using thermocouples connected to the spheres (Collier et al., 2004; Scott et al., 2004). The cooling process of such hot spheres can be easily traced and recorded in the CFD–DEM simulations, as shown in Fig. 14A. The predicted temperature is comparable with the measured one. The cooling curves of nine hot spheres are slightly different due to different local fluid flow and particle structures.

The comparison of the HTC–$U$ relationship between the simulated and the measured was made (Zhou et al., 2009). In physical experiments, Collier et al. (2004) and Scott et al. (2004) used different materials to examine the HTCs of hot spheres and found that there is a general tendency that the HTC of the hot sphere increases first with gas superficial velocity in the fixed bed ($U_g < U_{mf}$), and then remains constant, independent of the gas

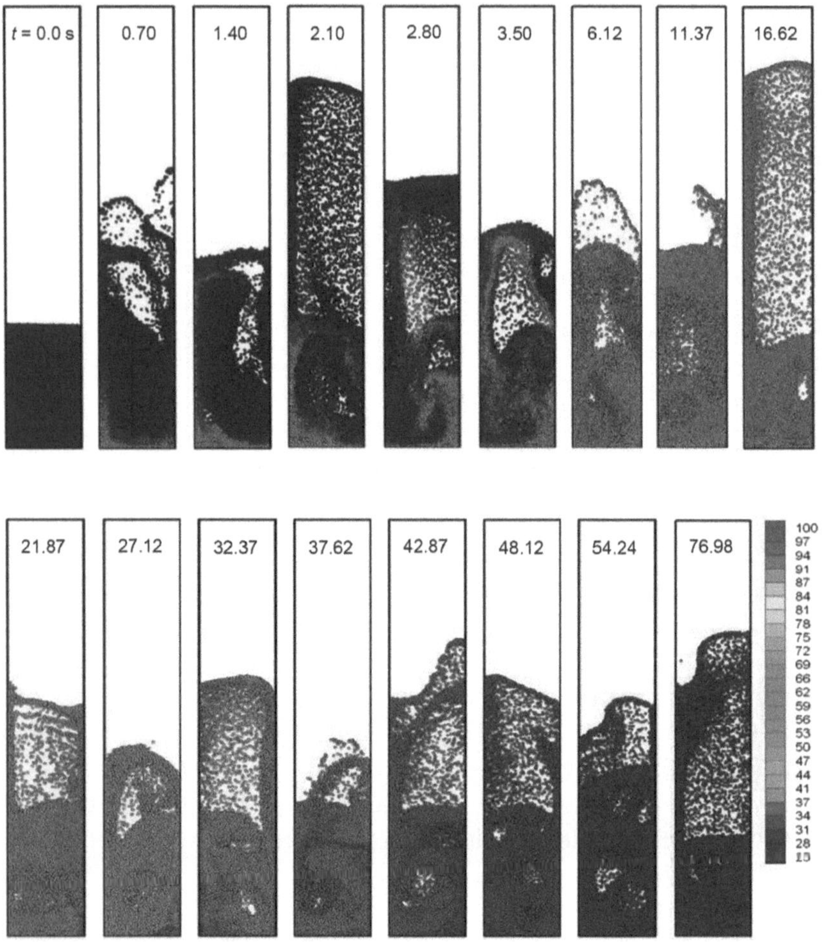


**Figure 13** Snapshots showing the heating process of a fluidized bed by hot gas (1.2 m/s, 100 °C) uniformly introduced from the bottom (bed width × thickness × height = 90 mm × 24 mm × 176 mm, $d_p = 3$ mm, $\rho_p = 420$ kg/m$^3$) (Zhou et al., 2009).

superficial velocities for fluidized beds ($U_g > U_{mf}$). The CFD–DEM simulation results also exhibit such a feature (Fig. 14B). For packed beds, the time-averaged HTC increases with gas superficial velocity and reaches its maximum at around $U_g = U_{mf}$. After the bed is fluidized, the HTC is almost constant in a large range.

The HTC–$U$ relationship is affected significantly by the thermal conductivity of bed particles (Zhou et al., 2009). The higher the value of $k_s$, the higher the HTC of hot spheres (Fig. 15). For example, when $k_s = 30$ W/(m K), the predicted HTC in the fixed bed ($U_g/U_{mf} < 1$) is so

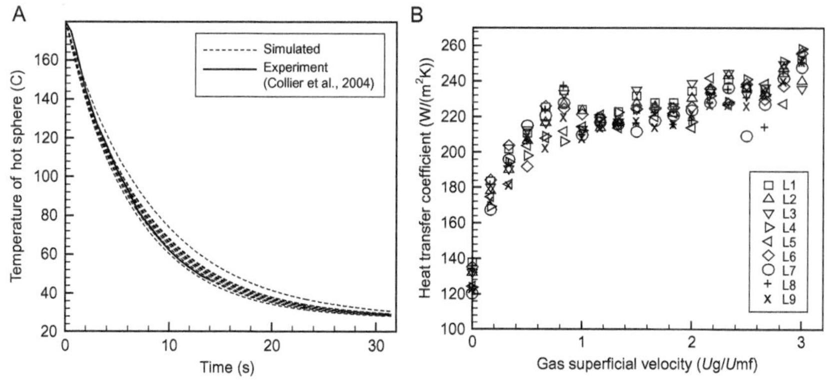


**Figure 14** (A) Temperature evolution of nine hot spheres when gas superficial velocity is 0.42 m/s and (B) time-averaged heat transfer coefficients of the nine hot spheres as a function of gas superficial velocity (Zhou et al., 2009).

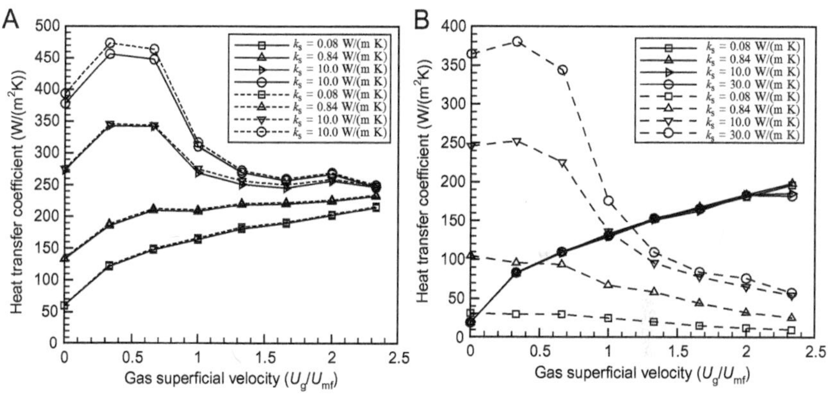


**Figure 15** Time-averaged heat transfer coefficients of one hot sphere: (A) total HTC calculated by different equations and (B) convective HTC (solid line) and conductive HTC (dashed-line) for different thermal conductivities (Zhou et al., 2009).

high that the HTC–$U_g$ relationship shown in Fig. 14B is totally changed. The HTC decreases with $U_g$ in the fixed bed, then may reach a constant HTC in the fluidized bed. But when thermal conductivity of particles is low, the HTC always increases with $U_g$, independent of bed state (Fig. 15A). Figure 15B further explains the variation trend of HTC with $U_g$. Generally, the convective HTC increases with $U_g$; but conductive HTC decreases with U. For a proper particle thermal conductivity, i.e., 0.84 W/(m K), the two contributions (convective HTC and conductive HTC) could compensate each other, then the total HTC is nearly constant after the bed is fluidized. So HTC independence of $U_g$ is valid under this condition. However, if particle thermal conductivity is too low or too high, the relationship of HTC and $U_g$ can be different, as illustrated in Fig. 15A.

The proposed CFD–DEM model can be used to analyze the sub-mechanisms for conduction, as shown in Fig. 3. The relative contributions by these heat transfer paths were quantified (Zhou et al., 2009). For example, when $k_s = 0.08$ W/(m K), particle–fluid–particle conduction always contributes more than particle–particle contact, but both vary with gas superficial velocity (Fig. 16A). For particle–fluid–particle conduction, particle–fluid–particle heat transfer with two contacting particles is far more important than that with two noncontacting particles in the fixed bed. Zhou et al. (2009) explained that it is because the hot sphere contacts about six particles when $U_g < U_{mf}$. But such a feature changes in the fluidized bed ($U_g > U_{mf}$), where particle–fluid–particle conduction between noncontacting particles is relatively more important. This is because most of particle–particle contacts with an overlap are gradually destroyed with increasing gas superficial velocity, which significantly reduces the contribution by the particle–fluid–particle conduction between two contacting particles. However, particle–particle conduction through the contacting area becomes more important with an increase of particle thermal conductivity. The percentage of its contribution is up to 42% in the fixed bed when $k_s = 30$ W/(m K), then reduces to around 15% in the fluidized bed (Fig. 16B). Correspondingly, the contribution percentage by particle–fluid–particle heat transfer is lower, but the variation with $U_g$ is similar to that for $k_s = 0.08$ W/(m K).

It should be noted that a fluid bed has many particles. A limited number of hot spheres cannot fully represent the averaged thermal behavior of all particles in a bed. Thus, Zhou et al. (2009) further examined the HTCs of all the particles and found that the features are similar to those observed for hot spheres (Fig. 17). The similarity illustrates that the hot sphere approach can, at least partially, represent the general features of particle thermal behavior in a particle–fluid bed. Overall, the particles in a uniformly fluidized bed behave similarly. But a particle may behave differently from another at a given time. The probability density distributions of time-averaged HTCs due to particle–fluid convection and particle conduction are obtained, respectively (Fig. 18). The convective HTC in the packed bed varies in a small range due to the stable particle structure. Then, the distribution curve moves to the right as $U$ increases, indicating the increase of convective HTC. The distribution curve also becomes wider. It is explained that, in a fluidized bed, clusters and bubbles can be formed, and the local flow

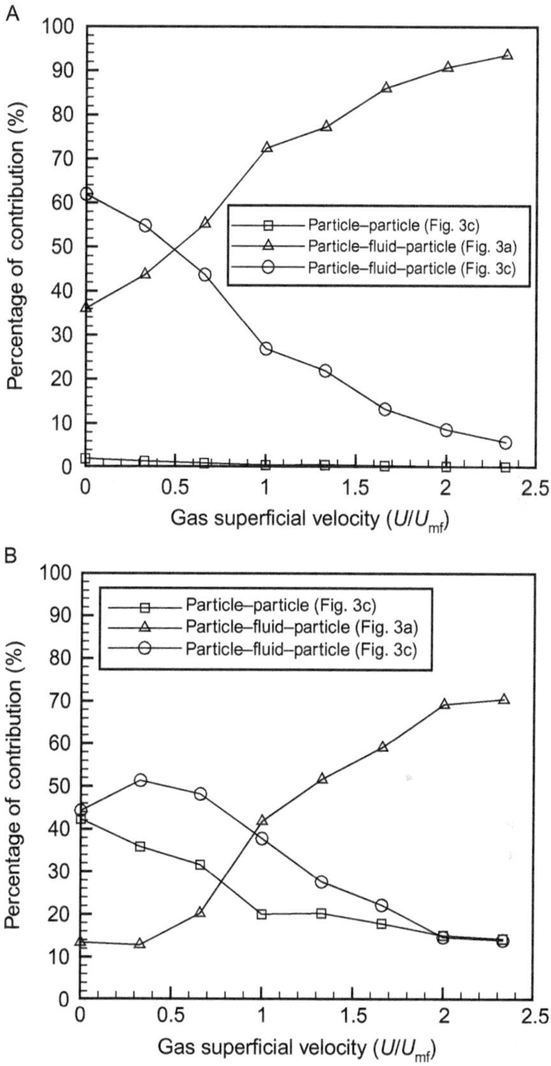


**Figure 16** Contributions to conduction heat transfer by different heat transfer mechanisms when (A) $k_s = 0.08$ W/(m K) and (B) $k_s = 30$ W/(m K) (Zhou et al., 2009).

structures surrounding particles vary in a large range. The density distribution of time-averaged HTCs by conduction shows that it has a wider distribution in a fixed bed (curves 1, 2, and 3) (Fig. 18B), indicating different local packing structures of particles. But curves 1 and 2 are similar, because statistically, the two bed packing structures are similar and do not vary much even if $U_g$ is different. When $U_g > U_{mf}$ (e.g., $U_g = 2.0U_{mf}$), the distribution

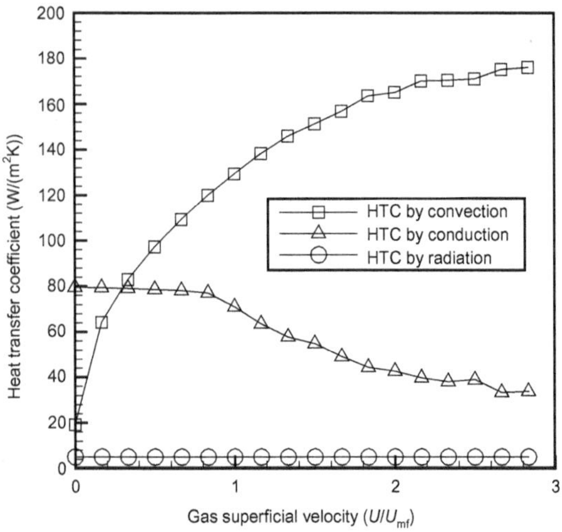


**Figure 17** Bed-averaged convective, conductive, and radiative heat transfer coefficients as a function of gas superficial velocity (Zhou et al., 2009).

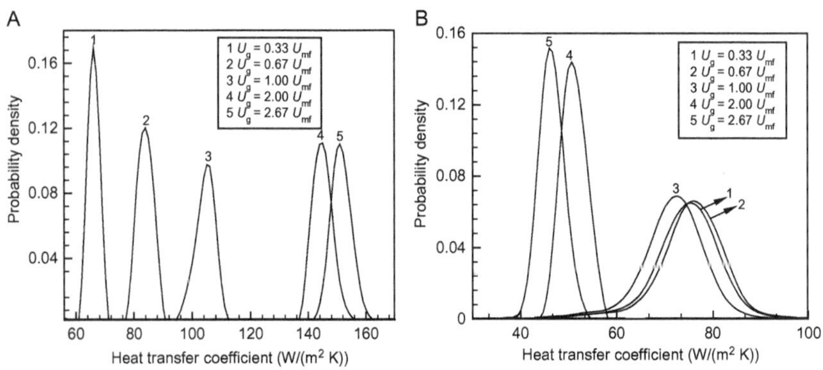


**Figure 18** Probability density distributions of time-averaged heat transfer coefficients of particles at different gas superficial velocities: (A) fluid convection and (B) particle conduction (Zhou et al., 2009).

curve moves to the left, indicating the heat transfer due to interparticle conduction is reduced. The bed particles occasionally collide and contact each other. Statistically, the number of collisions and contacts are similar in fully fluidized beds and not affected significantly by gas superficial velocities. Those features are consistent with those observed using the hot sphere approach. It confirms that hot sphere approach can represent the thermal behavior of all bed particles to some degree.

The particle thermal behavior in a fluidized bed is affected by bed temperature. Zhou et al. (2009) carried out a simulation at a high temperature of 1000 °C. It illustrated that the radiative HTC reaches 300 W/($m^2$ K), significantly larger than that for the case of hot gas at 100 °C (around 5 W/($m^2$ K)). The convective and radiative HTCs do not remain constant during the bed heating due to the variation of gas properties with temperature. The conductive HTC is not affected much by the bed temperature. This is because the conductive HTC is quite small in the fluidized bed, and only related to the gas and particle thermal conductivities.

### 3.2.2 Fine Particles

Powders are classified into four groups by the properties of fluidizing medium and particles (Geldart, 1973). Particle size is one of the important properties. For particles of small sizes, some forces other than gravitational force becomes significant. The van der Waals force is an important one besides electrostatic force and capillary force for wet particles (Castellanos, 2005). Geldart A particles can display three flow regimes at different gas velocities including packed bed, expanded bed, and fluidized bed. It can be expected that fluidized Geldart A particles would demonstrate new characteristics in these flow regimes because of different contact status (Hou et al., 2012c).

The combined CFD–DEM approach was extended to investigate the effects of some important parameters closely related to the van der Waals force such as particle size and Hamaker constant (Hou et al., 2012a). The heat transfer characteristics of cohesive particles were demonstrated in three flow regimes in Fig. 19. It revealed that the convective heat transfer is dominant for large particles while the conductive heat transfer becomes important with the decrease of particle size. This is mainly attributed to the increase of surface area per unit volume. Two transitional points with the increase of Hamaker constant were found in the variation of heat fluxes by convective and conductive heat transfer modes as shown in Fig. 20.

A macroscopic heating rate was derived based on particle scale simulations for the prediction of temperature at a bed scale. The heating rate as a function of Hamaker constant, particle diameter, and inlet gas velocity were obtained as shown in Fig. 21 (Hou et al., 2012a). The predicted trends agree well with general understanding and experimental observations of the effects of material properties and operating conditions.

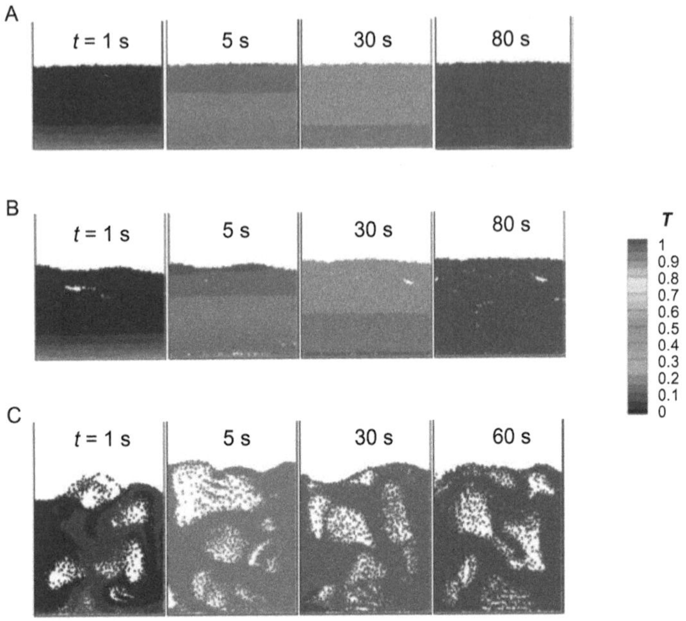


**Figure 19** Snapshots showing heating process of particles by hot air uniformly injected at the bed bottom in different flow regimes (bed width × thickness × height = 6 mm × 0.4 mm × 20 mm, $d_p = 0.1$ mm, $\rho_p = 1440$ kg/m$^3$): (A) packed bed ($U_g/U_{mf} = 0.5$), (B) expanded bed ($U_g/U_{mf} = 1.4$), and (C) fluidized bed ($U_g/U_{mf} = 6.0$). Particles are colored by their dimensionless temperatures. The dimensionless temperature $T$ of individual particles is obtained according to $(T_i - T_0)/(T_{in} - T_0)$, where $T_0$ is the initial bed temperature and $T_{in}$ is the air temperature at the inlet. *Reprinted from Hou et al. (2012a) with permission from ACS.*

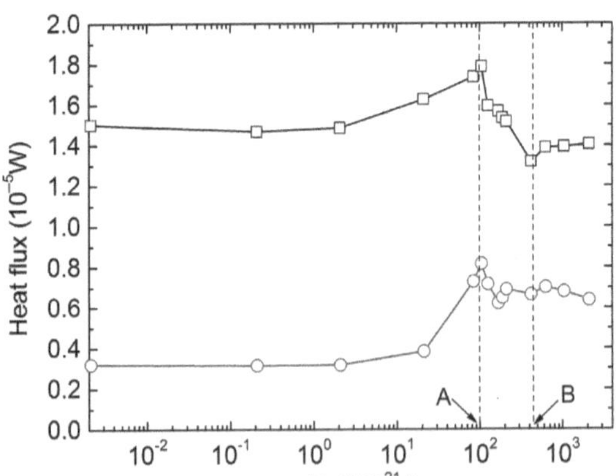


**Figure 20** Heat fluxes as a function of $H_a$ at $U_g/U_{mf} = 6$ (□, convective and ○, conductive). Transitional points are qualitatively denoted as Points A and B. *Reprinted from Hou et al. (2012a) with permission from ACS.*

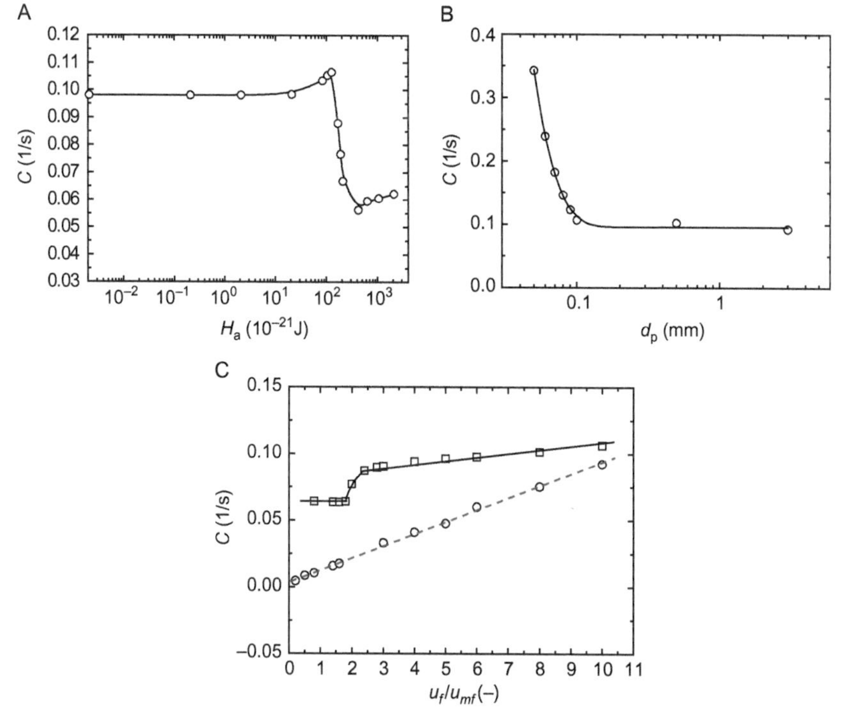


**Figure 21** (A) Lumped parameter $C$ as a function of the Hamaker constant $H_a$ when $d_p=0.1$ mm and $U_g/U_{mf}=6$, (B) $C$ as a function of $d_p$ when $U_g/U_{mf}=10.0$ and $H_a=4.2\times10^{-21}$J, and (C) predicted $C$ at different velocity ratio $U_g/U_{mf}$ when $H_a=4.2\times10^{-21}$J for group A and B powders: □, $d_p=0.1$ mm; ○, $d_p=3$ mm; and lines showing the trends. *Reprinted from Hou et al. (2012a) with permission from ACS.*

### *3.2.3 Heat Transfer Between a Fluidized Bed and Insert Tubes*

Immersed surfaces such as horizontal/vertical tubes, fins, and water walls are usually adopted in a fluidized bed to control the heat addition or extraction (Chen, 1998). Understanding the flow and heat transfer mechanisms is important to achieve its optimal design and control (Chen et al., 2005). The relation of the HTC of a tube and gas–solid flow characteristic in the vicinity of the tube such as particle residence time and porosity has been investigated experimentally using heat transfer probe and positron emission particle tracking method or an optical probe (Kim et al., 2003; Masoumifard et al., 2008; Wong and Seville, 2006). Alternatively, the CFD–DEM approach has been used to study the flow and heat transfer in fluidization with an immersed tube in the literature (Di Maio et al., 2009; Wong and Seville, 2006; Zhao et al., 2009). They obtained comparable prediction of HTC with experimental results at a low temperature. These studies show the applicability of the proposed CFD–DEM approach to a fluidized bed with an immersed tube. However, some important aspects are not considered in these studies such as the difference between two- and three-dimensional settings, the variation of fluid properties with temperature, detailed conductive heat transfer mechanisms, and the contribution of radiative heat transfer at a high temperature.

Recently, Hou et al. (2010, 2012b) used the proposed CFD–DEM model to investigate the heat transfer in gas fluidization with an immersed horizontal tube in a three-dimensional bed. The simulation conditions are similar to the experimental investigations by Wong and Seville (2006) except for the bed geometry. The predicted result of 0.27 m/s for minimum fluidization velocity ($U_{mf}$) is consistent with those experimental measurements (Chandran and Chen, 1982; Wong and Seville, 2006). Figure 22 shows the snapshots of flow patterns obtained from the CFD–DEM simulation. The bubbling fluidized bed behavior is significantly affected by the horizontal tube. Two main features were identified: defluidized region in the downstream and the air film in the upstream (Glass and Harrison, 1964; Hou et al., 2012b; Rong et al., 1999; Wong and Seville, 2006). Particles with small velocities tend to stay on the tube in the downstream and form the defluidized region intermittently. The thickness of the air film below the tube changes with time. There is no air film, and the upstream section is fully filled with particles at some time intervals (e.g., $t=1.3$ and 3.0 s in Fig. 22).

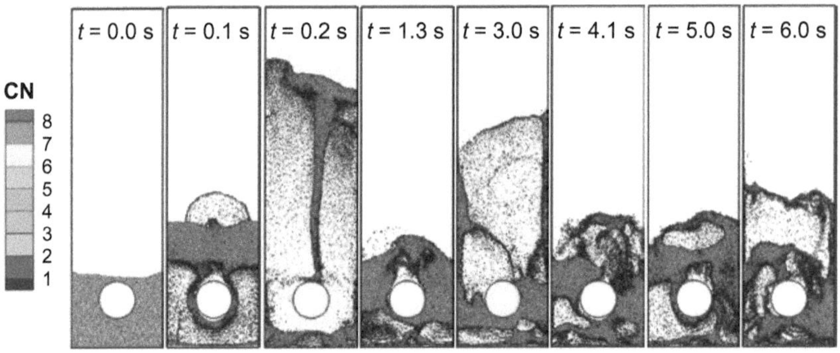


**Figure 22** Snapshots of solid flow pattern colored by coordination number of individual particles when $u_{exc}=0.50$ m/s (bed width × thickness × height = 36 mm × 2.4 mm × 460.8 mm, $d_p=0.6$ mm, $\rho_p=2600$ kg/m$^3$, tube diameter = 24 mm) (Hou et al., 2012b). The tube material is copper.

The tube exchanges heat with its surrounding particles and fluid. The local HTC has a distribution closely related to these observed flow patterns. The distribution and magnitude of HTC are two factors commonly used to describe the heat transfer in such a system (Botterill et al., 1984; Schmidt and Renz, 2005; Wong and Seville, 2006). The effects of the gas velocity and the tube position are examined, showing consistent results with those reported in the literature (Botterill et al., 1984; Kim et al., 2003; Fig. 23). The local HTC is high at sides of the tube around 90° and 270°, while it is low at the upstream and downstream of the tube around 0° and 180°. With the increase of gas velocity, the local HTC increases first and then decreases (Fig. 23A). The local HTC is also affected by tube positions and increases with the increase of tube level within the bed static height as shown in Fig. 23B.

The heat is mainly transferred through convection between gas and particles and between gas and the tube, and conduction among particles and between particles and the tube at low temperature. As an example, Fig. 24A shows the total heat fluxes through convection and conduction (the radiative heat flux is quite small at low temperatures). The convective and conductive heat fluxes vary temporally. Their percentages show that the convective heat transfer is dominant with a percentage over 90%. They are closely related to the microstructure around the tube, which can be indexed by the average porosity around the tube and by the contact number (CN) between the tube and the particles. The porosity and CN vary temporally depending on the complicated interactions between the particles and the tube and between the particles and fluid, which determine the flow pattern.

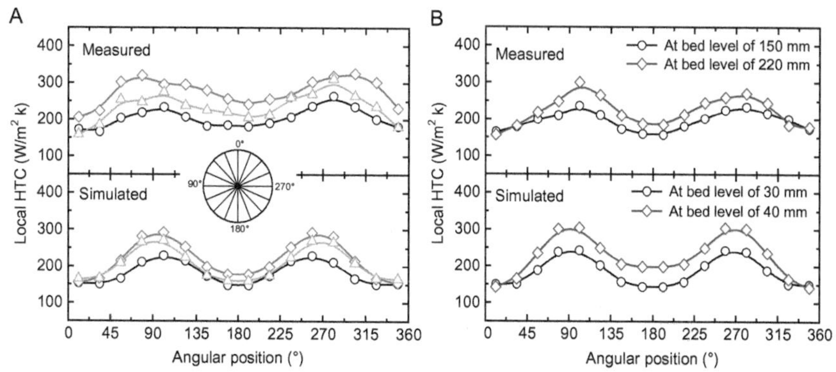


**Figure 23** Comparison of local HTCs between the simulated (Hou et al., 2012b) and the measured (Wong and Seville, 2006): (A) local HTC distribution for different excess gas velocities ($u_{exc}$) (○, 0.08 m/s; ◇, 0.50 m/s; and ▽, 0.80 m/s); and (B) local HTC with different tube positions when $u_{exc}=0.20$ m/s.

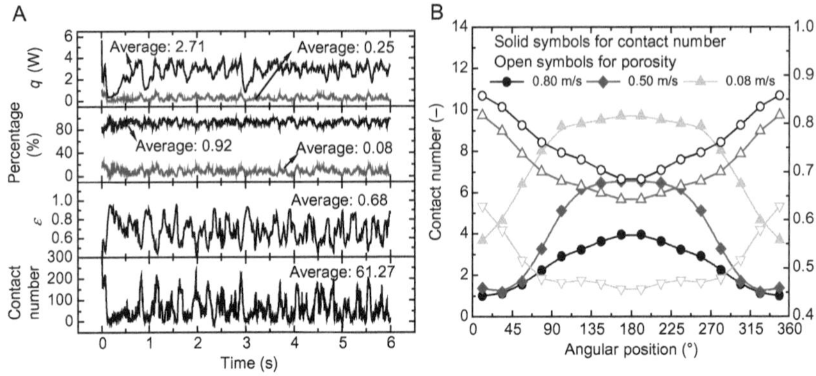


**Figure 24** (A) Overall convective and conductive heat fluxes ($q$) and their percentages (—, convection; ----, conduction), overall contact number (CN), and overall porosity ($\varepsilon$) as a function of time, where $u_{exc} = 0.40$ m/s (the overall heat flux and CN are the sum of the corresponding values of each section and the overall porosity are the averaged value of all the sections) and (B) local porosity and CN with different $u_{exc}$ (Hou et al., 2012b).

Generally, a region with a larger CN corresponds to a defluidized region with a smaller porosity in the vicinity of the tube. Otherwise, it corresponds to a passing bubble where the porosity is larger and the CN is smaller. Figure 24B further shows the distribution of local porosity and CN. It can be seen that local porosity is larger in downstream sections and lower in upstream sections while local CN has an opposite distribution.

The heat transfer between an immersed tube and a fluidized bed depends on many factors, such as the contacts of particles with the tube, porosity, and gas flow around the tube. These factors are affected by many variables related to operational conditions. Gas velocity is one of the most important parameters in affecting the heat transfer, which can be seen in Fig. 23. With the increase of $u_{exc}$ from 0.08 to 0.50 m/s, the overall HTC increases. However, when the $u_{exc}$ is further increased from 0.50 to 0.80 m/s, the HTC decreases. The effect of particle thermal conductivity $k_s$ on the local HTC was also examined and shown in Fig. 25A (Hou et al., 2012b). The local HTC increases with the increase of $k_s$ from 1.10 to 100 W/(m K). However, such an increase is not significant for $k_s$ from 100 to 300 W/(m K). The variation of percentages of different heat transfer modes with $k_s$ is further shown in Fig. 25B. When $k_s$ is lower than 100 W/(m K), the conductive heat transfer increases with the increase of $k_s$, while the convective heat transfer decreases. Further increase of $k_s$ has no significant effects.

The heat transfer by radiation is important in the considered system because the increase of environmental temperature of the tube and its

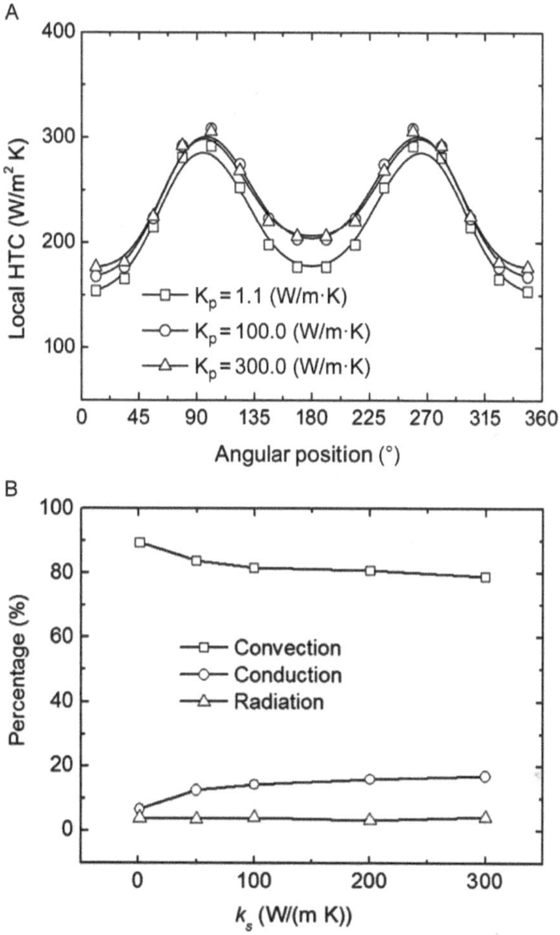


**Figure 25** Effect of $k_s$ on: (A) local HTC and (B) percentages of different heat fluxes, where $u_{exc} = 0.50$ m/s (Hou et al., 2012b).

significance has already been pointed out in the literature (see, for example, Chen et al., 2005; Mathur and Saxena, 1987). The effect of the tube temperature ($T_s$) on heat transfer characteristic was investigated in terms of the local HTC distribution and the heat fluxes by different heat transfer modes (Hou et al., 2012b). Figure 26A shows that the local HTC increase with the increase of $T_s$. The increased trend of HTC agrees well with the results of the experiments (Botterill et al., 1984). The increase of gas thermal conductivity with temperature is one of the main reasons for the increase of HTC (Zhou et al., 2009). This manifests the importance of using the temperature-related correlations of fluid properties. Variations of the heat fluxes with tube

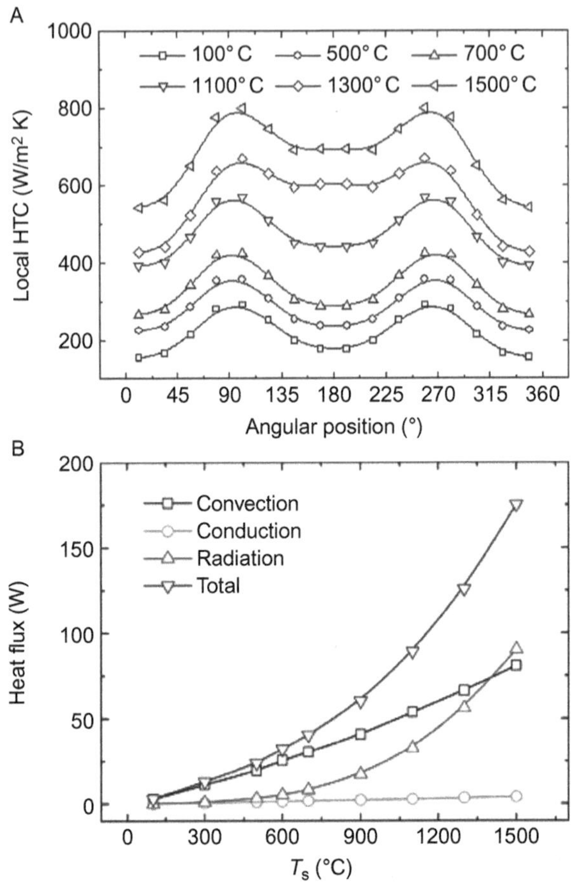


**Figure 26** Heat transfer behavior at high tube temperatures: (A) variations of local HTC with different $T_s$, where $u_{exc}=0.50$ m/s; and (B) convective, conductive, radiative, and total heat fluxes as a function of $T_s$, where $u_{exc}=0.50$ m/s (Hou et al., 2012b).

temperature $T_s$ are shown in Fig. 26B. The conductive heat flux changes insignificantly. The convective heat flux increases linearly, while the radiative heat fluxes increases exponentially with the increase of the $T_s$. Because of the increase of $T_s$, the difference between the environmental temperature ($T_e$) and the bed temperature ($T_b$) increases. The radiative heat flux increases more quickly than the convective heat flux according to the fourth power law of the temperature difference. The radiative heat flux becomes larger than that of conductive heat flux around $T_s=300$ °C and then, larger than that of the convective heat flux around $T_s=1200$ °C. These show that the radiation is an important heat transfer mode with high tube temperatures.

Using commercial software Fluent and user defined function, the heat transfer between a fluidized bed and tubes in a three-dimensional model was investigated (Wahyudi et al., 2013). No significant difference in the averaged HTC between inline and staggered arrangements of tube arrays was found at a moderate superficial gas velocity within the range of $1 \sim 2U_{\mathrm{mf}}$. This finding should be further examined with a large range of gas velocity.

The effects of tube bundle settings and materials properties were investigated using an in-house code (Hou et al., 2015b). The effect of material properties was illustrated by considering cohesive and noncohesive powders with different particle sizes. The contributions of different heat transfer mechanisms were discussed at two tube temperatures. It was found that conductive heat transfer between a fluidized bed and a tube is dominant for small cohesive particles while convective heat transfer is dominant for large noncohesive particles. The uniformity of particle velocity and temperature fields was analyzed. They vary with material properties and gas velocity in a complicated manner. The effect of tube array settings was examined in terms of two geometrical parameters for both in-line and staggered settings. Complicated gas–solid flow (Fig. 27) and heat transfer characteristics (Fig. 28) were observed. The link between macroscopic observations and microscopic information such as local porosity and CN between fluidized particles and tubes was revealed.

## 4. CONCLUSIONS AND FUTURE WORK

The combined CFD–DEM approach, incorporated with heat transfer models of convection, conduction, and radiation, has been developed and can be used in the study of heat transfer in packed and fluidized beds. It has various advantages over the conventional experimental techniques and continuum simulation approaches. For example, the detailed conductive heat transfer between particles can be examined and the factors such as local particle–fluid structure and materials properties in determining heat transfer can be investigated.

Notably, the approach has taken into account almost all the known heat transfer mechanisms including particle–fluid convection, particle–particle conduction, and radiative heat transfer between solid particles and surrounding environment. It can be tested through various comparisons of the predicted and measured results. The approach has a good capability in describing heat transfer in packed and fluidized beds at a particle scale.

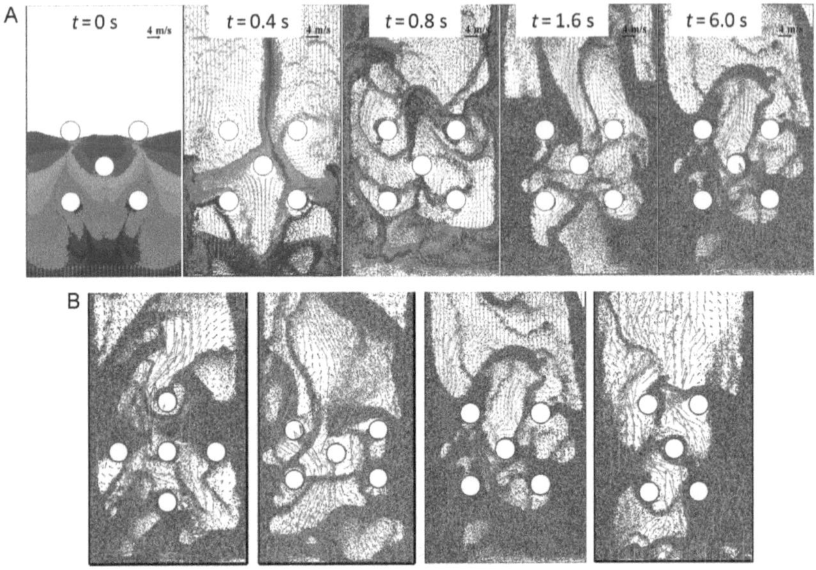


**Figure 27** Gas–solid flow patterns in fluidized beds with a tube array: (A) for different times when $\alpha=45°$ and (B) for different settings when $t=6.0$ s. The settings include a square one ($\alpha=0$) and three triangular ones ($\alpha=30°$, 45°, and 60°) from the left to right (bed width × thickness × height = 10 mm × 0.4 mm × 128 mm $d_p=0.1$ mm, $\rho_p=1440$ kg/m$^3$, tube diameter = 40 mm). The tube material is copper. *Reprinted from Hou et al. (2015b) with permission from Elsevier.*

On one hand, it has been used to investigate the complicated variation of different heat transfer mechanisms in different flow regimes of fluidized beds with coarse or fine particles. It can, for example, examine heat transfer between insert tube/tubes and fluidized beds as well. The effects of some pertinent variables can be quantified, and the underlying heat transfer mechanisms are identified. On the other hand, the approach can predict some macroscopic parameters based on the detailed heat transfer information at a particle scale, which is useful for the process design, control, and optimization. This can be seen from the studies of various complicated, coupled flow, and heat transfer systems including, for example, dryers (Mahmoudi et al., 2014; Tatemoto and Sawada, 2012; Weigler et al., 2012), fluid bed reactors (Bruchmuller et al., 2012; Kaneko et al., 1999; Wu et al., 2010), blast furnace (Yang et al., 2015a,c), and COREX ironmaking process (Hou et al., 2015a).

The CFD–DEM approach will be popular for the investigation of heat transfer in various particulate systems at a particle scale, particularly with the rapid development of discrete particle simulation techniques and computer

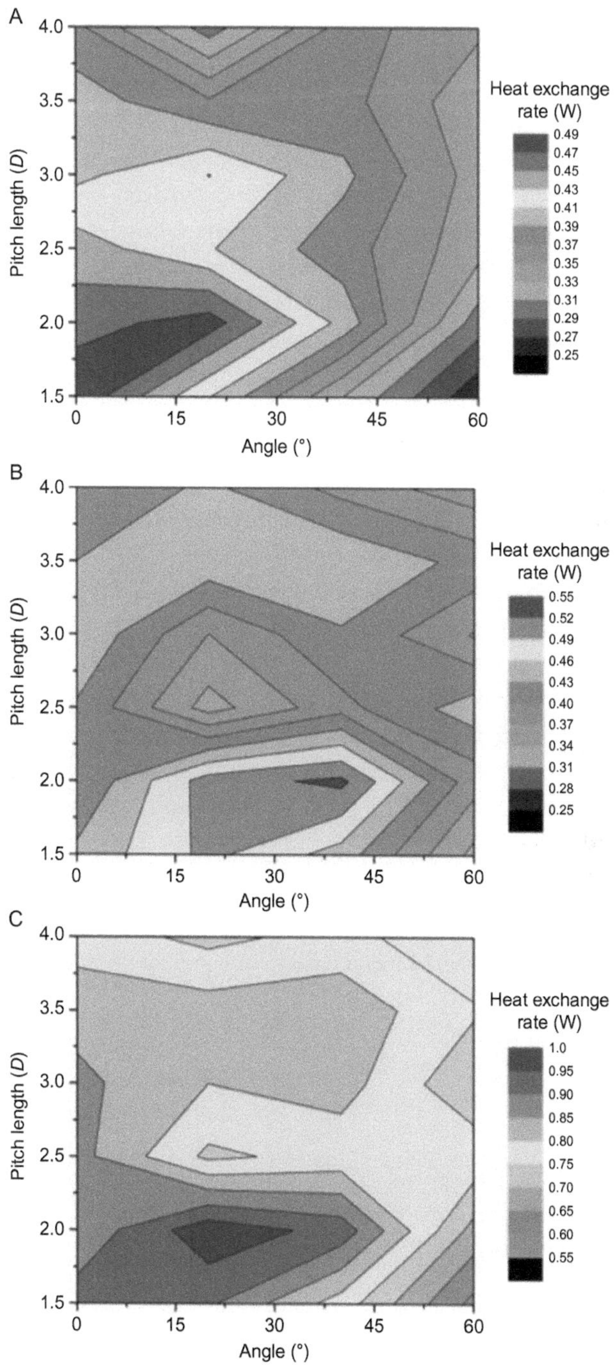


**Figure 28** Heat exchange rates as a function of pitch length and angle: (A) convection, (B) conduction, and (C) total. *Reprinted from Hou et al. (2015b) with permission from Elsevier.*

technology. However, it should be noted that the approach is still a state-of-art simulation technique. There are different treatments of various heat transfer mechanisms and further efforts to be made at different time and length scales are highlighted below:

- Subparticle/particle scale: To develop more general theories and models to better describe the heat transfer within a particle, and between a particle and fluid by various simulation techniques to overcome the existing limitations such as the need for Biot number smaller than 0.1 or the use of representative properties of a particle, generating a more concrete basis for particle scale simulation of coupled flow and heat transfer systems.
- Cluster/mesoscale: To develop a general theory to link the discrete and continuum approaches, so that particle scale heat transfer information, generated from DEM-based simulation, can be quantified in terms of (macroscopic) energy conservation equations, constitutive relations, and boundary conditions that can be implemented in continuum-based process modeling of thermochemical behaviors.
- Application: To develop more robust models and efficient computer codes by using advanced computer techniques so that the capability of particle scale simulation can be extended, say, from two- to three-dimensional and/or from simple spherical to complicated nonspherical particle system involving not only multiphase flow but also heat and mass transfer and chemical reactions, which is important to transfer the present phenomenon simulation to process simulation and hence meet real engineering needs.

## ACKNOWLEDGMENT

The authors are grateful to the Australian Research Council for the financial support.

## REFERENCES

Agarwal PK: Transport phenomena in multi-particle systems—IV. Heat transfer to a large freely moving particle in gas fluidized bed of smaller particles, *Chem Eng Sci* 46:1115–1127, 1991.

Amritkar A, Deb S, Tafti D: Efficient parallel CFD-DEM simulations using OpenMP, *J Comput Phys* 256:501–519, 2014.

Anderson TB, Jackson R: Fluid mechanical description of fluidized beds-equations of motion, *Ind Eng Chem Fund* 6:527–539, 1967.

Argento C, Bouvard D: Modeling the effective thermal conductivity of random packing of spheres through densification, *Int J Heat Mass Transfer* 39:1343–1350, 1996.

Baskakov AP, Filippovskii NF, Munts VA, Ashikhmin AA: Temperature of particles heated in a fluidized bed of inert material, *J Eng Phys (English Translation of Inzhenerno-Fizicheskii Zhurnal)* 52:574–578, 1987.

Batchelor GK, O'Brien RW: Thermal or electrical conduction through a granular material, *Proc R Soc Lond Ser A* 355:313–333, 1977.
Botterill JSM: *Fluid-bed heat transfer,* New York, 1975, Academic Press London.
Botterill JSM, Teoman Y, Yuregir KR: Factors affecting heat transfer between gas-fluidized beds and immersed surfaces, *Powder Technol* 39:177–189, 1984.
Bruchmuller J, van Wachem BGM, Gu S, Luo KH, Brown RC: Modeling the thermochemical degradation of biomass inside a fast pyrolysis fluidized bed reactor, *AIChE J* 58:3030–3042, 2012.
Castellanos A: The relationship between attractive interparticle forces and bulk behaviour in dry and uncharged fine powders, *Adv Phys* 54:263–376, 2005.
Chandran R, Chen JC: Bed-surface contact dynamics for horizontal tubes in fluidized beds, *AIChE J* 28:907–914, 1982.
Chen JC: Heat transfer in fluidized beds. In Wen-Ching Y, editor: *Fluidization, solids handling, and processing,* Westwood, NJ, 1998, William Andrew Publishing, pp 153–208.
Chen JC: Surface contact—its significance for multiphase heat transfer: diverse examples, *J Heat Trans-T ASME* 125:549–566, 2003.
Chen JC, Grace JR, Golriz MR: Heat transfer in fluidized beds: design methods, *Powder Technol* 150:123–132, 2005.
Cheng GJ: *Structural evaluation of the effective thermal conductivity of packed beds,* Sydney, Australia, 2003, The University of New South Wales.
Cheng GJ, Yu AB: Particle scale evaluation of the effective thermal conductivity from the structure of a packed bed: radiation heat transfer, *Ind Eng Chem Res* 52:12202–12211, 2013.
Cheng GJ, Yu AB, Zulli P: Evaluation of effective thermal conductivity from the structure of a packed bed, *Chem Eng Sci* 54:4199–4209, 1999.
Collier AP, Hayhurst AN, Richardson JL, Scott SA: The heat transfer coefficient between a particle and a bed (packed or fluidised) of much larger particles, *Chem Eng Sci* 59:4613–4620, 2004.
Cundall PA, Strack ODL: A discrete numerical model for granular assemblies, *Géotechnique* 29:47–65, 1979.
Di Felice R: The voidage function for fluid-particle interaction systems, *Int J Multiphase Flow* 20:153–159, 1994.
Di Maio FP, Di Renzo A, Trevisan D: Comparison of heat transfer models in DEM-CFD simulations of fluidized beds with an immersed probe, *Powder Technol* 193:257–265, 2009.
Enwald H, Peirano E, Almstedt AE: Eulerian two-phase flow theory applied to fluidization, *Int J Multiphase Flow* 22:21–66, 1996.
Ergun S: Fluid flow through packed columns, *Chem Eng Prog* 48:89–94, 1952.
Feng YT, Han K: An accurate evaluation of geometric view factors for modelling radiative heat transfer in randomly packed beds of equally sized spheres, *Int J Heat Mass Transfer* 55:6374–6383, 2012.
Feng YQ, Yu AB: Assessment of model formulations in the discrete particle simulation of gas-solid flow, *Ind Eng Chem Res* 43:8378–8390, 2004.
Feng YQ, Yu AB: Microdynamic modelling and analysis of the mixing and segregation of binary mixtures of particles in gas fluidization, *Chem Eng Sci* 62:256–268, 2007.
Feng YT, Han K, Li CF, Owen DRJ: Discrete thermal element modelling of heat conduction in particle systems: basic formulations, *J Comput Phys* 227:5072–5089, 2008.
Feng YT, Han K, Owen DRJ: Discrete thermal element modelling of heat conduction in particle systems: pipe-network model and transient analysis, *Powder Technol* 193:248–256, 2009.
Finney JL: Random packings and the structure of simple liquids. I. The geometry of random close packing, *Proc R Soc Lond Ser A* 319:479–493, 1970.

Fjellerup J, Henriksen U, Jensen AD, Jensen PA, Glarborg P: Heat transfer in a fixed bed of straw char, *Energy Fuel* 17:1251–1258, 2003.

Gan JQ: *Discrete particle simulation of packing, fluidization and heat transfer of ellipsoids*, Melbourne, Australia, 2015, Monash University.

Geldart D: Types of gas fluidization, *Powder Technol* 7:285–292, 1973.

Gidaspow D: *Multiphase flow and fluidization,* San Diego, 1994, Academic Press.

Glass DH, Harrison D: Flow patterns near a solid obstacle in a fluidized bed, *Chem Eng Sci* 19:1001–1002, 1964.

Gunn DJ: Transfer of heat or mass to particles in fixed and fluidised beds, *Int J Heat Mass Transfer* 21:467–476, 1978.

Hamaker HC: The London-van der Waals attraction between spherical particles, *Physica* 4:1058–1072, 1937.

Holman JP: *Heat transfer,* New York, 1981, McGraw-Hill Company.

Hou QF, Zhou ZY, Yu AB: Investigation of heat transfer in bubbling fluidization with an immersed tube. In Guo LJ, Joseph DD, Matsumoto Y, Sommerfeld M, Wang YS, editors: 6th international symposium on multiphase flow, heat mass transfer and energy conversion. AIP conference proceedings, Melville, 2010, American Institute of Physics, pp 355–360.

Hou QF, Zhou ZY, Yu AB: Computational study of the effects of material properties on heat transfer in gas fluidization, *Ind Eng Chem Res* 51:11572–11586, 2012a.

Hou QF, Zhou ZY, Yu AB: Computational study of the heat transfer in bubbling fluidized beds with a horizontal tube, *AIChE J* 58:1422–1434, 2012b.

Hou QF, Zhou ZY, Yu AB: Micromechanical modeling and analysis of different flow regimes in gas fluidization, *Chem Eng Sci* 84:449–468, 2012c.

Hou QF, Zhou ZY, Yu AB: Computational study of heat transfer in gas fluidization. In *Powders and grains 2013: proceedings of the 7th international conference on micromechanics of granular media, 1542*, 2013, pp 1114–1117.

Hou QF, Li J, Yu AB: CFD-DEM study of heat transfer in the reduction shaft of Corex, *Steel Res Int* 86:626–635, 2015.

Hou QF, Zhou ZY, Yu AB: Gas-solid flow and heat transfer in fluidized beds with tubes: effects of material properties and tube array settings, *Powder Technol*, 2015. http://dx.doi.org/10.1016/j.powtec.2015.03.028. In Press.

Incropera FP, Dewitt DP: *Fundamentals of heat and mass transfer,* New York, 2002, John Wiley & Sons.

Ishii M: *Thermo-fluid dynamic theory of two-phase flow. Collection de la Direction des Etudes et Recherches d'Electricite de France,* Paris, 1975, Eyrolles.

Kaneko Y, Shiojima T, Horio M: DEM simulation of fluidized beds for gas-phase olefin polymerization, *Chem Eng Sci* 54:5809–5821, 1999.

Khraisha YH: Thermal conductivity of oil shale particles in a packed bed, *Energ Sources* 24:613–623, 2002.

Kim SW, Ahn JY, Kim SD, Lee DH: Heat transfer and bubble characteristics in a fluidized bed with immersed horizontal tube bundle, *Int J Heat Mass Transfer* 46:399–409, 2003.

Kishore N, Gu S: Momentum and heat transfer phenomena of spheroid particles at moderate Reynolds and Prandtl numbers, *Int J Heat Mass Transfer* 54:2595–2601, 2011.

Kobayashi M, Maekwa H, Nakamura H, Kondou Y: Calculation of the mean thermal conductivity of heterogeneous solid mixture with the Voronoi-polyhedron element method, *Trans Jpn Soc Mech Eng B* 57:1795–1801, 1991.

Kunii D, Levenspiel O: *Fluidization engineering,* Boston, 1991, Butterworth-Heinemann.

Kunii D, Smith JM: Heat transfer characteristics of porous rocks, *AIChE J* 6:71–78, 1960.

Langston PA, Tüzün U, Heyes DM: Continuous potential discrete particle simulations of stress and velocity fields in hoppers: transition from fluid to granular flow, *Chem Eng Sci* 49:1259–1275, 1994.

Langston PA, Tüzün U, Heyes DM: Discrete element simulation of internal-stress and flow-fields in funnel flow hoppers, *Powder Technol* 85:153–169, 1995.
Launder BE, Spalding DB: The numerical computation of turbulent flows, *Comput Method Appl Mech Eng* 3:269–289, 1974.
Li JT, Mason DJ: A computational investigation of transient heat transfer in pneumatic transport of granular particles, *Powder Technol* 112:273–282, 2000.
Li JT, Mason DJ: Application of the discrete element modelling in air drying of particulate solids, *Drying Technol* 20:255–282, 2002.
Mahmoudi AH, Hoffmann F, Peters B: Application of XDEM as a novel approach to predict drying of a packed bed, *Int J Therm Sci* 75:65–75, 2014.
Masoumifard N, Mostoufi N, Hamidi A-A, Sotudeh-Gharebagh R: Investigation of heat transfer between a horizontal tube and gas-solid fluidized bed, *Int J Heat Fluid Flow* 29:1504–1511, 2008.
Mathur A, Saxena SC: Total and radiative heat transfer to an immersed surface in a gas-fluidized bed, *AIChE J* 33:1124–1135, 1987.
Mickley HS, Fairbanks DF: Mechanism of heat transfer to fluidized beds, *AIChE J* 1:374–384, 1955.
Molerus O, Wirth KE: *Heat transfer in fluidized beds,* London, 1997, Chapman and Hall.
Moreira MFP, Thomeo JC, Freire JT: Analysis of the heat transfer in a packed bed with cocurrent gas-liquid upflow, *Ind Eng Chem Res* 44:4142–4146, 2005.
Parker DJ, Broadbent CJ, Fowles P, Hawkesworth MR, McNeil P: Positron emission particle tracking—a technique for studying flow within engineering equipment, *Nucl Instrum Methods Phys Res Sect A* 326:592–607, 1993.
Parmar MS, Hayhurst AN: The heat transfer coefficient for a freely moving sphere in a bubbling fluidised bed, *Chem Eng Sci* 57:3485–3494, 2002.
Patankar SV: *Numerical heat transfer and fluid flow,* New York, 1980, Hemisphere.
Patil AV, Peters E, Sutkar VS, Deen NG, Kuipers JAM: A study of heat transfer in fluidized beds using an integrated DIA/PIV/IR technique, *Chem Eng J* 259:90–106, 2015.
Peters B: Measurements and application of a discrete particle model (DPM) to simulate combustion of a packed bed of individual fuel particles, *Combust Flame* 131:132–146, 2002.
Prins W, Draijer W, van Swaaij WPM: Heat transfer to immersed spheres fixed or freely moving in a gas-fluidized bed. In 20th proceedings of the international centre for heat and mass transfer. Heat and mass transfer in fixed and fluidized beds. Washington, 1985, Hemisphere, pp 317–331.
Rong DG, Horio M: DEM simulation of char combustion in a fluidized bed. In Second international conference on CFD in the minerals and process industries. Melbourne, Australia, 1999, CD-ROM, pp 65–70.
Rong DG, Mikami T, Horio M: Particle and bubble movements around tubes immersed in fluidized beds—a numerical study, *Chem Eng Sci* 54:5737–5754, 1999.
Rong LW, Dong KJ, Yu AB: Lattice-Boltzmann simulation of fluid flow through packed beds of uniform spheres: effect of porosity, *Chem Eng Sci* 99:44–58, 2013.
Rong LW, Dong KJ, Yu AB: Lattice-Boltzmann simulation of fluid flow through packed beds of spheres: effect of particle size distribution, *Chem Eng Sci* 116:508–523, 2014.
Rong LW, Zhou ZY, Yu AB: Lattice-Boltzmann simulation of fluid flow through packed beds of uniform ellipsoids, *Powder Technol* 285:146–156, 2015.
Schmidt A, Renz U: Numerical prediction of heat transfer between a bubbling fluidized bed and an immersed tube bundle, *Heat Mass Transfer* 41:257–270, 2005.
Scott SA, Davidson JF, Dennis JS, Hayhurst AN: Heat transfer to a single sphere immersed in beds of particles supplied by gas at rates above and below minimum fluidization, *Ind Eng Chem Res* 43:5632–5644, 2004.

Sparrow EM, Abraham JP, Tong JCK: Archival correlations for average heat transfer coefficients for non-circular and circular cylinders and for spheres in cross-flow, *Int J Heat Mass Transfer* 47:5285–5296, 2004.

Sun J, Chen MM: A theoretical analysis of heat transfer due to particle impact, *Int J Heat Mass Transfer* 31:969–975, 1988.

Tatemoto Y, Sawada T: Numerical analysis of drying characteristics of wet material immersed in fluidized bed of inert particles, *Drying Technol* 30:979–988, 2012.

Tsuji Y, Tanaka T, Ishida T: Lagrangian numerical simulation of plug flow of cohesionless particles in a horizontal pipe, *Powder Technol* 71:239–250, 1992.

Tsuji Y, Kawaguchi T, Tanaka T: Discrete particle simulation of two-dimensional fluidized bed, *Powder Technol* 77:79–87, 1993.

Vargas WL, McCarthy JJ: Stress effects on the conductivity of particulate beds, *Chem Eng Sci* 57:3119–3131, 2002.

Verma LS, Shrotriya AK, Singh R, Chaudhary DR: Thermal conduction in two-phase materials with spherical and non-spherical inclusions, *J Phys D Appl Phys* 24:1729, 1991.

Vortmeyer D: Radiation in packed solids. In 6th international heat transfer conference, Toronto. 1978, pp 525–539.

Wadewitz A, Specht E: Limit value of the Nusselt number for particles of different shape, *Int J Heat Mass Transfer* 44:967–975, 2001.

Wahyudi H, Chu KW, Yu AB: Discrete particle simulation of heat transfer in pressurized fluidized bed with immersed cylinders. In *Powders and grains 2013: proceedings of the 7th international conference on micromechanics of granular media, vol. 1542*, 2013, pp 1118–1121.

Wakao N, Kaguei S: *Heat and mass transfer in packed beds*, New York, 1982, Gordon and Breach Science Publishers.

Weigler F, Scaar H, Mellmann J: Investigation of particle and air flows in a mixed-flow dryer, *Drying Technol* 30:1730–1741, 2012.

Wen CY, Yu YH: Mechanics of fluidization, *Chem Eng Prog Symp Ser* 162:100–111, 1966.

Wong YS, Seville JPK: Single-particle motion and heat transfer in fluidized beds, *AIChE J* 52:4099–4109, 2006.

Wu CN, Cheng Y, Ding YL, Jin Y: CFD-DEM simulation of gas-solid reacting flows in fluid catalytic cracking (FCC) process, *Chem Eng Sci* 65:542–549, 2010.

Xu BH, Yu AB: Numerical simulation of the gas-solid flow in a fluidized bed by combining discrete particle method with computational fluid dynamics, *Chem Eng Sci* 52:2785–2809, 1997.

Xu BH, Yu AB: Comments on the paper "Numerical simulation of the gas-solid flow in a fluidized bed by combining discrete particle method with computational fluid dynamics" *Chem Eng Sci* 53:2646–2647, 1998.

Xu BH, Yu AB, Chew SJ, Zulli P: Numerical simulation of the gas-solid flow in a bed with lateral gas blasting, *Powder Technol* 109:13–26, 2000.

Yagi S, Kunii D: Studies on effective thermal conductivities in packed beds, *AIChE J* 3:373–381, 1957.

Yang RY, Zou RP, Yu AB: Computer simulation of the packing of fine particles, *Phys Rev E* 62:3900–3908, 2000.

Yang RY, Zou RP, Yu AB: Voronoi tessellation of the packing of fine uniform spheres, *Phys Rev E* 65:041302, 2002.

Yang WJ, Zhou ZY, Pinson D, Yu AB: A new approach for studying softening and melting behavior of particles in a blast furnace cohesive zone, *Metall Mater Trans B* 46:977–992, 2015a.

Yang WJ, Zhou ZY, Yu AB: Particle scale studies of heat transfer in a moving bed, *Powder Technol* 281:99–111, 2015b.

Yang WJ, Zhou ZY, Yu AB, Pinson D: Particle scale simulation of softening–melting behaviour of multiple layers of particles in a blast furnace cohesive zone, *Powder Technol* 279:134–145, 2015c.
Zehner P, Schlunde EU: Thermal conductivity of packings at moderate temperatures, *Chem Ing Tech* 42:933–941, 1970.
Zehner P, Schlünder EU: Thermal conductivity of packed beds (in German), *Chem Ing Tech* 42:933–941, 1970.
Zhao YZ, Jiang MQ, Liu YL, Zheng JY: Particle-scale simulation of the flow and heat transfer behaviors in fluidized bed with immersed tube, *AIChE J* 55:3109–3124, 2009.
Zhou YC, Wright BD, Yang RY, Xu BH, Yu AB: Rolling friction in the dynamic simulation of sandpile formation, *Phys A* 269:536–553, 1999.
Zhou HS, Flamant G, Gauthier D, Flitris Y: Simulation of coal combustion in a bubbling fluidized bed by distinct element method, *Chem Eng Res Des* 81:1144–1149, 2003.
Zhou HS, Flamant G, Gauthier D: DEM-LES simulation of coal combustion in a bubbling fluidized bed part II: coal combustion at the particle level, *Chem Eng Sci* 59:4205–4215, 2004a.
Zhou YC, Yu AB, Stewart RL, Bridgwater J: Microdynamic analysis of the particle flow in a cylindrical bladed mixer, *Chem Eng Sci* 59:1343–1364, 2004b.
Zhou JH, Yu AB, Horio M: Finite element modeling of the transient heat conduction between colliding particles, *Chem Eng J* 139:510–516, 2008.
Zhou ZY, Yu AB, Zulli P: Particle scale study of heat transfer in packed and bubbling fluidized beds, *AIChE J* 55:868–884, 2009.
Zhou ZY, Kuang SB, Chu KW, Yu AB: Discrete particle simulation of particle-fluid flow: model formulations and their applicability, *J Fluid Mech* 661:482–510, 2010a.
Zhou ZY, Yu AB, Zulli P: A new computational method for studying heat transfer in fluid bed reactors, *Powder Technol* 197:102–110, 2010b.
Zhou ZY, Hou QF, Yu AB: Particle scale simulation of heat transfer in fluid bed reactors. In Belmiloudi A, editor: *Heat transfer—mathematical modelling, numerical methods and information technology*, Rijeka, Croatia, 2011a, InTech, pp 383–408.
Zhou ZY, Pinson D, Zou RP, Yu AB: Discrete particle simulation of gas fluidization of ellipsoidal particles, *Chem Eng Sci* 66:6128–6145, 2011b.
Zhu HP, Yu AB: Averaging method of granular materials, *Phys Rev E* 66:021302, 2002.
Zhu HP, Zhou ZY, Yang RY, Yu AB: Discrete particle simulation of particulate systems: theoretical developments, *Chem Eng Sci* 62:3378–3396, 2007.
Zhu HP, Zhou ZY, Yang RY, Yu AB: Discrete particle simulation of particulate systems: a review of major applications and findings, *Chem Eng Sci* 63:5728–5770, 2008.
Zhu HP, Hou QF, Zhou ZY, Yu AB: Averaging method of particulate systems and its application to particle-fluid flow in a fluidized bed, *Chin Sci Bull* 54:4309–4317, 2009.
Zhu HP, Zhou ZY, Hou QF, Yu AB: Linking discrete particle simulation to continuum process modelling for granular matter: theory and application, *Particuology* 9:342–357, 2011.

CHAPTER FIVE

# Mesoscale Transport Phenomena and Mechanisms in Gas–Liquid Reaction Systems

**Ning Yang**[1]

Center for Mesoscience and State Key Laboratory of Multiphase Complex Systems, Institute of Process Engineering, Chinese Academy of Sciences, Beijing, PR China

[1]Corresponding author: e-mail address: nyang@ipe.ac.cn

## Contents

1. Introduction 247
2. Understanding the Mesoscale Mechanisms Through the EMMS Model 251
3. Physical Interpretation of Macroscale Behavior from Mesoscale Perspective 255
4. Consolidated Understanding of the EMMS Model for gas–Liquid Systems 258
   4.1 The DBS Model—An Intrinsic Model for Structure Evolution 258
   4.2 Dual Effects of Liquid Viscosity and Surface Tension 260
   4.3 Intrinsic Similarity Between Gas–Liquid and Gas–Solid Systems 261
   4.4 Coupling with Mass Transfer and Reaction 264
   4.5 Further Explanation 268
5. Stability-Constrained Multifluid CFD Model 268
   5.1 Stability Condition as a Closure Law for CFD Models 268
   5.2 Simulation of Bubble Column Reactors 270
6. Conclusions and Future Issues 276
Acknowledgments 278
References 278

## Abstract

Mesoscale transport phenomena and mechanisms are essential to achieve a more fundamental understanding on the mass, momentum, and heat transfer in the classical study of transport phenomena and on the mixing, residence time distribution, and rate-limiting analysis in chemical reaction engineering, yet they are now beyond the scope of classical textbooks of chemical engineering. We highlight a heuristic mesoscale modeling approach starting from a conceptual energy-minimization multiscale (EMMS) model and ending at the stability-constrained multifluid computational fluid dynamics (CFD) model. While the stability condition determines the direction of system evolution, the stability-constrained CFD further describes the dynamics of structure evolution. We establish the dual-bubble-size model, an extended EMMS approach for gas–liquid systems. Stability condition is formulated as the minimization of the sum of two energy dissipations, reflecting the compromise of a liquid-dominant regime at which smaller

*Advances in Chemical Engineering*, Volume 46
ISSN 0065-2377
http://dx.doi.org/10.1016/bs.ache.2015.10.007

bubbles prevail and a gas-dominant regime favoring the existence of larger bubbles. It supplies a mesoscale constraint for conservation equations and a mesoscale perspective to understand the macroscale regime transition. The model calculation for gas–liquid and gas–solid systems demonstrates the intrinsic similarity of the two systems; the system evolution at macroscale is driven by stability conditions. Theoretically stability condition may offer closure laws for CFD simulation, leading to the stability-constrained multifluid CFD model. While direct integration is difficult, we propose various simplified approaches to derive the closure models for drag, bubble-induced turbulence and correction factors for coalescence rate in population balance equations. The stability-constrained multifluid CFD model shows much advantage over traditional closure models.

## NOMENCLATURE

$C_{D0,b}$ drag coefficient for a bubble in a quiescent liquid, dimensionless
$C_{D0,p}$ drag coefficient for a particle in a quiescent fluid, dimensionless
$C_{Db}$ drag coefficient for a bubble in a swarm, dimensionless
$C_{Dp}$ drag coefficient for a particle in multiparticle systems, dimensionless
$c_f$ coefficient of surface area increase, $c_f = f_{BV}^{2/3} + (1 - f_{BV})^{2/3} - 1$, dimensionless
$d_b$ bubble diameter, m
$d_L$ bubble diameter of large bubbles, m
$d_S$ bubble diameter of small bubbles, m
$D_T$ column diameter, m
$Eo$ Eötvos number, dimensionless
$f_b$ volume fraction of gas phase, dimensionless
$f_{BV}$ breakup ratio of daughter bubble to its mother bubble, dimensionless
$f_L$ volume fraction of large bubbles, dimensionless
$f_S$ volume fraction of small bubbles, dimensionless
$g$ gravitational acceleration, m/s$^2$
$l_E$ the entrance length in Taitel's model, m
$Mo$ Morton number, dimensionless
$n_b$ number density of bubbles, 1/m$^3$
$N_{break}$ rate of energy consumption due to bubble breakage and coalescence per unit mass, m$^2$/s$^3$
$N_{st}$ rate of energy dissipation for suspending and transporting particles per unit mass, m$^2$/s$^3$
$N_{surf}$ rate of energy dissipation due to bubble oscillation per unit mass, m$^2$/s$^3$
$N_T$ total rate of energy dissipation
$N_{turb}$ rate of energy dissipation in turbulent liquid phase per unit mass, m$^2$/s$^3$
$P_b$ bubble breakup probability, dimensionless
$U_g$ superficial gas velocity, m/s
$U_{g,L}$ superficial gas velocity for large bubbles, m/s
$U_{g,S}$ superficial gas velocity for small bubbles, m/s
$U_{g,trans}$ transition gas velocity, m/s
$U_l$ superficial liquid velocity, m/s
$W_{st}$ rate of energy dissipation for suspending and transporting particles per unit volume, m$^2$/s$^3$

## GREEK LETTERS

$\varepsilon$ voidage, dimensionless
$\lambda$ character size of eddy, m
$\mu$ viscosity, Pa s
$\rho$ density, kg/m$^3$
$\sigma$ surface tension, N/m
$\omega$ collision frequency, 1/s

## ABBREVIATIONS

**DBS** double bubble size
**EMMS** energy-minimization multiscale
**SBS** single bubble size

# 1. INTRODUCTION

The increasing tempo of change in human life, society, economics, and environment creates a correspondingly urgent need for scientists and engineers seeking for new perspectives for traditional problems some of which may be long-standing, or posing new questions and offering new answers. For example, although new catalysts and chemical technologies can be invented or patented in laboratory every year, process scale-up from laboratory to industrial application remains a troublesome or challenging issue. Successful cases are limited and risky, relying on the empirical correlations and the engineers whose knowledge and experience are acquainted through the long-term case study of previous well-established processes. It is generally acknowledged that the main technical problem, among others, is how to create an ideal transport environment for reactions and separation, and hence chemical reactions could be compatible with their carrier, i.e., the fluid flow, mass, and heat transfer in multiphase reactors. A new angle to achieve a fundamental understanding and then seek efficient solutions of these classical problems is to reveal the mystery on mesoscales, i.e., the mesoscale transport phenomena and mechanisms relevant to bubbles, droplets, and particles (Li, 2015; Li and Huang, 2014; Li et al, 2013). Actually mesoscale problems are essential to a more fundamental understanding of momentum, mass, and heat transfer in the classical study of transport phenomena, and to the mixing, residence time distribution, and rate-limiting analysis in the chemical reaction engineering, yet they are beyond the scope of those classical textbooks of chemical engineering.

Mesoscale is indeed a level-dependent concept in multiphase systems (Fig. 1). For example, the mesoscales at the interfacial level and the reactor level may focus on different phenomena and mechanisms. At the interfacial level, the concentration distributions of the absorbed species on some catalyst particle surfaces form complex structures denoted "chemical turbulence" such as propagating and standing waves or rotating spirals, and the temporal variation of reaction rate may become oscillatory or even chaotic (Ertl, 1991). At the interface of bubbles or droplets, the so-called interfacial turbulence, a manifestation of hydrodynamic instability, arises from the gradient of interfacial tension. At the reactor level, particle clusters, bubble swarms, or liquid vortices generate more complicated multiphase turbulence than that in single-phase systems. Here, the traditional terminology "turbulence" can be used to signify the fluctuation or variation of parameters on the large scales due to the complex behavior at smaller scales, though it may not necessarily be a perfect description of mesoscales since the complexity and characteristics at the mesoscales in multiphase reaction

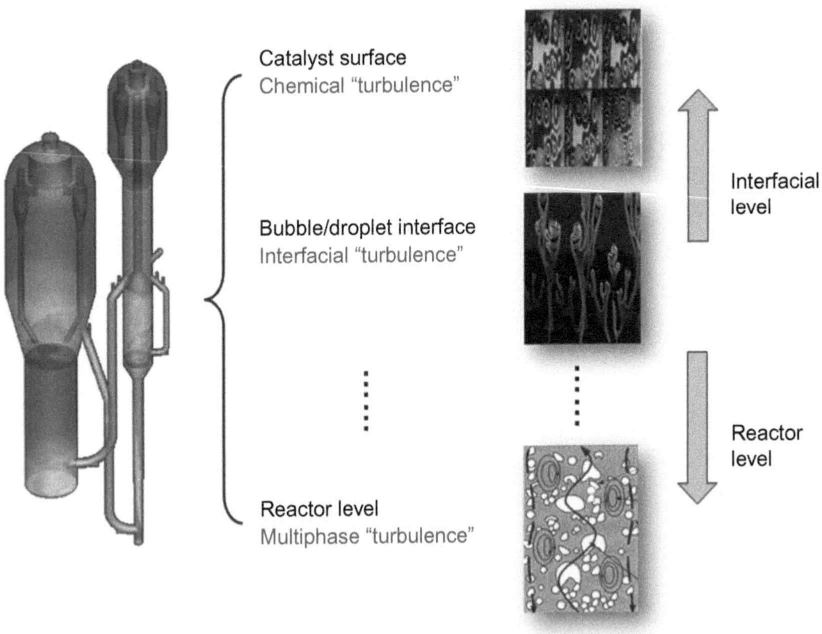


**Figure 1** Mesoscales in multiphase reactors: a level-dependent concept.

systems are beyond the capability of classical turbulence theory. These mesoscale problems at different levels may occur simultaneously in multiphase reactors, and here we only confine our attention to the mesoscales at the reactor level other than the material or interfacial or system levels.

Mesoscales in multiphase reactors are also dependent on reactor types and operating conditions (Fig. 2): the heterogeneous distribution, evolution, and transmission of strong force chain in granular matter influence the resulting macroscopic stress in rotating drums or fixed beds (Sun et al, 2009). The dynamic formation and dissolution of particle clusters greatly influence the local mixing and fluid dynamics in circulating fluidized beds. In bubble column reactor systems, the mesoscales may involve bubble swarms, bubble breakup and coalescence, bubble wake, and liquid vortices, generating far more complex coherent structure than that of gas–solid systems (Chen et al, 1994). It should be pointed out that the above descriptions about mesoscales should be more essentially termed as mesoscale phenomena than mesoscale mechanisms. There is much confusion between these two concepts which are sometimes used without any distinction in literature. While concentrating on the first concept is common, it is challenging to discover the mesoscale mechanisms behind the mesoscale phenomena.

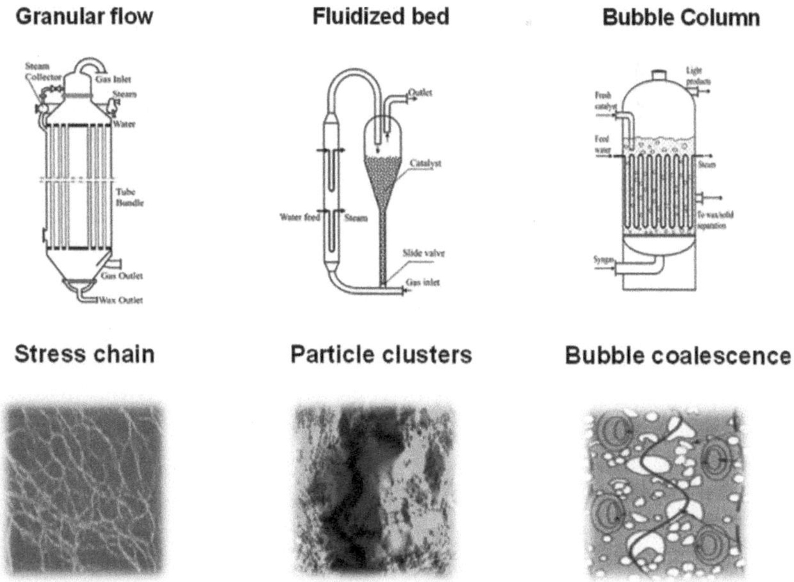


**Figure 2** Mesoscales in multiphase reactors: dependence on reactor types and operating conditions.

There are a couple of open questions about mesoscales for multiphase reactor systems. What happens at mesoscales and what are the dominant mechanisms behind mesoscale phenomena? Is it possible to establish a stability condition to physically reflect the mesoscale mechanisms and how can we describe them mathematically? How could these new findings or theory help to understand the macroscale behavior of multiphase systems, or be incorporated into current framework of reactor engineering models or computational fluid dynamics (CFD)? These traditional models either oversimplify the transport mechanisms for ease of engineering applications or smooth out the mesoscale heterogeneity and physics due to the intrinsic assumptions or ensemble averaging of conservation equations and jump conditions. Rather than a rigorous and complete analysis of the above open questions, this chapter attempts to present a new perspective and then a preliminary solution for these questions. We utilize a heuristic method to resolve the energy dissipation and structure in the gas–liquid systems of bubble columns, and then the dominant mechanisms and their compromise can be analyzed through a stability condition related to two bubble groups. We propose an energy-minimization multiscale (EMMS) model for gas–liquid systems, i.e., the dual-bubble-size (DBS) model (Yang et al, 2007, 2010). It can be used to understand and interpret the evolution of macroscale structures, viz., the regime transition, and integrated into CFD to predict the effect of mesoscale structures.

This chapter is organized into the following parts which are essentially correlated with each other based on our previous publications. The interested readers can be referred to the publications for further details of each section. Section 2 introduces the basic concepts of the EMMS approach in gas–liquid systems, and the conservation equations closed by the stability condition are given so that the structure parameters in the model can be obtained by solving the nonlinear optimization problem. The stability condition physically reflects the compromise between two different dominant mechanisms relevant to small bubbles and larger bubbles, supplementing the missing laws in multiphase complex systems in addition to the mass and momentum conservation equations. Section 3 further gives the solutions of the EMMS model for gas–liquid bubble column systems. The solutions of the structure parameters represent the macroscale structure variation and regime transition. The latter could be physically interpreted from a mesoscale perspective. Section 4 aims to enhance the readers' understanding of this conceptual approach: Section 4.1 compares the DBS, triple-bubble-size

(TBS), and multiple-bubble-size (MBS) models and demonstrates that all the model predictions would be reduced to that of DBS model, implying the intrinsic compromising mechanisms relevant to small bubbles and large bubbles in gas–liquid systems. Section 4.2 shows the reasonableness of this model in predicting the dual effects of viscosity and surface tension on regime transition. Section 4.3 compares the EMMS models for gas–solid and gas–liquid systems and points out that there indeed exists some intrinsic similarity between these two systems though the stability condition and mesoscale structures are system dependent. The stability condition also drives the system evolution in addition to those conservation laws. Section 4.4 introduces a new reactor model for bubble columns combining the EMMS model and the species transport equations. While the above sections deal with a zero-dimensional conceptual model, Section 5 integrates the model with the CFD approach, and the EMMS model is further used to derive the constitutive models for drag, bubble-induced turbulence and the kernel functions of bubble coalescence and breakup in population balance equations (PBE). The new models eliminate the empirical fitting parameters in the CFD models and improve the simulation accuracy.

## 2. UNDERSTANDING THE MESOSCALE MECHANISMS THROUGH THE EMMS MODEL

Mesoscale structures in gas–solid systems are generally in the form of particle clusters, strands, or streamers which have been extensively corroborated in experimental research. CFD simulation using the EMMS drag model also proves the significance of incorporating mesoscale structures in CFD (Jiradilok et al, 2006; Shah et al, 2011; Wang and Li, 2007; Yang et al, 2003, 2004): the solid entrainment rate is in good agreement with experiments, the dynamic formation of dissolution of particle clusters can be captured, and the coexistence of the dense bottom and the dilute top can be predicted. These phenomena could not be predicted by traditional multiphase CFD models in which the drag coefficient model is generally derived from the analysis of homogeneous systems and the mesoscale structures are neglected.

Gas–liquid systems are however more complex than gas–solid systems. Here, the gas–liquid system of interest is a swarm of dispersed bubbles in continuous liquid. Rising bubbles are introduced through a gas distributor at bottom, momentum is transferred through the interactions of gas and liquid, and energy is dissipated through shear-induced or bubble-induced turbulence. Bubbles may deform with shape oscillation or suffer from breakup or coalescence with increasing gas flow rate, adding new degrees of freedom to an already complex system. The system is also characterized by diversified coherent two-phase structures involving liquid vortical structures and circulation cells (Groen et al, 1996; Joshi et al, 2002). Although there are a large number of experimental analyses on flow structure in literature, it is difficult to distinguish the dominant mechanisms and the scales at which these phenomena occur since these structures coexist simultaneously and are tangled together.

The heuristic methodology we utilize is illustrated in Fig. 3. There are two kinds of resolution in the model, viz., the resolution of structure and dominant mechanisms. First, it is reasonable to resolve the flow structure into small bubbles, large bubbles, and liquid since there are plenty of reports in literature about the coexistence of small and large bubbles or a bimodal bubble size distribution (De Swart et al, 1996). This structure resolution is different from the spatial–temporal resolution of fluid domain in CFD models, and the significance lies in that the two bubble classes (TBCs) are actually relevant to the two kinds of dominant mechanisms. We will also show later that further resolution with more bubble classes in the model is not necessary and the MBSs are reduced to the TBCs. We use the following structure parameters for the description of the TBCs: bubble diameters

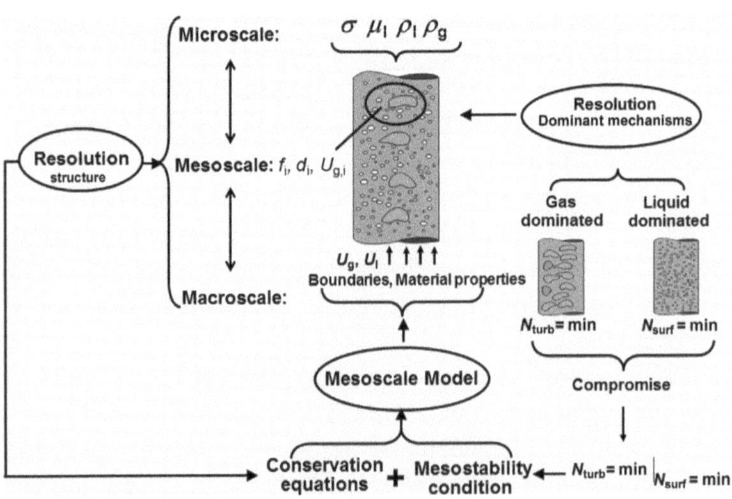


**Figure 3** EMMS approach for gas–liquid systems.

($d_S$, $d_L$), volume fraction ($f_S$, $f_L$), and superficial gas velocities ($U_{g,S}$, $U_{g,L}$). The subscripts S and L denote the small and large bubbles, respectively, and one should note that the TBCs are symmetrical in the model equations and we do not distinguish the small and large bubbles specially. Then we can establish three conservation equations:

$$f_S \rho_l g = \frac{f_S}{\pi/6 \cdot d_S^3} \cdot C_{DS} \frac{\pi}{4} d_S^2 \cdot \frac{1}{2} \rho_l \left( \frac{U_{g,S}}{f_S} - \frac{U_l}{1 - f_b} \right)^2 \tag{1}$$

$$f_L \rho_l g = \frac{f_L}{\pi/6 \cdot d_L^3} \cdot C_{DL} \frac{\pi}{4} d_L^2 \cdot \frac{1}{2} \rho_l \left( \frac{U_{g,L}}{f_L} - \frac{U_l}{1 - f_b} \right)^2 \tag{2}$$

$$U_{g,S} + U_{g,L} = U_g \tag{3}$$

The correlations of Grace et al (1976) and Clift et al (1978) are employed for the drag coefficients:

$$C_{Db} = C_{D0,b}(1 - f_b)^4 \tag{4}$$

$$C_{D0,b} = \frac{4}{3} \frac{g d_b}{U_T^2} \frac{\rho_l - \rho_g}{\rho_l} \tag{5}$$

$$U_T = \frac{\mu_l}{\rho_l d_b} M^{-0.149} (J - 0.857) \tag{6}$$

$$M = \frac{\mu_l^4 g \left( \rho_l - \rho_g \right)}{\rho_l^2 \sigma^3} \tag{7}$$

$$J = \begin{cases} 0.94 H^{0.757} & (2 < H \le 59.3) \\ 3.42 H^{0.441} & (H > 59.3) \end{cases} \tag{8}$$

$$H = \frac{4}{3} Eo \cdot M^{-0.149} \left( \frac{\mu_l}{\mu_{\mathrm{ref}}} \right)^{-0.14} \tag{9}$$

$$Eo = \frac{g \left( \rho_l - \rho_g \right) d_b^2}{\sigma} \tag{10}$$

Apparently the three conservation equations cannot be solved for the six structure variables, and an additional constraint, i.e., the stability condition reflecting the compromise between different dominant mechanisms, is required to close the model equations.

Then the energy dissipation is resolved for the system. The total energy fed into the system per unit mass of liquid $N_T$ can be calculated from the rate of work done by the drag force on unit mass of liquid:

$$N_{\mathrm{T}} = \frac{\left[(1-f_{\mathrm{b}})\rho_{\mathrm{l}} - \rho_{\mathrm{g}}\right] g \sum_i n_i V_i u_{\mathrm{slip},i}}{(1-f_{\mathrm{b}})\rho_{\mathrm{l}}} \approx \left[U_g - \frac{f_{\mathrm{b}} U_{\mathrm{l}}}{1-f_{\mathrm{b}}}\right] g \tag{11}$$

where $n_i$ denotes the number density of bubbles of $i$th class, $V_i$ the volume of $i$th class, and $u_{\mathrm{slip},i}$ the interstitial slip velocity. It becomes $U_g g$ for bubble columns. $N_{\mathrm{T}}$ can be resolved into three portions: $N_{\mathrm{turb}}$ denotes the energy dissipated through the energy cascade process of liquid turbulence, $N_{\mathrm{surf}}$ stands for the energy dissipation due to the slip of liquid along bubble surfaces and bubble shape oscillation, and $N_{\mathrm{break}}$ represents the energy stimulated from the interaction of turbulence eddies with bubbles and then stored temporarily as surface energy generated from bubble breakage and finally released to the liquid bulk phase during bubble coalescence. Compared to the first two parts which are directly dissipated through the interaction of bubbles and liquid and the liquid turbulence, the latter occurs at relatively larger spatial and temporal scales. Hence, the latter is regarded as a kind of mesoscale energy dissipation compared to the first two parts. The mesoscale dissipation is used to sustain the formation and evolution of mesoscale structures and serves as a buffer for energy dissipation. Each bubble class has its own $N_{\mathrm{surf}}$ and $N_{\mathrm{break}}$, whereas $N_{\mathrm{turb}}$ is a property of surrounding liquid shared by the TBCs.

The dominant mechanisms are then represented by the tendencies of energy dissipation toward extremes. $N_{\mathrm{surf}} \to \min$ represents a liquid-dominant regime, and in this case larger bubbles are likely to break into smaller bubbles and hence smaller bubbles prevail in the system. $N_{\mathrm{turb}} \to \min$ represents a gas-dominant regime to favor the coalescence of smaller bubbles and the formation of larger bubbles. The system would be governed by the compromise of these two dominant mechanisms when a dynamic equilibrium of breakup and coalescence is established, and this can be formulated as a stability condition:

$$N_{\mathrm{st}} = N_{\mathrm{surf,S}} + N_{\mathrm{surf,L}} + N_{\mathrm{turb}} \to \min \tag{12}$$

The stability condition for gas–liquid systems was formulated by Ge et al (2007) and Zhao (2006) and then extended to the DBS model by Yang et al (2007). More details of the model can be found in our previous publications (Chen et al, 2009a; Ge et al, 2007; Wang et al, 2012; Yang et al, 2007, 2010; Zhao, 2006).

A rough description of $N_{\mathrm{surf}}$ is the difference between the drag of rigid solid particles and that of bubbles:

$$N_{\mathrm{surf}} = \left[1 - \frac{C_{\mathrm{D,p}}}{C_{\mathrm{D,b}}}\right] N_{\mathrm{T}} \tag{13}$$

and $N_{\mathrm{break}}$, the surface energy generated from bubble breakup, can be calculated from the classical statistical theory of isotropic turbulence

$$N_{\mathrm{break}} = \int_{\lambda_{\min}}^{d_b} \int_0^{0.5} \frac{\omega(d_b, \lambda)}{(1 - f_b)\rho_1 + f_b \rho_g} \cdot P_b(d_b, \lambda, f_{\mathrm{BV}}) \cdot c_f \pi d_b^2 \sigma \cdot df_{\mathrm{BV}} d\lambda \tag{14}$$

where $\omega(d_b, \lambda)$ is the collisional frequency of bubbles and eddies, and $P_b(d_b, \lambda, f_{\mathrm{BV}})$ is the breakage probability (Luo and Svendsen, 1996). The stability condition is then used to close the conservation equations, i.e., Eqs. (1)–(3) as a nonlinear optimization problem. Hence, we establish an EMMS model for gas–liquid systems, i.e., the DBS model. With given superficial gas velocity $U_g$ and physical properties of gas and liquid, the six structure variables can be obtained for the system.

# 3. PHYSICAL INTERPRETATION OF MACROSCALE BEHAVIOR FROM MESOSCALE PERSPECTIVE

In the EMMS model for gas–solid systems, choking is identified as a jump between two branches of the stable solution of the EMMS model, each of which is of the same minimum value of $N_{\mathrm{st}}$ (energy consumption for suspending and transporting particles per unit mass) and corresponds to the macroscale regime of dense fluidization and dilute transport, respectively (Ge and Li, 2002). Chen et al (2012) further gives a 3D contour plot of $N_{\mathrm{st}}$ within the space of two structure parameters where the roles of the three different levels of constraints, i.e., the conservation equations, the correlation of mesoscale structures (cluster diameter), and the stability condition, can be clarified. On the other hand, the choking can also be reproduced by the integrated model of CFD and EMMS drag (Wang et al, 2007; Yang et al, 2003, 2004): when the solid inventory in the circulating fluidized bed system increases within a certain range at a given superficial gas velocity $U_g$, the simulated output solid mass flux $G_s$ is almost invariable, indicating that the entrainment capability of the system reaches the state of saturation. In this case, the voidage at the dilute top and dense bottom region does not change, and the system can be self-adapted by moving the inflection point of the S-shaped axial voidage profile. Physically this self-adapting capability on macroscale structure can be attributed to the cluster formation at mesoscales. These macroscale phenomena cannot be reproduced by traditional two-fluid CFD models with empirical drag laws unless the EMMS drag model is integrated. These studies offer a new mesoscale perspective to understand the macroscale regime transition for gas–solid systems.

Similarly, we could supply a physical interpretation of the regime transition in gas–liquid systems. A large number of experimental analysis or empirical correlations have been developed to identify the regime transition in bubble columns (Chen et al, 1994; Ruthiya et al, 2005; Ruzicka et al, 2001; Shaikh and Al-Dahhan, 2007), and a fundamental understanding of these macroscale behaviors is still lacking. With increasing the superficial gas velocity, the whole spectrum is generally grouped into three regimes. Gas holdup first increases linearly with superficial gas velocity at the homogeneous regime when the size of bubbles is uniform, and then the linearity is broken down, marking the beginning of the transitional regime. The gas holdup increases gradually to form a plateau or shoulder until it attains a maximum and then falls down. When the gas flow rate exceeds a critical value, the gas holdup increases once again and the macroscale structure of two-phase flow reaches the fully developed heterogeneous regime (Zahradnik and Fialova, 1996). In this regime, there is a population of large and highly nonuniform bubbles with a strong tendency to coalesce.

Figure 4 illustrates the EMMS model prediction of gas holdup. Clearly observed is a jump change after a gradual increase of gas holdup. The jump change corresponds to the regime shift from the homogeneous and transitional regimes to the fully developed heterogeneous regime (Camarasa et al, 1999; Zahradnik and Fialova, 1996). The two contour plots in Fig. 4 illustrate the iso surface of $N_{surf}+N_{turb}$ in the 3D space of free structure parameters at the two superficial gas velocities around the jump change of gas holdup. There are two local minima of $N_{surf}+N_{turb}$ in the contour plots. The global minimum lies in the left one and shifts to the right one when the superficial gas velocity is beyond a critical value. Theoretically there are two local minima of the same value of $N_{surf}+N_{turb}$ at a critical velocity which could be captured by higher resolution of the space of structure parameters. This is very analogous to the model calculation of gas–solid systems, which will be further analyzed in the succeeding section. The jump change of global minimum of $N_{surf}+N_{turb}$ gives rise to a dramatic variation of structure parameters and finally leads to the jump change of gas holdup, and hence the macroscale regime transition occurs. We believe that this can provide a physical understanding and interpretation of macroscale regime transition from a mesoscale perspective.

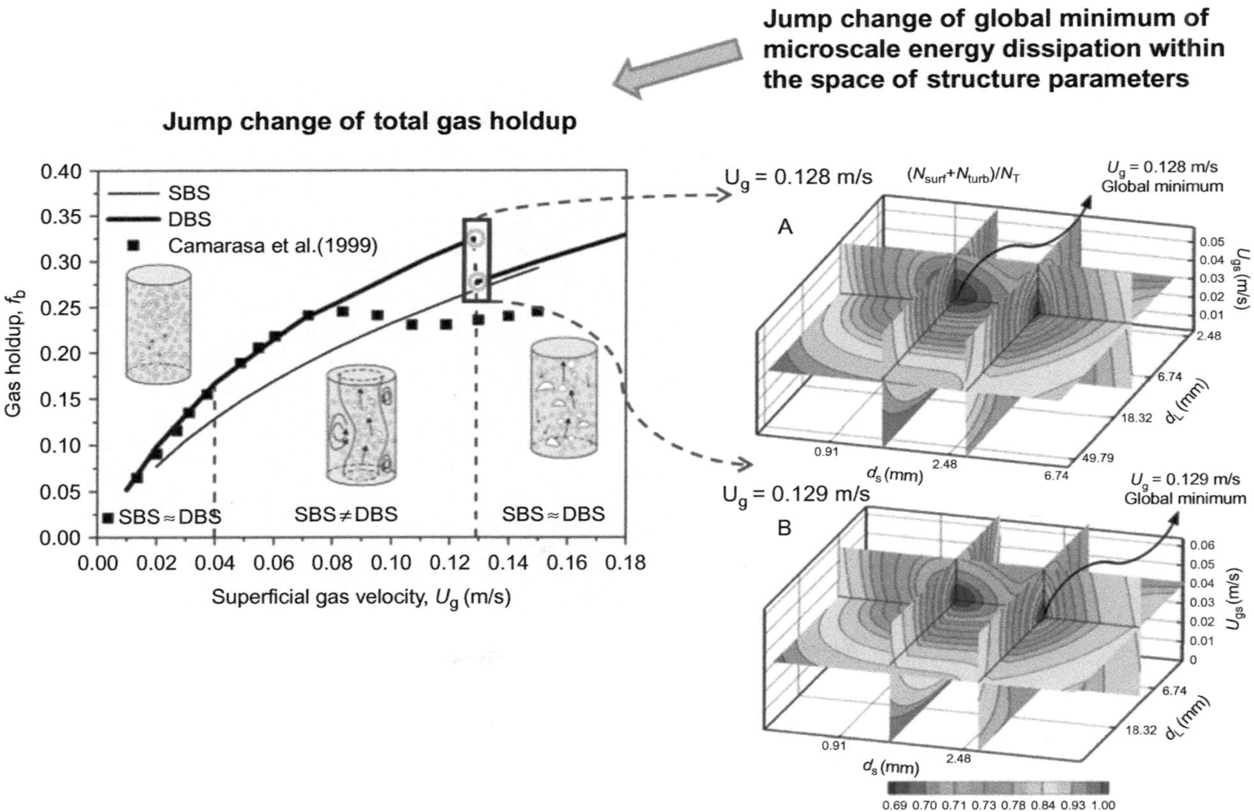


**Figure 4** Understanding of macroscale regime transition (left: gas holdup; right: contour plot of $N_{st}$ in the 3D space of structure parameters). *Adapted from Yang et al (2007) with permission from Elsevier.*

# 4. CONSOLIDATED UNDERSTANDING OF THE EMMS MODEL FOR GAS–LIQUID SYSTEMS

## 4.1 The DBS Model—An Intrinsic Model for Structure Evolution

Figure 4 also compares the single-bubble-size (SBS) model and DBS model. In the former case, only one bubble class is introduced in the resolution of structure and energy dissipation, and hence there are only three structure parameters and two conservation equations. The three energy dissipation terms also only hold for one bubble class. The SBS model calculation shows only a monotonous increase of gas holdup and there is no jump change. This can be understood since the mesoscale mechanism, i.e., the compromise between the two dominant mechanisms pertinent to the TBCs, cannot be reflected in the SBS model, and therefore, the capability of reflecting the structure heterogeneity and evolution at macroscale is lost.

We also find that the calculation of DBS model using the Tomiyama drag correlation (1998) other than the Grace correlation shows also a monotonous increase of gas holdup and cannot generate the jump change (Chen et al, 2009a). In this case, an interesting point is that the energy dissipation $N_{surf}$ for small bubbles approaches zero, implying that one of the TBCs vanishes and the DBS model is more like the SBS model. This problem is related to the rough correlation of $N_{surf}$ because it should have been a nonzero value for small bubbles according to Clift et al (1978). It can be inferred from the analysis above that it is essentially the compromising mechanism of the two energy dissipation terms for the TBCs which leads to the jump change.

One may think that the TBS or even MBS models should be more suitable than the DBS model since there is indeed a relatively wide distribution of bubble size in bubble columns. We use a simulated annealing method to solve the TBS and MBS model equations involving MBSs ($i=3$ or 10):

$$f_i \rho_l g = \frac{f_i}{\pi/6 \cdot d_i^3} \cdot C_{D,i} \frac{\pi}{4} d_i^2 \cdot \frac{1}{2} \rho_l \left( \frac{U_{g,i}}{f_i} - \frac{U_l}{1-f_b} \right)^2 \tag{15}$$

$$\sum_i U_{g,i} = U_g \tag{16}$$

$$N_{st} = \sum_{i=S,L} \left( N_{surf,i} + N_{turb,i} \right) \rightarrow \min \tag{17}$$

and find that the TBS and MBS model prediction is reduced to that of DBS model (Wang et al, 2012), as illustrated in Figs. 5 and 6. Two characteristic bubble classes are distinct in the model calculation of structure parameters even if more bubble classes are introduced. This offers another evidence that the gas–liquid flow is essentially governed by the compromise of two dominant mechanisms pertinent to the TBCs. In this sense, the DBS model can

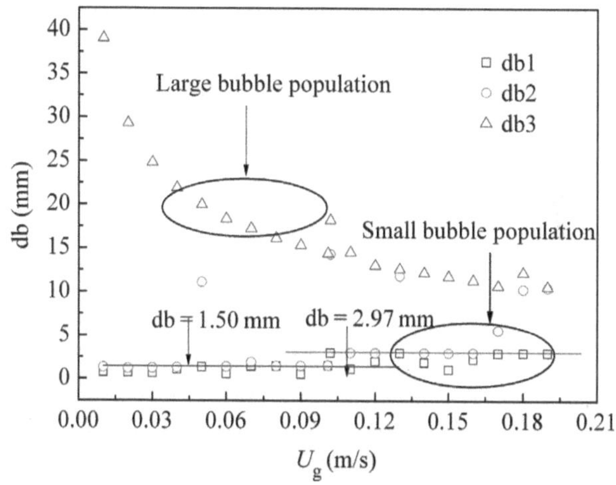


**Figure 5** Bubble size distribution in the TBS model. *Reprinted from Wang et al (2012) with permission from American Chemical Society.*

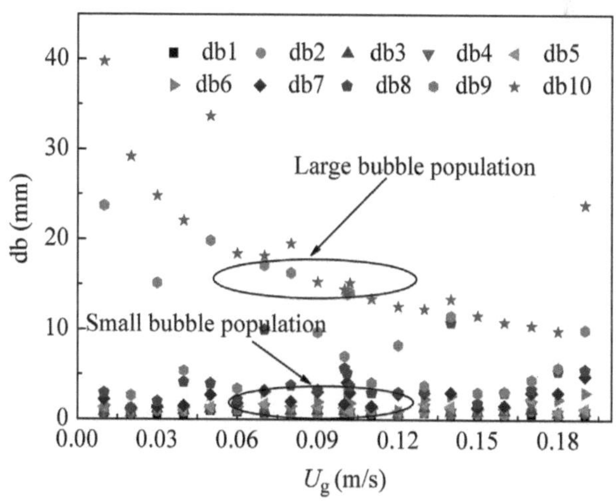


**Figure 6** Bubble size distribution in the MBS model. *Reprinted from Wang et al (2012) with permission from American Chemical Society.*

be regarded as an intrinsic model reflecting the compromising mechanisms in the system and thereby is able to describe the system evolution and gas–liquid interaction.

## 4.2 Dual Effects of Liquid Viscosity and Surface Tension

The DBS model can predict the dual effects of liquid viscosity and surface tension on regime transition. Ruzicka et al (2003) report that the moderate viscosity (3–22 mPa s) destabilizes the homogeneous regime and advances the regime transition, and the low viscosity (1–3 mPa s) could stabilize the homogeneous regime. It was attributed to the larger drag force to reduce bubble rise velocity at lower viscosity and the tendency of bubble coalescence and size polydispersity over drag reduction at higher viscosity. Interestingly, this experimental finding can be reproduced by the move of jump change in the DBS model calculation, as illustrated in Fig. 7. Slight increase in liquid viscosity from 1.0 to 3.0 mPa s moves the jump change to higher superficial gas velocities, stabilizing the homogeneous regime. By contrast, further increase from 3.5 to 8.0 mPa s of liquid viscosity turns the jump change toward a reverse direction, i.e., lower superficial gas velocities, implying the destabilization of the homogeneous regime. Figure 8 further delineates the dual effects of liquid viscosity and the critical velocity for regime transition first increases and then decreases.

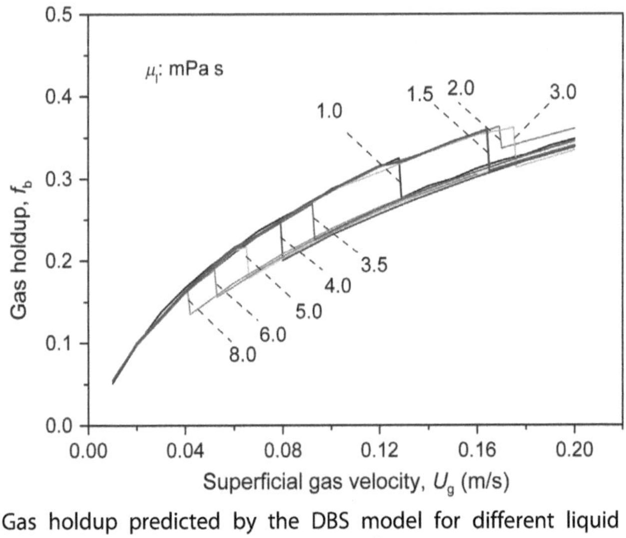


**Figure 7** Gas holdup predicted by the DBS model for different liquid viscosities. *Reprinted from Yang et al (2010) with permission from Elsevier.*

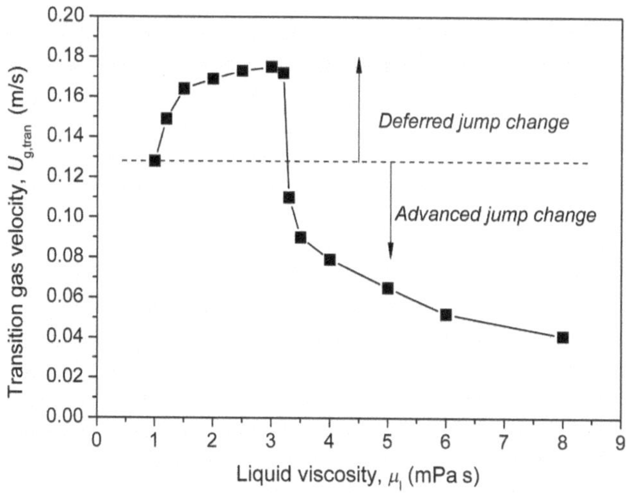


**Figure 8** Transition velocity as a function of liquid viscosity. *Reprinted from Yang et al (2010) with permission from Elsevier.*

Surface tension is also an important parameter which affects the microscale and mesoscale behaviors of bubble breakup and coalescence and then the macroscale regime transition. Surface tension appears in the correlation of drag coefficient $C_{Db}$, i.e., Eq. (10), and the breakup rate in Eq. (14) for calculating $N_{break}$ in the DBS model. Although the adsorption of surfactant on bubble surface is also another critical mesoscale problem (interfacial level), we do not consider and analyze this effect at the reactor level. We find that our model can also reasonably support the experimental findings about the dual effects of surface tension on regime transition as reported by Ruzicka et al (2008). Figure 9 indicates that the increase of surfactant concentration from 20 to 40 mN/m moves the jump change to higher superficial gas velocities, stabilizing the homogeneous regime. By contrast, further increase in surfactant concentration from 40 to 90 mN/m moves the jump change to lower superficial gas velocities and hence destabilizes the homogeneous regime. Figure 10 further delineates the dual effects of surface tension and the critical velocity for regime transition first increases and then decreases (Yang et al, 2010).

## 4.3 Intrinsic Similarity Between Gas–Liquid and Gas–Solid Systems

Since we have already established the EMMS models for gas–solid and gas–liquid systems, we are in a position to analyze the intrinsic similarity between

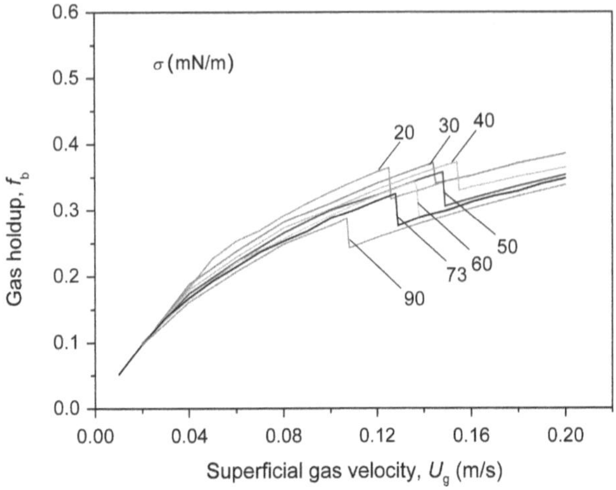


**Figure 9** Gas holdup predicted by the DBS model for different surface tension. *Reprinted from Yang et al (2010) with permission from Elsevier.*

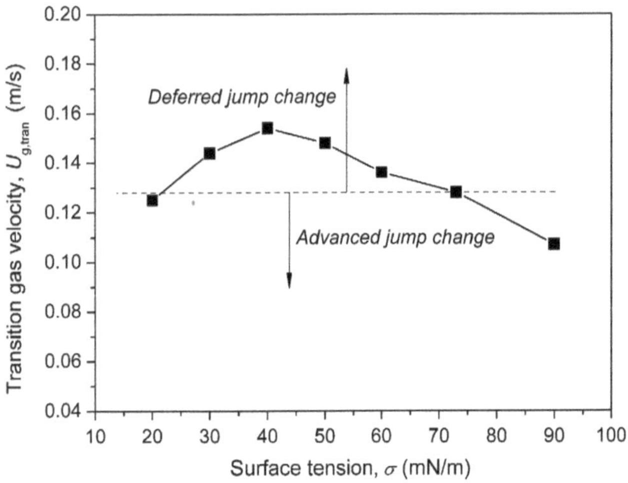


**Figure 10** Transition velocity as a function of surface tension. *Reprinted from Yang et al (2010) with permission from Elsevier.*

gas–solid and gas–liquid systems by using the EMMS approach (Chen et al, 2012). First, the model solution spaces for the two systems can be depicted through a unified numerical solution strategy, so that we are able to better understand the model hierarchies in the EMMS approach for gas–solid systems and the roles of different constraints in addition to the simplified conservation equations, i.e., the constraints relevant to mesoscale structures (cluster diameter correlation) and mesoscale mechanisms (stability condition). Second, the common characteristics of gas–solid and gas–liquid systems can be found by comparing the model solutions for the two systems, though the structural parameters and stability criteria are specific in each system. In both cases, there are two local minima of $N_{st}$ simultaneously existing in the solution space of structure parameters. The two local minima reflect the compromise of two different dominant mechanisms and can share an equal value at a critical condition of operating conditions. The position of global minimum of $N_{st}$ in the space of structure parameters can migrate from one to the other when the operating condition is around a critical value. As a result, structure parameters such as voidage or gas holdup exhibit a jump change due to this migration, leading to the dramatic structure variation and hence regime transition of these systems. This demonstrates that it is the stability condition that drives the macroscale structure variation and system evolution, which may be the intrinsic similarity of gas–solid and gas–liquid systems.

Figure 11 depicts the corresponding relationship between the mesoscale mechanism (stability condition) and regime transition (voidage curve) for gas–solid systems. A common feature is the bifurcation caused by the jump change. In Fig. 11, the two local minima of $N_{st}$ correspond to two different branches, and one of them would be a pseudo solution at certain state. The branch 1-1 corresponds to one of the local minimum of $N_{st}$ when the voidage of dense phase $\epsilon_c$ is equal to the voidage at minimum fluidization $\epsilon_{mf}$. The corresponding branch $1'$-$1'$ in the voidage curve represents the dense fluidization state. The branch 2-2 corresponds to the local minimum of $N_{st}$ when $\epsilon_c$ has a larger value and the corresponding branch $2'$-$2'$ represents the dilute transport state. The stability condition requires the global minimum of $N_{st}$ and shifts the real solution from branch 1-1 to branch 2-2, leading to a corresponding jump of voidage from branch $1'$-$1'$ to $2'$-$2'$.

Figure 12 depicts the similar mechanism for gas–liquid systems. The solid line denotes the real solution and the broken line denotes the pseudo solution. The branch 1-1 represents one of the local minimum of $N_{st}$ and the corresponding branch $1'$-$1'$ the homogeneous regime. The branch 2-2 represents another local minimum of $N_{st}$ and the corresponding branch $2'$-$2'$ the heterogeneous regime. The voidage or gas holdup is a function of structure parameters. The relationship and the bifurcations of stability condition and structure parameters coexist in gas–solid and gas–liquid systems. This implies that the stability condition, though system-dependent in its formulation, is the driving force for structure evolution.

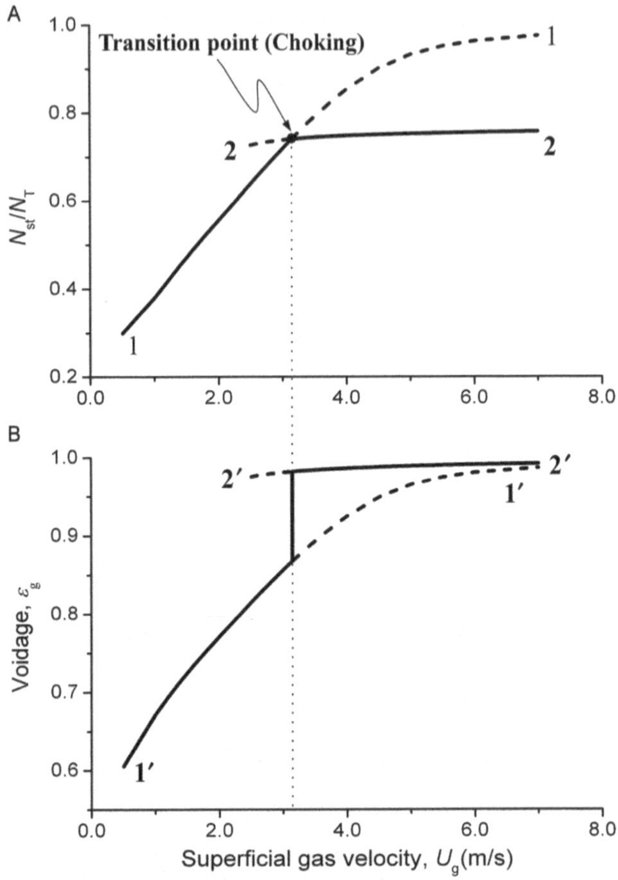


**Figure 11** Relationship between structure bifurcation and stability condition in gas–solid systems. *Reprinted from Chen et al (2012) with permission from Elsevier.*

## 4.4 Coupling with Mass Transfer and Reaction

A number of reaction engineering models in literature have already utilized the concepts of TBC or single-bubble-class (SBC) together with the axial dispersion models to couple the hydrodynamics, mass transfer, and reaction in bubble column reactors. While some researchers claim that there is less difference in the SBC and TBC model predictions, others believe that the TBC model prediction is in better agreement with experiments. We find that the difference is actually relevant to the submodels for hydrodynamics, mass transfer, and reaction kinetic as well as gas contraction, and in particular the gas holdup model and whether the system is limited by reaction or mass transfer. Then a new reactor model is developed to replace the empirical

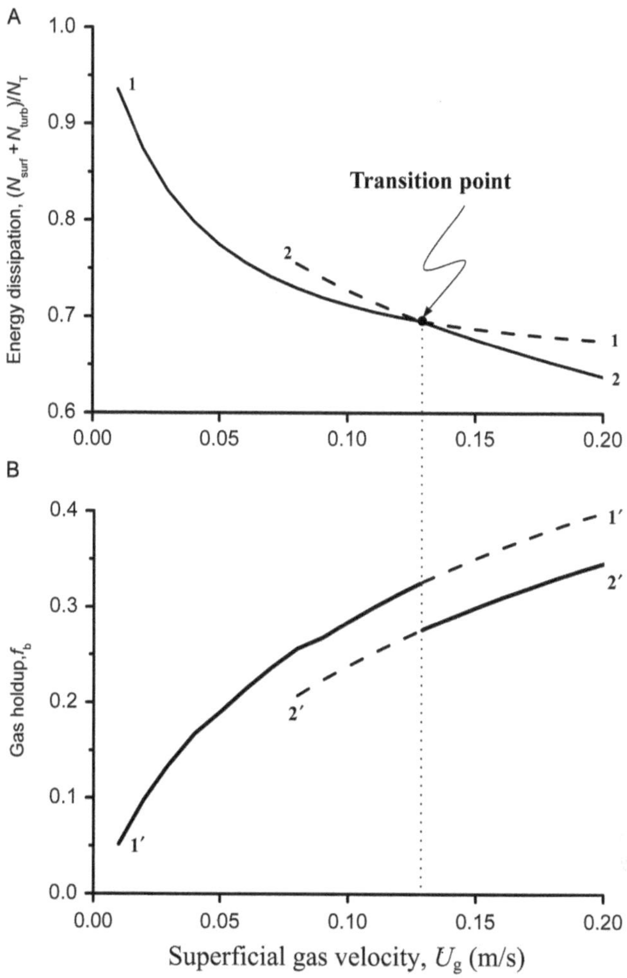


**Figure 12** Relationship between structure bifurcation and stability condition in gas–liquid systems. *Reprinted from Chen et al (2012) with permission from Elsevier.*

correlations for gas holdup by using the DBS model of the EMMS approach. The empirical contraction factor is eliminated by introducing a gas state equation and overall mass balance. The exchange between small and large bubbles is considered in the species transport equations, and the model frameworks are illustrated in Figs. 13 and 14. The terms in Fig. 13 represent the different transport mechanisms in the transport equations of species concentration for liquid and the TBCs. Figure 14 shows the coupling of hydrodynamic model with species transport equations and gas contraction model. More discussions can be found in Jiang et al (2015).

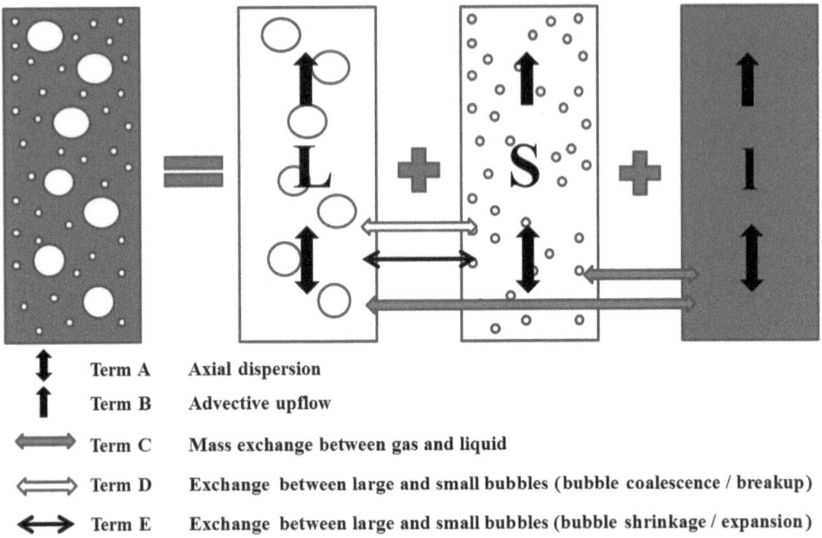


**Figure 13** A reactor model based on the EMMS approach. *Reprinted from Jiang et al (2015) with permission from Elsevier.*

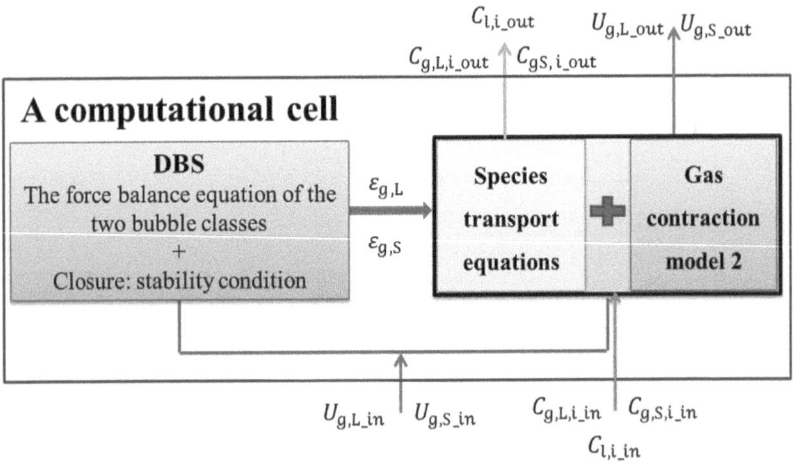


**Figure 14** Schematic diagram of the new reactor model framework. *Reprinted from Jiang et al (2015) with permission from Elsevier.*

The model calculation for Fischer–Tropsch synthesis indicates that the exchange between small and large bubbles can be neglected for reaction-limited systems but becomes important for mass transfer limited systems. The gas holdup and syngas conversion predicted by the new reactor model

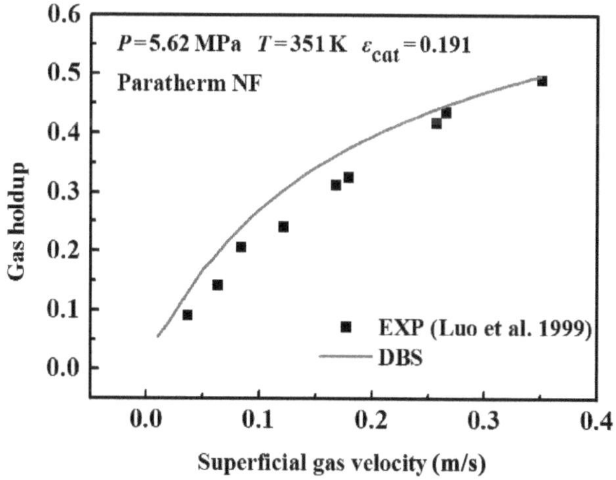


**Figure 15** Comparison of model calculation of gas holdup and experiments. *Reprinted from Jiang et al (2015) with permission from Elsevier.*

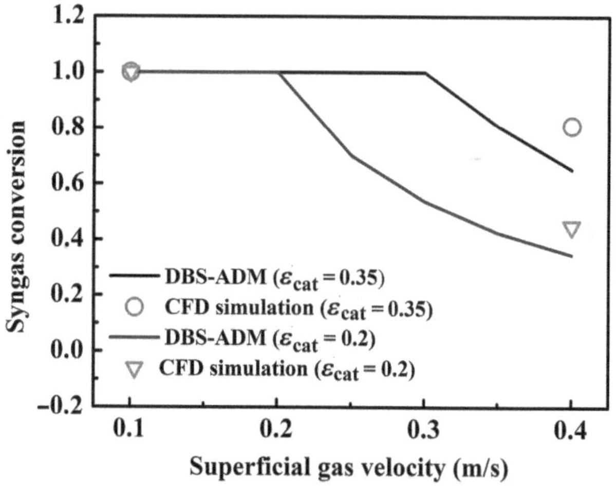


**Figure 16** Comparison of syngas conversion for different catalyst volume fractions. *Reprinted from Jiang et al (2015) with permission from Elsevier.*

are in reasonable agreement with experiments and CFD simulation (Figs. 15 and 16). Here, we demonstrate that the compromise of the two dominant mechanisms reflected by the TBCs also plays important roles in coupling hydrodynamics with mass transfer and reactions.

## 4.5 Further Explanation

It should be pointed out that the DBS model or the EMMS model for gas–liquid systems is only a zero-dimensional conceptual model for now. It does not consider the boundary conditions such as gas distributor, column geometry, wall condition, or the microscale or mesoscale behavior at the interfacial level. Hence, it may not be appropriate to expect close consistency between specific experimental data and the model calculation of structure parameters. For example, the shoulder or plateau on the gas holdup curve may be a gradual change rather than a jump change (Fig. 4), depending on the type of gas distributor and liquid properties. In fact, the types of gas distributor or sparger also affect the structure of gas–liquid flow, gas holdup, and regime formation as reported in literature (Camarasa et al, 1999; Ruzicka et al, 2001, Zahradnik and Fialova, 1996). More uniform gas aeration at inlet inhibits the bubble coalescence, and the homogeneous regime can be held to higher superficial gas velocities and gas holdup. In this case, the transition point may become steeper (Mudde et al, 2009). We may suppose that the gradual change may evolve into a jump change if a perfectly uniform aeration is imposed. In practice, the jump change is dampened to a gradual or smooth one because there is always some disturbance arising from the nonuniform gas aeration. The disturbance and its amplification may affect the bubble coalescence or breakup at mesoscales. In this sense, the DBS model reflects some inherent or intrinsic characteristics of the system evolution. Integration of the DBS model into CFD simulation should provide a more complete solution to these practical problems.

# 5. STABILITY-CONSTRAINED MULTIFLUID CFD MODEL

## 5.1 Stability Condition as a Closure Law for CFD Models

CFD simulation has been widely used to predict the multiphase flow in fluidized bed or bubble column reactors. There are always some issues on the accuracy and reliability of CFD simulation, especially for higher superficial gas velocities and higher gas holdup at which industrial-scale reactors generally operate, though a number of simulations have been reported to be able to achieve good agreement with lab-scale experiments. This difficulty is seemingly relevant to a number of factors, i.e., closure models for stress–strain relationship and interphase forces, wall and boundary conditions, numerical methods, grid resolution, etc. For example, CFD simulation of bubble columns has been reported to be sensitive to drag force, lift, turbulence models, discretization schemes, boundary conditions, and grid resolution (Monahan et al, 2005). Current CFD models are strictly derived by some time- or ensemble-averaged conservation equations for mass and momentum, and the unresolved mesoscale structures are however overlooked and hidden in the closure laws or constitutive equations. These structures cannot be resolved unless the effects of mesoscale structure and phenomena or mesoscale mechanisms are integrated by some seamless or simplified methods. Empirical correlations for the interaction of particles, bubbles, or droplets with the surrounding continuous fluid are usually employed in current CFD model framework. The mesoscale structures like particle aggregation in gas–solid systems or mesoscale mechanisms, viz., the compromise of two dominant mechanisms relevant to the TBC, are seldom taken into account in CFD modeling. But the mesoscale mechanisms beyond transport equations may be critical.

An ideal solution is to directly integrate the stability condition reflecting mesoscale mechanisms into Eulerian–Eulerian two-fluid models through the following conceptual model:

$$\frac{\partial(\varepsilon_k\rho_k)}{\partial t} + \nabla\cdot(\varepsilon_k\rho_k\mathbf{u}_k) = \Gamma_k \tag{18}$$

$$\frac{\partial(\varepsilon_k\rho_k\mathbf{u}_k)}{\partial t} + \nabla\cdot(\varepsilon_k\rho_k\mathbf{u}_k\mathbf{u}_k) = -\varepsilon_k\nabla P + \mu_{k,\,\mathrm{eff}}\varepsilon_k\left[\nabla\mathbf{u}_k + (\nabla\mathbf{u}_k)^T\right] + \varepsilon_k\rho_k\mathbf{g} + \mathbf{F}_\mathrm{D} \tag{19}$$

both of which are subject to the stability condition:

$$N_\mathrm{st} \rightarrow \min \tag{20}$$

Here, the stability condition reflects the compromise of dominant mechanisms at mesoscales and can be used to describe the direction of structure evolution of the system. By contrast, the transport equations for mass and momentum, i.e., Eqs. (18) and (19), describe the dynamics of structure evolution. The relationship of these two approaches is more or less like the thermodynamics and chemical kinetics. Compared to the traditional CFD models, this integrated approach can be termed the stability-constrained multifluid (SCMF) CFD model.

However, the direct integration is technically difficult. $N_\mathrm{st}$ is generally expressed as a function of structure parameters which is complicated to formulate and to be incorporated into the current two-fluid models. Moreover, the nonlinear optimization problem has to be solved for each time and space step of the averaged Navier–Stokes equations, which cannot be affordable for current computational resources. In literature, some two-average approach for conservation equations can be established to separate the particle-scale and mesoscale structure parameters and phase interactions (Zhang and VanderHeyden, 2002), and the mesoscopic model could be derived and formulated to separate the mesoscale closure, yet the direct solution is usually intractable (Fox, 2012).

A simplified way is to decouple the two approaches: first, the stability condition serves as the close law for the simplified algebraic conservation equations, as described by the EMMS model for gas–liquid and gas–solid systems. The nonlinear optimization problem can therefore be solved to obtain the global or local structure parameters which are then be used to derive the closure law or correlations for the drag, bubble-induced turbulence and even the correction factors for the kernel functions of bubble coalescence and breakup for PBEs.

This idea is first practiced by Yang et al (2003, 2004) deriving an EMMS drag model for gas–solid systems. For gas–liquid systems, the EMMS drag model is formulated as the correlations for the ratio of drag coefficient to bubble diameter $C_D/d_b$ (Chen et al, 2009b; Xiao et al, 2013; Xu et al, 2015; Yang et al, 2011). We notice that it is this lumped parameter other than the drag coefficient $C_D$ or bubble diameter $d_b$ which determines the averaged drag coefficient $\beta$ in CFD simulation. Hence, once we employ this drag correlations derived from the EMMS model, we do not need to specify the bubble diameter or correction factors for drag coefficient. These parameters may have to be adjusted in the CFD simulation without using the EMMS drag.

## 5.2 Simulation of Bubble Column Reactors

There are three types of closure models in CFD simulation of gas–liquid flow in bubble columns, i.e., drag force, bubble-induced turbulence, and kernel functions of bubble breakup and coalescence. We will show how we utilize the EMMS approach to derive new models and integrate them into CFD simulation.

Figure 17 illustrates the total gas holdup in a bubble column as a function of superficial gas velocity for two different bubble column systems. The traditional drag models either underestimate or overestimate the total gas holdup, showing a monotonous increase of gas holdup. By contrast, the CFD simulation with the EMMS drag is in conformity with experiments. We also depict the calculation of EMMS (DBS) model which is regarded

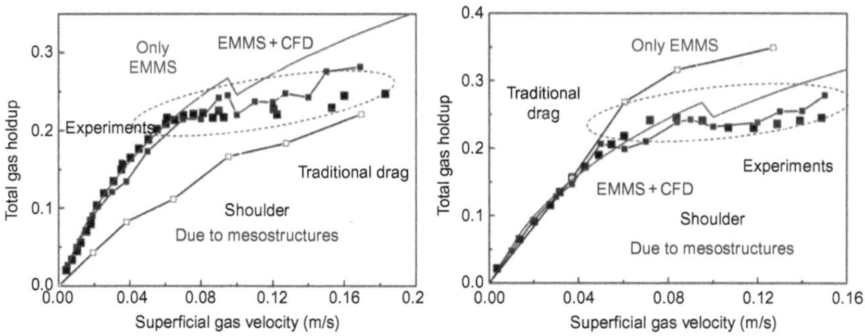


**Figure 17** Comparison of experiments of Hills (1974, left figure) and Camarasa et al (1999, right figure), EMMS model calculation, CFD using EMMS drag, and CFD using traditional drag model. *Adapted from Xiao et al (2013) with permission from Elsevier.*

as a zero-dimensional conceptual model in the preceding section. It is interesting to see that the jump change generated by the EMMS model itself is smoothed down in the integrated CFD-EMMS simulation and the calculation is in more accordance with experiments for bubble columns of different gas distributors.

The radial profile of gas holdup is illustrated in Fig. 18. The EMMS drag shows much better performance than Schiller–Naumann or Ishii–Zuber models. In general, the Schiller–Naumann correlation is recommended for particles, whereas the Ishii–Zuber correlation for bubbles or droplets in commercial CFD package. Olmos et al (2003) and Yang et al (2011) report that the correction factors have to be used for current drag correlations in order to acquire the accurate prediction. Actually the suitable correction factor may vary with operating conditions. By using the EMMS drag model for the lumped parameter $C_D/d_b$, Yang et al (2011) and Xiao et al (2013) demonstrate that such correction factors are not required and the new drag model is able to achieve good agreement with experiments at different superficial gas velocities. The difference in the radial profile of normalized liquid axial velocity among different model predictions is small though the EMMS drag model is slightly better. The streamline predicted by the EMMS drag indicates that the flow structure is relatively homogeneous at lower superficial gas velocity, and the liquid vortices can be observed at higher superficial gas velocity. In the latter case, the bubble column can be separated into several local circulation cells which promote the local mixing. This prediction conforms to the description of coherent structure and multiple circulation cells of Groen et al (1996) and Joshi et al (2002). Figure 19 illustrates the dynamic structure (contour plot of gas holdup)

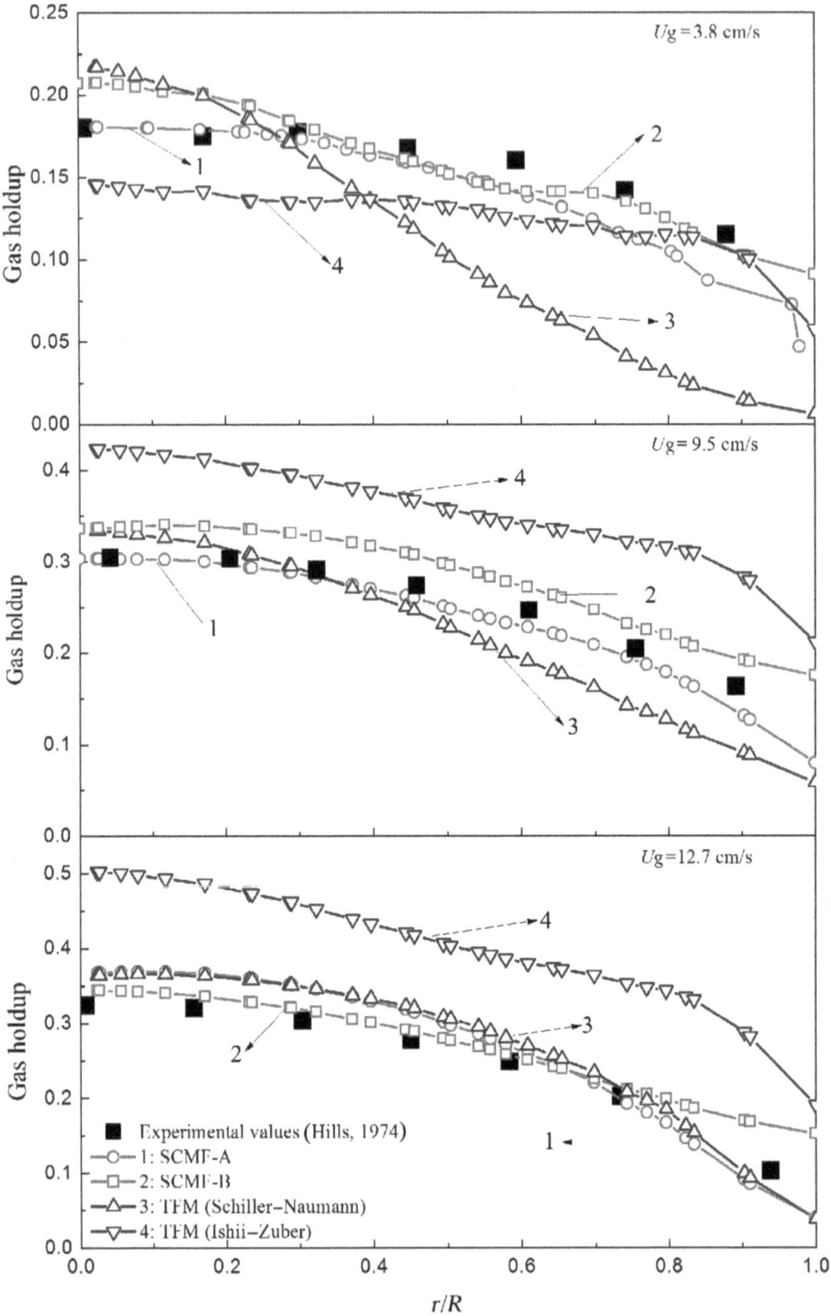


**Figure 18** Radial profile of gas holdup (SCMF-A and SCMF-B denote two types of EMMS drag, see Xiao et al, 2013). *Adapted from Xiao et al (2013) with permission from Elsevier.*

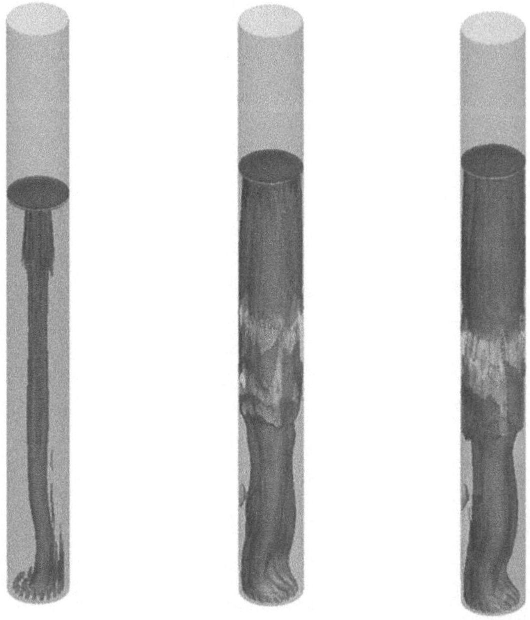


**Figure 19** Dynamic flow structure predicted by the CFD simulation using the EMMS drag.

predicted by the CFD simulation using the EMMS drag model. The flow structure at lower gas flow rate is relatively homogeneous and the gas rises up directly. By contrast, there is strong interaction between the different layers of gas along the radial and axial directions at higher flow rates.

The simulation of internal-loop airlift reactor is more complicated in that complex mesoscales may exist in different sections of the reactor separated by a draft tube. Commercial CFD package using the traditional Schiller–Naumann drag model predicts a lower gas holdup in the riser and the absence of gas in the downcomer. We extend the EMMS drag model into the internal-loop airlift reactor systems (Xu et al, 2015). The prediction illustrated in Figs. 20 and 21 demonstrate that the gas exists in both the riser and downcomer when the riser superficial gas velocity increases, and the EMMS (DBS) drag is in good agreement with experiments. By using chaotic analysis of pressure fluctuation signals, Fu et al (2007) report that bubbles in the downcomer may be absent, remain stationary, or be circulated into the riser, depending on the reactor geometry, bubble size distribution, and operating conditions. Further study about the drag in conjunction with two-phase

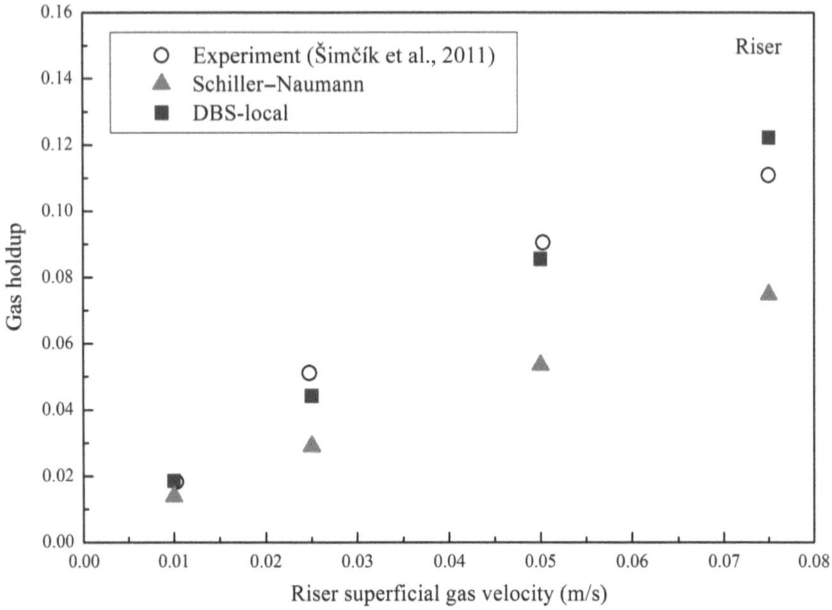


**Figure 20** Gas holdup in the riser. *Reprinted from Xu et al (2015) with permission from Elsevier.*

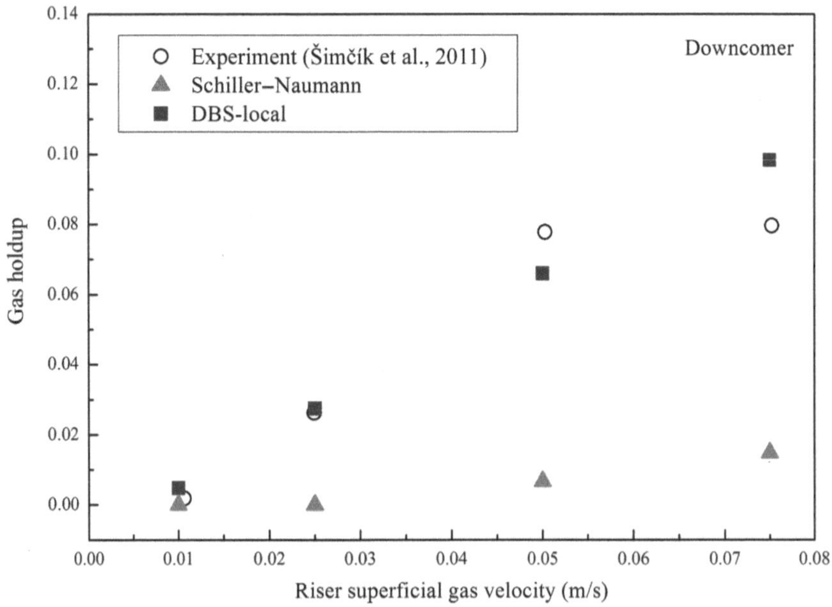


**Figure 21** Gas holdup in the downcomer. *Reprinted from Xu et al (2015) with permission from Elsevier.*

turbulence and population balance model is required to reflect these experimental findings.

CFD simulation of gas–liquid systems is more complicated than that of gas–solid systems because the turbulence interaction between two phases and the kernel functions of bubble coalescence and breakup in PBE also affect the simulation. While some researchers believe that the bubble-induced turbulence is physically important by experimental analysis (Cui and Fan, 2004), others find that the influence of the correlation for bubble-induced turbulence is marginal in their CFD simulation (Laborde-Boutet et al, 2009). The issue is actually because the bubble-induced turbulence is also dependent on the drag model used in CFD simulation. We integrate the EMMS drag into the transport equations for turbulent kinetic energy and energy dissipation and demonstrate that the EMMS drag is necessary to correctly predict the local gas holdup and axial liquid velocity over a relatively wider range of superficial gas velocities. RNG $k$–$\varepsilon$ model is recommended in literature but we find that it does not have the absolute advantage since the interphase between phases is critical in the turbulence model for the CFD simulation of gas–liquid flow. When Re-Normalisation Group (RNG) $k$–$\varepsilon$ model is used along with the Schiller–Naumann drag model, the simulation could not generate good results. In this sense, the drag model is the primary factor compared to the single-phase turbulence models for CFD simulation of gas–liquid flow (Xu et al, 2015).

The kernel functions of bubble breakup and coalescence are required to supply the source term in PBE to predict the bubble size distribution. These kernel functions are generally some phenomenological models together with some derivation using statistical analysis and classical theory of isotropic turbulence. PBE has been coupled with CFD in literature and the predicted bubble size agrees well with the experiments at low superficial gas velocity less than 0.01 m/s or small gas volume fraction. The bubble size is usually overpredicted at relatively higher superficial gas velocity or gas volume fraction because the coalescence rate is always overpredicted. Hence, correction factors are used by some studies, either as a constant or as a function of gas holdup or Stokes number. However, these correction factors are empirical and only work well for limited operating conditions or specified kernel functions.

With the EMMS model, we are able to obtain the energy dissipation $N_{break}$, that is, the energy generated during bubble breakage and finally released to the liquid bulk phase during bubble coalescence. This mesoscale energy dissipation can be used to determine the correction factor in the

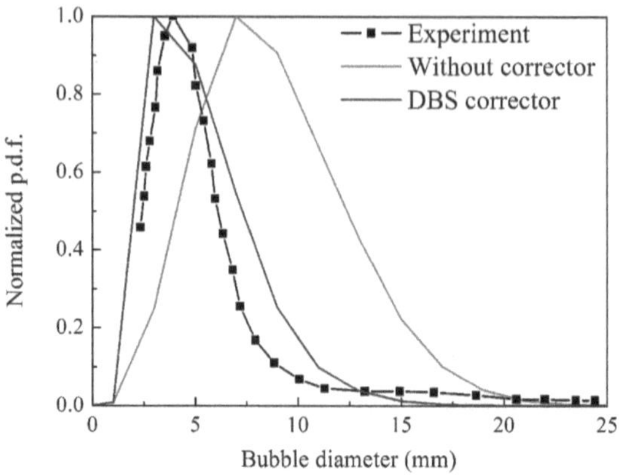


**Figure 22** CFD–PBE simulation with the EMMS correction factor for coalescence rate.

coalescence kernel function. Figure 22 indicates that the CFD–PBE prediction using the new corrector model is in reasonable agreement with experiments, whereas the simulation without corrector overpredicts the bubble size. Although the above simulation needs further study and validation, it shows the reasonableness and prospect of the EMMS approach in supplying the closure law for CFD simulation.

## 6. CONCLUSIONS AND FUTURE ISSUES

Mesoscale transport phenomena and mechanisms indeed offer new angles to achieve a more fundamental understanding on the traditional problems in chemical engineering:

**(1)** Despite the complexity of mesoscale structures and mechanisms, we highlight a heuristic mesoscale modeling approach starting from a zero-dimensional conceptual model (EMMS model) and ending at the SCMF CFD model. While the stability condition determines the direction of structure evolution of the system, the stability-constrained CFD model further describes the dynamics of structure evolution. The relationship between the two approaches is more or less like that of thermodynamics and chemical kinetics.

**(2)** By resolving the structures and dominant mechanisms, it is possible to establish a stability condition reflecting the compromise between different dominant mechanisms for multiphase reaction systems, and the stability condition supplies a mesoscale constraint in addition to mass and momentum conservation equations. The calculation of EMMS model demonstrates that it is the stability condition that drives the structure variation and system evolution at macroscales, which may be the intrinsic similarity of gas–solid and gas–liquid systems.

**(3)** We establish the DBS model, an extended EMMS approach for gas–liquid systems. The total energy dissipation in gas–liquid systems can be resolved into three parts, i.e., $N_{surf}$, $N_{turb}$, and $N_{break}$. The former two are directly dissipated at microscales. The third reflects a kind of mesoscale energy dissipation, i.e., the energy stimulated from the interaction of turbulence eddies with bubbles and then stored temporarily as surface energy generated from bubble breakage and finally released to the liquid bulk phase during bubble coalescence. The mesoscale dissipation is used to sustain the formation and evolution of mesoscale structures and serves as a buffer for energy dissipation.

**(4)** Stability condition is formulated as the minimization of energy directly dissipated at microscales ($N_{surf}+N_{turb} \to \min$) or equivalently the maximization of energy consumption at mesoscales ($N_{break} \to \max$). It reflects the compromise between two dominant mechanisms: $N_{surf} \to \min$ represents a liquid-dominant regime and in this case larger bubbles are likely to break into those smaller bubbles which prevail in the system. $N_{turb} \to \min$ represents a gas-dominant regime to favor the coalescence of smaller bubbles and the existence of larger bubbles. Stability condition in the DBS model is used to close the simplified conservation equations, and hence the structure parameters can be obtained.

**(5)** Stability condition provides a mesoscale perspective to understand the regime transition at macroscale. The transition can be interpreted as the jump change of the global minimum point of microscale energy dissipation in the 3D space of structure parameters.

**(6)** Theoretically stability condition may offer closure laws for CFD simulation, leading to SCMF CFD model. While the direct integration is difficult, we propose various simplified approaches to derive the closure models for drag, bubble-induced turbulence, and the correction factors for the kernel functions of bubble coalescence and breakup for PBEs. The SCMF CFD model shows much advantage over current closure models.

Despite these explorations and knowledge on mesoscales, mesoscale transport phenomena and mechanisms would continue to be a challenge for multiphase reaction systems. For example, there are currently no general guidelines to identify the dominant mechanisms and formulate the stability condition for different specific systems. Although we establish mesoscale models for gas–liquid and gas–solid systems, the presence of particles in gas–liquid–solid systems and the internals used to intensify the mixing in multiphase reactors have great influence on mesoscale structures, which requires further investigation. The adsorption of surfactant on bubble or droplet surface constitutes another mesoscale problem at the interfacial level, and it is necessary to bridge this mesoscale problem and the mesoscale at the reactor level before we could completely understand the complexity of multiphase reaction systems.

## ACKNOWLEDGMENTS

The author gained valuable insight into mesoscale modeling approach for multiphase systems through extensive collaboration with the colleagues and students in the EMMS group (http://www.emms.cn). The author also wishes to thank the long-term support from the National Natural Science Foundation of China (21222603, 91434121), Ministry of Science and Technology of China (2013BAC12B01), and Strategic Priority Research Program of Chinese Academy of Sciences (XDA07080301).

## REFERENCES

Camarasa E, Vial C, Poncin S, et al: Influence of coalescence behaviour of the liquid and of gas sparging on hydrodynamics and bubble characteristics in a bubble column, *Chem Eng Proc* 38:329–344, 1999.

Chen RC, Reese J, Fan LS: Flow structure in a three-dimensional bubble column and three-phase fluidized bed, *AIChE J* 40:1093–1104, 1994.

Chen J, Yang N, Ge W, et al: Modeling of regime transition in bubble columns with stability condition, *Ind Eng Chem Res* 48:290–301, 2009a.

Chen J, Yang N, Ge W, et al: Computational fluid dynamics simulation of regime transition in bubble columns incorporating the dual-bubble-size model, *Ind Eng Chem Res* 48:8172–8179, 2009b.

Chen J, Yang N, Ge W, et al: Stability-driven structure evolution: exploring the intrinsic similarity between gas-solid and gas-liquid systems, *Chinese J Chem Eng* 20(1):167–177, 2012.

Clift R, Grace JR, Weber ME: *Bubbles, drops, and particles,* London, 1978, Academic Press.

Cui Z, Fan LS: Turbulence energy distributions in bubbling gas-liquid and gas-liquid-solid flow systems, *Chem Eng Sci* 59:1755–1766, 2004.

De Swart JWA, van Vliet RE, Krishna R: Size, structure and dynamics of "large" bubbles in a two-dimensional slurry bubble column, *Chem Eng Sci* 51:4619–4629, 1996.

Ertl G: Oscillatory kinetics and spatio-temporal self-organization in reactions at solid surfaces, *Science* 254:1750–1755, 1991.

Fox RO: Large-eddy-simulation tools for multiphase flow, *Annu Rev Fluid Mech* 44:47–76, 2012.

Fu CC, Fan LS, Wu WT: Flow regime transitions in an internal-loop airlift reactor, *Chem Eng Technol* 30:1077–1082, 2007.

Ge W, Li J: Physical mapping of fluidization regimes—the EMMS approach, *Chem Eng Sci* 57:3993–4004, 2002.

Ge W, Chen F, Gao J, et al: Analytical multi-scale method for multi-phase complex systems in process engineering—bridging reductionism and holism, *Chem Eng Sci* 62:3346–3377, 2007.

Grace JR, Wairegi T, Nguyen TH: Shapes and velocities of single drops and bubbles moving freely through immiscible liquids, *Trans IChemE* 54:167–173, 1976.

Groen JS, Oldeman RGC, Mudde RF, et al: Coherent structures and axial dispersion in bubble column reactors, *Chem Eng Sci* 51:2511–2520, 1996.

Hills JH: Radial non-uniformity of velocity and voidage in a bubble column, *Trans Inst Chem Eng* 52:1–9, 1974.

Jiang X, Yang N, Zhu J, et al: On the single and two-bubble class models for bubble column reactors, *Chem Eng Sci* 123:514–526, 2015.

Jiradilok V, Gidaspow D, Damronglerd S, et al: Kinetic theory based CFD simulation of turbulent fluidization of FCC particles in a riser, *Chem Eng Sci* 61:5544–5559, 2006.

Joshi JB, Vitankar VS, Kulkarni AA, et al: Coherent flow structures in bubble column reactors, *Chem Eng Sci* 57:3157–3183, 2002.

Laborde-Boutet C, Larachi F, Dromard N, et al: CFD simulation of bubble column flows: investigations on turbulence models in RANS approach, *Chem Eng Sci* 64:4399–4413, 2009.

Li J: Approaching virtual process engineering with exploring mesoscience, *Chem Eng J* 278:541–555, 2015.

Li J, Huang W: *Towards mesoscience: the principle of compromise in competition,* Berlin, 2014, Springer.

Li J, Ge W, Wang W, et al: *From multiscale methodology to meso-science—a chemical engineering perspective,* Berlin, 2013, Springer.

Luo H, Svendsen HF: Theoretical model for drop and bubble breakup in turbulent dispersions, *AIChE J* 42(5):1225–1233, 1996.

Luo XK, Lee DJ, Lau R, et al: Maximum stable bubble size and gas holdup in high-pressure slurry bubble columns, *AIChE J* 45:665–680, 1999.

Monahan SM, Vitankar VS, Fox RO: CFD predictions for flow-regime transitions in bubble columns, *AIChE J* 51:1897–1923, 2005.

Mudde RF, Harteveld WK, van den Akker HEA: Uniform flow in bubble columns, *Ind Eng Chem Res* 48(1):148–158, 2009.

Olmos E, Gentric C, Midoux N: Numerical description of flow regime transitions in bubble column reactors by a multiple gas phase model, *Chem Eng Sci* 58:2113–2121, 2003.

Ruthiya KC, Chilekar VP, Warnier MJF, et al: Detecting regime transitions in slurry bubble columns using pressure time series, *AIChE J* 51:1951–1965, 2005.

Ruzicka MC, Zahradnik J, Drahos J, Thomas NH: Homogeneous–heterogeneous regime transition in bubble columns, *Chem Eng Sci* 56:4609–4626, 2001.

Ruzicka MC, Drahos J, Mena PC, Teixeira JA: Effect of viscosity on homogeneous-heterogeneous flow regime transition in bubble columns, *Chem Eng J* 96:15–22, 2003.

Ruzicka MC, Vecer MM, Orvalho S, et al: Effect of surfactant on homogeneous regime stability in bubble column, *Chem Eng Sci* 63:951–967, 2008.

Shah MT, Utikar RP, Tade MO, et al: Hydrodynamics of an FCC riser using energy minimization multiscale drag model, *Chem Eng J* 168:812–821, 2011.

Shaikh A, Al-Dahhan MH: A review on flow regime transition in bubble columns, *Int J Chem React Eng* 57:1–68, 2007.

Šimčík M, Mota A, Ruzicka MC, et al: CFD simulation and experimental measurement of gas holdup and liquid interstitial velocity in internal loop airlift reactor, *Chem Eng Sci* 66:3268–3279, 2011.

Sun Q, Wang G, Hu K: Some open problems in granular matter mechanics, *Prog Nat Sci* 19:523–529, 2009.

Tomiyama A: Struggle with computational fluid dynamics, *Multiphase Sci Technol* 10:369–405, 1998.
Wang W, Li J: Simulation of gas-solid two-phase flow by a multi-scale CFD approach: extension of the EMMS model to the sub-grid scale level, *Chem Eng Sci* 62:208–231, 2007.
Wang W, Lu B, Li J: Choking and flow regime transitions: simulation by a multi-scale CFD approach, *Chem Eng Sci* 62(3):814–819, 2007.
Wang Y, Xiao Q, Yang N, et al: In-depth exploration of the dual-bubble-size model for bubble columns, *Ind Eng Chem Res* 51(4):2077–2083, 2012.
Xiao Q, Yang N, Li J: Stability-constrained multi-fluid CFD models for gas–liquid flow in bubble columns, *Chem Eng Sci* 100:279–292, 2013.
Xu T, Jiang X, Yang N, et al: CFD simulation of internal-loop airlift reactor using EMMS drag mode, *Particuology* 19:24–132, 2015.
Yang N, Wang W, Ge W, et al: CFD simulation of concurrent-up gas-solid flow in circulating fluidized beds with structure-dependent drag coefficient, *Chem Eng J* 96:71–80, 2003.
Yang N, Wang W, Ge W, et al: Simulation of heterogeneous structure in a circulating fluidized-bed riser by combining the two-fluid model with the EMMS approach, *Ind Eng Chem Res* 43:5548–5561, 2004.
Yang N, Chen J, Zhao H, et al: Explorations on the multi-scale flow structure and stability condition in bubble columns, *Chem Eng Sci* 62:6978–6991, 2007.
Yang N, Chen J, Ge W, et al: A conceptual model for analyzing the stability condition and regime transition in bubble columns, *Chem Eng Sci* 65(1):517–526, 2010.
Yang N, Wu Z, Chen J, et al: Multi-scale analysis of gas–liquid interaction and CFD simulation of gas–liquid flow in bubble columns, *Chem Eng Sci* 66(14):3212–3222, 2011.
Zahradnik J, Fialova M: The effect of bubbling regime on gas and liquid phase mixing in bubble column reactors, *Chem Eng Sci* 51:2491–2500, 1996.
Zhang DZ, VanderHeyden WB: The effects of mesoscale structures on the macroscopic momentum equations for two-phase flows, *Int J Multiphase Flow* 28:805–822, 2002.
Zhao H: *Multi-scale modeling of gas-liquid (slurry) reactors,* Ph.D. Thesis, Beijing, 2006, Chinese Academy of Sciences.

CHAPTER SIX

# Mesoscale Flow Structures and Fluid–Particle Interactions

**Harry E.A. Van den Akker**[1]
Bernal Institute and Department of Mechanical, Aeronautical and Biomedical Engineering, University of Limerick, Limerick, Ireland
Department of Chemical Engineering, Delft University of Technology, Delft, The Netherlands
[1]Corresponding author: e-mail address: Harry.VanDenAkker@ul.ie

## Contents

1. Introduction 282
2. Evidence of Mesoscale Coherent Structures 284
3. The Origin of Mesoscale Structures 293
   3.1 Applying Extremum Principles 294
   3.2 Applying Linear Stability Analyses 296
   3.3 Running CFD Simulations 298
   3.4 An Intermediate Conclusion 302
4. The Fundamentals of Fluid–Particle Interaction 303
5. The Steady-State Drag Force 308
6. The Effect of Acceleration: Added Mass and Basset Force 310
7. The Lift Force 313
   7.1 The Saffman Lift Force 315
   7.2 The Magnus Lift Force 319
8. The Impact of Turbulence 320
9. Multiple Particle Systems 324
10. Conventional CFD Simulations 328
   10.1 The Euler–Lagrange Approach 328
   10.2 The Euler–Euler Approach 330
   10.3 Computational Results from the Literature 332
11. Direct Numerical Simulations 335
12. Conclusions 340
References 342

## Abstract

During the last two decades, the insight has gradually grown—thanks to advances in both experimental techniques and computational simulation tools—that many dispersed two-phase flows are dominated by dynamic mesoscale coherent structures in which particles, bubbles, or drops organize themselves. Evidence from experiments, hydrodynamic analyses, and computational simulations on the presence and dynamics of such structures, strands, and clusters is presented. Their origin being less well

*Advances in Chemical Engineering*, Volume 46
ISSN 0065-2377
http://dx.doi.org/10.1016/bs.ache.2015.10.010

understood, the general consensus is that fluid–particle interaction forces play a dominant role in bringing and keeping the dispersed-phase particles together. This chapter then revisits the topic of the fluid–particle interaction in all of its aspects and constituents such as single-particle steady-state drag, added mass, Basset force, lift, and particle rotation, while also the effects of particle and fluid acceleration and carrier phase turbulence are reviewed. The common practice of linearly adding the various correlations obtained for very specific canonical cases is to be rejected for other flow conditions than just creeping flow. Particular attention is paid to the way the pertinent fluid–particle interaction correlations are utilized in computational simulations of both the Euler–Lagrange (point-particle tracking) and Euler–Euler (two-fluid) type, where a distinction is made between Reynolds-averaged Navier–Stokes-based simulations and large-eddy simulations. Tracking a particle immersed in a 3D turbulent fluid in the presence of other particles by means of the steady-state drag force is a dubious approach. Undoubtedly, the best representation of the fluid–particle interaction can be obtained from periodic-box direct numerical simulations.

## 1. INTRODUCTION

Two-phase flow in one way or another already has been a topic in classical fluid mechanics (or hydrodynamics) for some 300 years—although it all started with the motion of single particles. As a result, a wealth of data and concepts are available in the literature. In the context of this introduction, some typical references to the oldest literature suffice to illustrate the early interest. In 1710, Sir Isaac Newton dropped spheres from inside the dome of St. Paul's cathedral in London to find out about gravity and the resistance of the (stagnant) air; Eiffel did the same from the Eiffel Tower in Paris in 1912; and Lunnon did it in a mine shaft in 1928. In 1851, Stokes derived an expression, nowadays known as Stokes law, for the frictional force exerted on a small sphere by a viscous fluid, while also in 1851 Pernolet carried out a systematic study on the effect of shape on the fall velocity of particles. The impact of particle rotation on the flight of a particle was investigated by Magnus in 1852 and by Lord Rayleigh in 1857.

In 1888, Basset revisited the slow motion of a single spherical particle in a stagnant fluid under the action of gravity—later on followed by treatises by Boussinesq in 1903 and by Oseen in 1927. In 1895, the British mathematician Sir Horace Lamb published the first edition of his classical textbook *Hydrodynamics* in which, among other topics, he considered the motion of a sphere and a cylinder in a potential flow. In the 1930s, Prandtl carried out his flow visualization studies in the Kaiser Wilhelm Institut in Göttingen (Germany), among other things on the flow patterns about cylinders under conditions not amenable to the mathematical tools available in those days. Yet, all the above studies were limited to fluid–particle interaction in the case of single bodies.

The strong development of process industries after the Second World War triggered a vivid interest in multiparticle systems. An example is fluid catalytic cracking of crude oil, where fluidized beds were throughout thanks to their great advantage of an intense degree of mixing; yet, most processes in fluidized beds were described in 1D terms. Richardson and Zaki (1954) studied sedimentation and fluidization of solid particles and presented a correlation for the vertical slip velocity of particles in a swarm as a function of the volume fraction of the carrier phase. Davidson and Harrison (1963) proposed a 1D two-phase flow model for gas bubbles rushing in plug flow through the dense phase of a fluidized bed. In the fields of stirred tank reactors and column-type gas–liquid reactors, strongly simplified models were exploited as well. Zwietering (1958) studied solids suspension in stirred vessels and just proposed the still widely used empirical correlation for the just suspended impeller speed that carries his name. Westerterp et al (1963) correlated interfacial area in agitated gas–liquid contactors mainly to agitation rate and tank to impeller ratio provided the agitation rate was above a certain minimum value. None of these authors studied the hydrodynamics of these two-phase flow systems in detail: the tools for doing this were simply not available.

Wallis (1969) in his treatise *One-Dimensional Two-Phase Flow* dealt with 1D (co-current or countercurrent) flow of two phases, almost tacitly assuming a homogeneous phase distribution across a pipe, channel, or vessel, and a uniform and usually constant bubble, droplet, or particle size. The concept of the 1D drift flux model introduced by Zuber and Findlay (1965) and Wallis (1969) became rather popular among chemical engineers (see, e.g., Rietema, 1982; Wesselingh, 1987). In another 1D approach, Barnea and Mizrahi (1973, 1975) derived a slip velocity—voidage correlation on the basis of some elementary physics-based concepts which is an improvement over the Richardson and Zaki (1954) relationship, particularly at lower particle concentrations. Somewhat similarly, Brucato et al (1998) focused on the net averaged sedimentation rate of particles settling in a turbulent medium.

In several papers on bubble columns (Beek, 1965; De Nevers, 1968; Freedman and Davidson, 1969; Hills, 1974; Rietema and Ottengraf, 1970; Whalley and Davidson, 1974), a variety of models for a circulatory flow mode of the liquid in aerated systems were proposed, based on experiments in contactors of varying geometry and varying types of aeration. Wijffels and Rietema (1972a, 1972b) derived a model for a steady-state single overall circulation in liquid–liquid spray columns, while Van den Akker and Rietema (1979a, 1979b) reported that a positive temperature gradient breaks up this single overall circulation and reduces the axial mixing. Most of this gas–liquid and liquid–liquid column research was aimed at deriving a simple 1D flow model with axial dispersion, on the analogy of the Taylor dispersion concept in laminar pipe flow.

All the above approaches, however, imply a gross simplification of the strongly chaotic 3D flow and transport processes in real-life two-phase flow reactors. They all echo the moderate experimental and modeling capabilities available in those decades of the twentieth century. In reality, and certainly under operating conditions usual in industry, 1D uniform multiphase flow is more the exception than the rule. Both gas–solid and gas–liquid flows are typified by a wealth of transient coherent structures exhibiting varying degrees of vorticity (Van den Akker, 1998). This will be highlighted in the next section in much greater detail.

Our understanding of the hydrodynamics of multiphase flows has progressed substantially in the recent three decades, thanks to the development of advanced experimental techniques, particularly laser Doppler anemometry (LDA), particle image velocimetry (PIV), computer-automated radioactive particle tracking (CARPT), and optical bubble probes. In addition, computational fluid dynamics (CFD) simulations allow for "inner" views in two-phase process equipment.

## 2. EVIDENCE OF MESOSCALE COHERENT STRUCTURES

In the 1960s–1980s, the hydrodynamics of industrial processes used to be interpreted in terms of steady-state models. Rietema and Ottengraf (1970) studied bubbling through viscous liquids in baffled bubble columns 22 cm in diameter. They were able to get a stable axisymmetric bubble street and to avoid cluster formation and coalescence at varying liquid viscosities by installing proper baffles. Van den Akker and Rietema (1979b) interpreted their experimental observations of multiple circulatory cells in thermally driven liquid–liquid spray columns in terms of a steady-state model (see Fig. 1), although visual observations did reveal some transient flow structures which were blamed to the use of many short glass QuickFit sections; such sections (which widen a bit at both ends) were used to allow insertion of thermocouples via the flanges between them. Joshi and Sharma (1979) and Zehner (1982, 1986) also reported multiple circulatory flow modes for bubble columns.

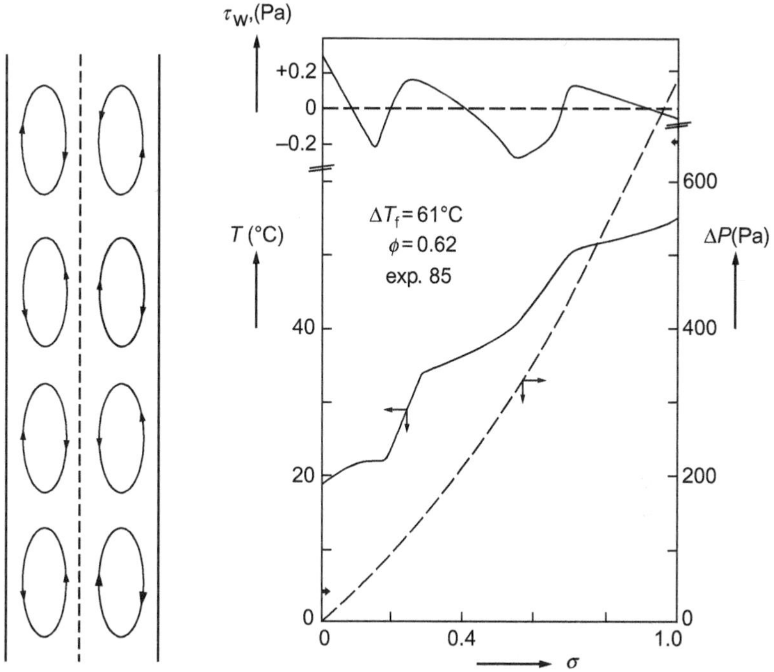


**Figure 1** The multiple circulation pattern in a liquid–liquid spray column operated with a positive temperature difference between top and bottom of the column, as observed experimentally by Van den Akker and Rietema (1979b). *Left: Sketch, reprinted from Rietema (1982) with permission from Elsevier; right: wall shear stress, temperature, and pressure profiles versus axial coordinate for the conditions shown, reprinted from Van den Akker and Rietema (1979b) with permission from Elsevier.*

Gradually, however, several researchers in the 1980s–1990s (Chen et al, 1994; Devanathan et al, 1995; Lin et al, 1996; Lübbert et al, 1982; Mudde et al, 1997a; Reese and Fan, 1994; Reese et al, 1993) observed that gas–liquid flows, certainly at higher aeration rates, exhibit a chaotic interplay of dynamic coherent, self-organizing mesoscale structures. In fact, Hills (1974) already reported the steady-state liquid circulation pattern to be the long-time averaging resultant of a much more complex flow field. A thorough review of the developments in those years can be found in Groen (2004) and in Mudde (2005). The parade in Fig. 2 summarizes the evolution of models, from very rudimentary to more sophisticated, for describing the circulatory and even chaotic flow behavior of bubble columns.

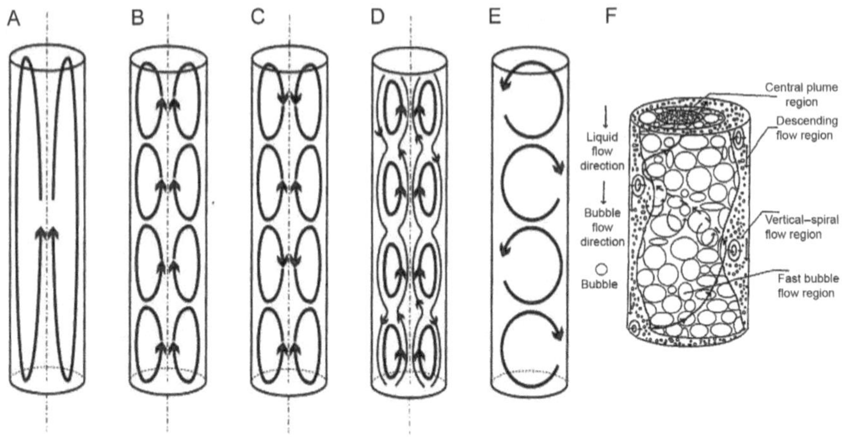


**Figure 2** Parade of circulation cell models developed over time: (A) a single overall circulation pattern, Rietema (1982); (B) the donut model of Joshi and Sharma (1979); (C) counter-rotating donuts as proposed by Van den Akker and Rietema (1982); (D) the interacting donut model due to Joshi and Sharma (1982); (E) the nonsymmetric circulation cell model of Zehner (1982, 1986); (F) the vortical model proposed by Chen et al (1994). *Panels (A)—(E): Adapted from Groen (2004); panel (F): Reprinted from Chen et al (1994) with permission from Wiley.*

LDA measurements by Groen et al (1995, 1996) and Mudde et al (1997b) demonstrated the strongly transient character of bubbly flows under conditions relevant to industrial processes. The meandering plume experiments by Becker et al (1994) and the sophisticated CARPT observations by the Dudukovic group at Washington University in St. Louis (Cassanello et al, 2001; Degaleesan et al, 2001) once more testified the chaotic flow behavior of bubble columns. Becker et al (1994) and Bauer and Eigenberger (1999) were among the first to computationally reproduce by means of CFD the dynamic flow patterns of bubble columns. More recently, Harteveld et al (2003) showed that these dynamic mesoscale structures may be provoked by the way the bubbles are supplied to the column. They were capable of keeping the uniform bubbly flow stable up to bubble holdups of some 55% by using a gas supply system consisting of 559 identical needles arranged in a well-thought layout (see also Mudde et al, 2009). All this leads to the conclusion that industrial bubble columns with a usually (very) inhomogeneous gas supply may exhibit a complex and chaotic flow behavior dominated by coherent mesoscale structures.

In the field of fluidization, Tsuo and Gidaspow (1990) were among the very first (see also Tsuji et al, 1998) to find cluster formation by means of

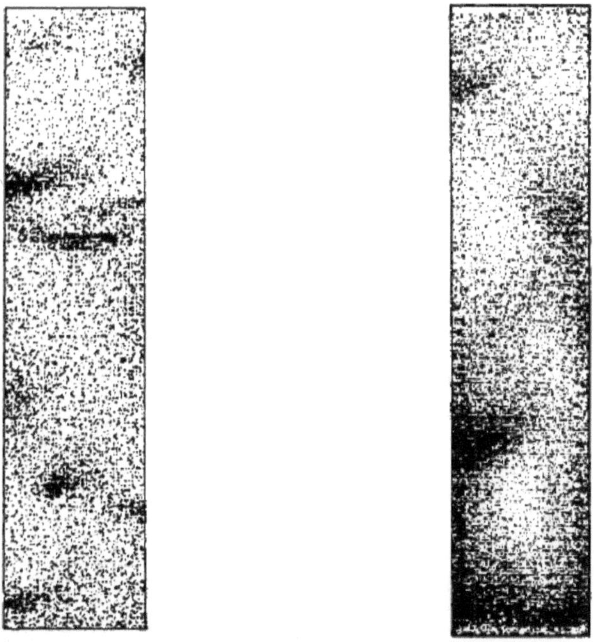

**Figure 3** Spatial distribution of particle density in a circulating fluidized bed at various operating conditions as calculated by Tsuo and Gidaspow (1990). *Reprinted from Tsuo and Gidaspow (1990) with permission from Wiley.*

CFD simulations exploiting the two-fluid model: see Fig. 3; their clusters moved downwards along the wall. Verloop et al (1993, 1995) performed measurements on the particulates behavior in the freeboard of the Delft University Fluidized Bed Combustor (49 cm in diameter, 6.5 m in height). Superimposed upon the overall upward flow, they observed downward particle motions mainly along the walls of the freeboard.

This is illustrated in Fig. 4 reproduced from Verloop (1994) and Van den Akker (1998). These authors were able to reproduce this behavior by means of CFD simulations in which they assumed bubbles bursting from the bed surface and entraining particles into the freeboard. Similarly, the overall dilute gas–solid flow in risers (used for catalytic cracking of crude oil) is often dominated by inhomogeneities usually described as strands or clusters (Fligner et al, 1994; Horio and Kuroki, 1994; Yonemura et al, 1995). Harris et al (2002) reviewed experimental literature data on cluster formation in riser flow. Mudde et al (1999) made observations in a turbulent fluidized bed 38.4 cm in diameter by means of a dual gamma-ray densitometry technique, illustrated by the measuring series of Fig. 5, and from such

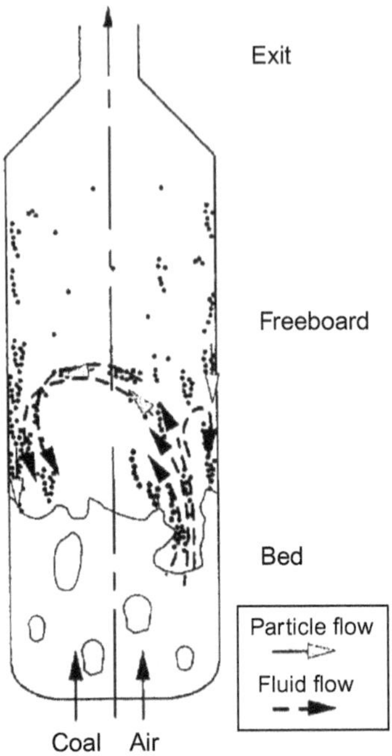


**Figure 4** A sketch of the gas–particle flow in the freeboard of a pressurized fluidized bed combustor. The broken arrows indicate the particle motions. *Reprinted from Van den Akker (1998) with permission from Elsevier.*

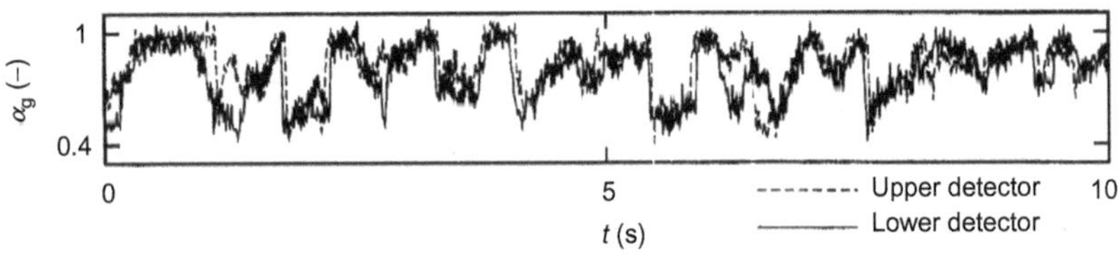


**Figure 5** Two time signals from two gamma-ray beams through the freeboard zone of a turbulent fluidized bed. At some moments in time the broken signals is leading, at other moments the same signal is lagging behind—interpreted as created by a different strand moving in the opposite direction. *Reprinted from Van den Akker (1998) with permission from Elsevier.*

fingerprints distinguished several types of coherent structures (see also Van den Akker, 1998).

Such an obviously inherent instability of dispersed two-phase flows has also been found in CFD simulations which in the last few decades have developed into an interesting research tool that allows to "measure" inside a flow field to a degree not feasible for experimental techniques. Lapin and

Lübbert (1994) were among the first to find the unsteadiness of gas–liquid flows by means of a mixed Eulerian–Lagrangian code developed in-house; the papers by Sokolichin and Eigenberger (1994) and Sokolichin et al (1997) illustrated this once more. More recent simulations of dispersed bubbly flows by Deen et al (2001), Bombardelli et al (2004), Zhang et al (2006), Dhotre et al (2009), and Niceno et al (2009) all exhibit the chaotic behavior of the bubbly plume and the liquid bulk. Wardle (2011) reported similar chaotic structures in the liquid–liquid mixture in an annular centrifugal contactor.

Tanaka et al (1995), in their simulation results, obtained with a direct simulation Monte Carlo (DSMC) method, discerned V-shaped strands and inverse V-shaped clusters interacting in a dynamic network in 2D upward gas–solid flows (see Fig. 6). Tsuji et al (1998), Helland et al (2000), Agrawal et al (2001), and Yan et al (2013) studied cluster formation in circulating fluidized beds and high-velocity vertical gas–solid flows by means of various numerical simulations. Li and Kwauk (2001) presented several snapshots of gas–solid two-phase systems revealing a so-called self-organization characterized by a dynamic alternation of solid-rich and gas-rich zones exhibiting strands and clusters (see Fig. 7). Clustering in particle-laden shear flows was also observed in direct numerical simulations (DNSs) reported by Gualteri et al (2013).

Gas–solid cyclones exhibit not only strongly unsteady flow behavior (Derksen et al, 2008) but also another type of self-organization: the particles collect on the inner walls of the cyclone in downward-driven ropes, as illustrated by the computational large-eddy simulations (LESs) results of Derksen which are confirmed by visual observations (Fig. 8).

Derksen (2003) carried out LESs for a solid suspension of a low solid volume fraction in a lab-scale turbulent stirred vessel. Figure 9 (reproduced from his paper) shows that the spatial distribution of the liquid–solid mixture

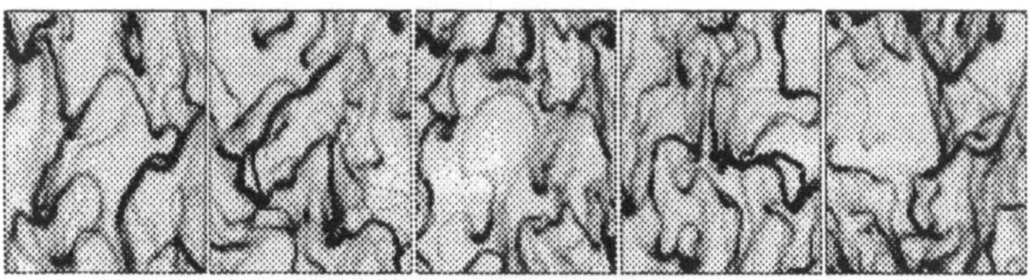


**Figure 6** Snapshots of the dynamic pattern of solid volume fractions indicative of particle clusters, in a gas–particle flow, as calculated by means of a DSMC by Tanaka et al (1995). *Reprinted from Tanaka et al (1995) with permission from the ASME.*

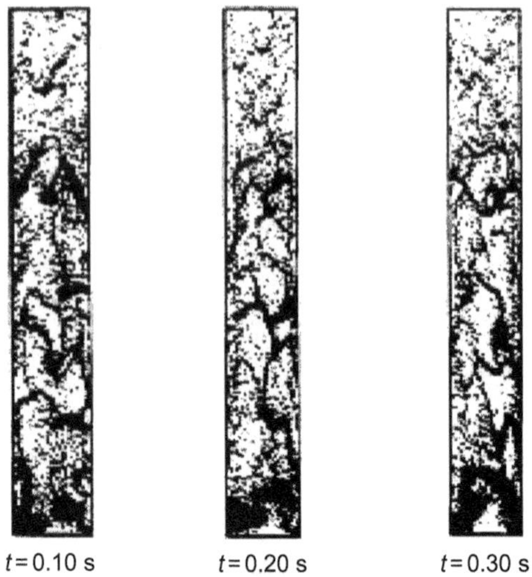


**Figure 7** Local heterogeneity and its dynamic change over time in a gas–solid two-phase system. *Reprinted from Li and Kwauk (2001) with permission from the ACS.*

**Figure 8** The phenomenon of roping in a gas–solid cyclone. Left : external view of a cyclone with two types of particles exhibiting roping as found in a two-way coupled large-eddy simulation by Derksen et al (2008)—figure not published before; right : picture of three gas–solid cyclones operated in parallel.

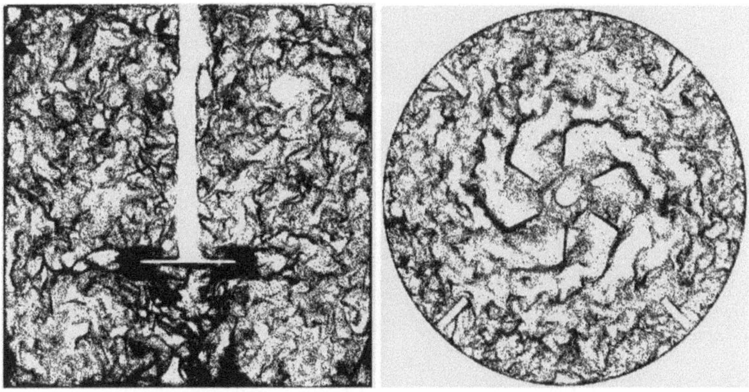

**Figure 9** Spatial distributionsLeft: in a vertical cross section; right: in a horizontal cross section at the level of the impeller—in a liquid–solid suspension in a stirred vessel, as found by particle tracking in an LES liquid flow field. *Reprinted from Derksen (2003) with permission from Wiley.*

is rather inhomogeneous and that the solid particles collect in strands at the outer rims of the eddies in the turbulent flow field created by the impeller. Centrifugal forces in the eddies are held responsible for the formation of these strands (see, e.g., Goto and Vassilicos, 2008). This is also confirmed by Fig. 10 and also suggested by Fig. 11 both of which result from simulations carried out in Toulouse. So far, these computational results have never been confirmed by experimental observations due to the visual or optical inaccessibility of such suspensions. Reversely, some authors do report that small bubbles collect in the intense vortex cores of Kolmogorov-scale turbulence (see Balachandar and Eaton, 2010; Goto and Vassilicos, 2008).

The inevitable conclusion after all the above experimental evidence and computational results is that—unless special precautions are being taken to avoid the creation of instabilities and/or to prevent instabilities from growing—dispersed two-phase flow is inherently unstable and as such is prone to the chaotic occurrence of coherent, often vortical, flow structures and strands. In the words of Radl and Sundaresan (2014): "Fluidized gas–particle systems are inherently unstable and they manifest structures on a wide range of length and time scales." Zones exhibiting strong differences in bubble, drop, or particle number density dynamically alternate and chaotically shake and mix the carrier phase in the flow domain. One may wonder about the origin of these hydrodynamic instabilities and about the driving force(s) and the physical mechanisms behind this chaotic

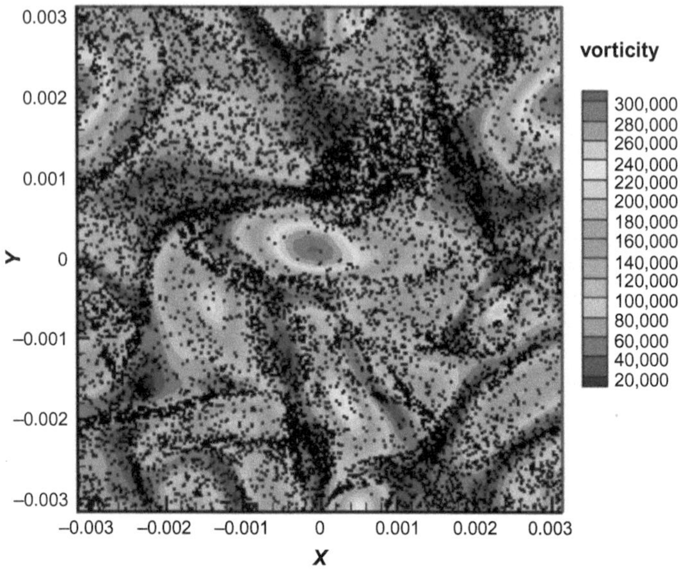


**Figure 10** Particles collecting in strands due to the action of vorticity, as found by M. Garcia. *Reproduced from the CERFACS Web site with permission: http://cerfacs.fr/gallery-of-images/.*

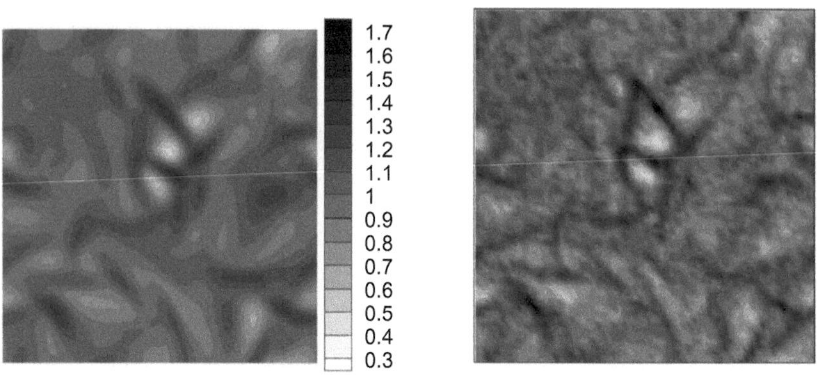


**Figure 11** Snapshots of spatial distributions of particle number density as found from a Eulerian simulation (left) and a Lagrangian simulation (right). *Reprinted from Kaufmann et al (2008) with permission from Elsevier.*

behavior. One may further raise the question whether, given the different values of the density difference between the phases, the formation of strands and structures in gas–liquid, liquid–liquid, gas–solid, and liquid–solid flows have much in common. Such questions are the topic of this chapter.

## 3. THE ORIGIN OF MESOSCALE STRUCTURES

Two-phase flow has more degrees of freedom, or a larger degree of undeterminedness, than single-phase flow: in many cases, the spatial distribution of the dispersed phase across the flow domain and its temporal evolution is not prescribed—neither in the real world of experiment, process equipment, and the natural environment, nor in the virtual world of computational simulations. The larger the flow domain, the more freedom two-phase flows have to self-organize and to redistribute the dispersed phase spatially and temporally.

The situation is comparable to, but also different from, turbulent single-phase flow. In a turbulent flow, the cascade of turbulent eddies is dynamically reacting to the overall power input at the macroscale by adapting the size of the "smallest" eddies named after Kolmogorov as a result of which the inertial subrange is reduced or extended. In two-phase flows, a spatial inhomogeneity created by purpose or just provoked by some perturbation may result in local differences in the average density of the two-phase mixture which may induce motions, often creating vorticity, in the mesoscale range giving rise to a cascade of turbulent, or turbulent-like, eddies. Two-phase flows then turn into typical multiscale phenomena where, compared to turbulent single-phase flows, the size of bubble, drop, or particle may play a complicating role. When the dispersed phase is a fluid, the situation may be even more complex as bubbles and drops may adapt their shape and even break up or coalesce in response to the forces exerted by the carrier phase. Breakup and coalescence of drops and bubbles are beyond the scope of this chapter.

The origin of mesoscale structures in dispersed multiphase flows may be investigated in three ways:

- By applying extremum principles: to see whether thermodynamically, or energetically, a state manifesting mesoscale structures is more favorable than the state of uniform homogeneous flow;
- By applying linear stability analyses: to see whether or not uniform homogeneous flow is stable against disturbances, in the same way as a stagnant fluid layer heated from below may become unstable when heated from below;
- By running CFD simulations: to see in a more mechanistic approach which effects may be held responsible for the clustering of particles, drops, and bubbles into coherent structures or strands; then questions

may come up with respect to the type of CFD simulations and the models used in these CFD simulations, in particular for describing the fluid–particle interaction.

These three approaches will now first be discussed in some detail; this leads to an intermediate conclusion with respect to the dominating role of fluid–particle interaction in the development of mesoscale structures. On this basis of this conviction, the main body of this chapter will be devoted to fluid–particle interaction in dispersed multiphase flows and a proper description of the forces involved.

## 3.1 Applying Extremum Principles

An intriguing question is whether or not the degree of spatial inhomogeneity and unsteadiness is governed by some general extremum principle. According to the Second Law of Thermodynamics, for an arbitrary isolated process, the entropy of the final state is larger than or equal to the entropy of the initial state (Clausius). An important question is how the Second Law translates to continuous-flow processes. In virtually all flows, energy is dissipated, i.e., entropy increases. The intriguing question is at which rate—and this is outside the reach of thermodynamics.

While Prigogine advocates the nonequilibrium thermodynamics doctrine that linear nonequilibrium systems obey to the principle of minimum entropy production (see, e.g., Li and Kwauk, 2001), other scientists (see Martyushev and Seleznev, 2006) claim the general validity of maximum entropy production, just the antipode of Prigogine's doctrine. This means there are two absolutely different principles, subject to different constraints and in terms of different variable parameters. Martyushev and Seleznev (2006) claim that maximum entropy production is more universal and governs the evolution of nonequilibrium dissipative systems. In this view, entropy not only tends to increase, but it will increase to the maximum value allowed by the imposed constraints. The above controversy (see also Ross et al, 2012) explains the confusion in the domain of multiphase flow research about the potential application of some extremum principle which anyhow is appealing as it suggests room for self-organization of two-phase flows at the mesoscale. In addition, applying an extremum principle would result in an additional equation that would allow solving for an additional variable in some two-phase flow model allowing for the occurrence of mesoscale clusters.

In addition, literature contains several heuristic empirical claims that—subject to certain provisos—the principle of minimum energy dissipation first (as early as 1868) recognized by Helmholtz may be applied to two-phase flow systems. Interesting discussions on the use of extremum principles and variational techniques as means for deriving additional equations which may govern the extent of spatial inhomogeneities can be found in some two-phase flow papers such as Rietema and Ottengraf (1970), Hyre and Glicksman (1997), and Li and Kwauk (2001). Generally, multiphase flows are typically nonlinear and nonequilibrium, and for such systems a generally accepted and universally valid variational principle is not available.

Rietema and Ottengraf (1970) applied the principle of minimum energy dissipation to their laminar bubble column to provide an additional equation that allowed finding the bubble street diameter which was an additional unknown in their model. Wijffels and Rietema (1972a) invoked the principle of a minimum pressure gradient. Energy dissipation also plays a dominating role in the models presented by Whalley and Davidson (1974) and Joshi and Sharma (1979). Variational techniques, related to Hamilton's principle of least action from classical mechanics and making use of Lagrangian multipliers, have been applied to bubbly flows by Geurst (1985, 1988) and to fluidized and sedimenting beds by Grbavcic et al (1991). The literature on circulating fluidized beds contains several more claims about minimum potential energy and minimum flow resistance, which were discussed by Hyre and Glicksman (1997). Their conclusion, however, runs as follows: "In light of past experimental evidence, there seems to be little reason to make the assumption that heterogeneous two-phase flows tend toward a state of minimum energy dissipation or minimum pressured drop. Such a hypothesis needs to be substantiated by detailed experimental proof before it can be considered for fluidized beds."

In spite of the above described controversy on extremum principles, researchers affiliated with the National Key Laboratory of Multi-Phase Complex Systems at the Beijing Institute of Process Engineering of the Chinese Academy of Sciences developed a complex Energy-Minimization Multi-Scale (EMMS) methodology and applied it to a variety of multiphase flow processes: see Li and Kwauk (2001), Wang (2009) and Wang et al (2010).

Finally, in a rather popular class of gas–liquid and liquid–liquid CFD simulations, the dynamics of the flexible and moving interfaces is described in terms of a so-called phase field. In a zone of finite thickness about the (sharp) interface between the two fluid phases, such a phase field models how an order parameter gradually changes its value from its value (e.g., +1) at one boundary of this zone to its value (e.g., −1) at the other boundary, the result being that the interface is now diffuse rather than sharp. In common versions of the phase field model, a free energy functional is formulated comprising expressions for the bulk free energy and for the interfacial free energy. In a variational approach, this free energy is assumed to strive for a minimum. In the Cahn–Hilliard concept (see, e.g., Lee et al, 2014), the interfacial free energy is described in terms of the spatial gradient of the above order parameter. The Cahn–Hilliard free energy approach is exploited in the context of both finite-volume techniques and lattice Boltzmann (LB) methods; in both simulation techniques, solving an additional equation increases the computational burden. In spite of its evident popularity in the realm of computational simulations, the application of a minimum free energy principle for the purpose of describing phase separation and interface dynamics may not have a very sound physical basis.

## 3.2 Applying Linear Stability Analyses

If, then at the current state of the art, no generally accepted guidance can be obtained from applying a universally valid extremum principle, it may be tempting to investigate whether or not uniform two-phase flow is hydrodynamically stable against perturbations. The background of such a query is that for many chemical engineering applications, plug flow of the dispersed phase through the carrier phase may be the preferred mode of operation. If uniform two-phase flow would not be stable against perturbations, this would explain why mesoscale structures develop.

Linear hydrodynamic stability theory was established in the field of free convection, starting with Lord Rayleigh. Pellew and Southwell (1940) present a nice review of stability analyses on buoyancy-driven convection. In the 1960s and 1970s, linear stability analyses on the basis of two-fluid models were rather popular among fluidization researchers who were looking for the conditions under which homogeneous fluidization is stable before it develops into heterogeneous (bubbling) fluidization: see, e.g., Jackson (1963), Anderson and Jackson (1968), Garg and Pritchett (1975), Mutsers and Rietema (1977), and Liu et al (2007). The technique was thoroughly explained by Liu et al (2008).

Van den Akker (1978) carried out a linear stability analysis for a liquid–liquid spray column operated with a positive temperature gradient. The leading idea was that local differences in mixture density would give rise to a multiple circulation pattern in a way similar to the multicellular natural convection mode in a vertical slot heated from one side (Elder, 1965; Lee and Korpela, 1983). The only way to find in the liquid–liquid spray column that a perturbation of a multiple circulation type would grow was by introducing a reduced drag coefficient for the radial component of the fluid–particle interaction force. And although a physical explanation for such a reduced radial drag coefficient was presented, this result did not make it into a journal paper as no confirmation of such an effect was found in the literature.

Linear stability analyses were also carried out for bubble columns by, e.g., Shnip et al (1992), Minev et al (1999), León-Becerril and Liné (2001), Sankaranarayanan and Sundaresan (2002), Lucas et al (2005, 2006), and Monahan and Fox (2007). The reader is also referred to the review paper by Joshi (2001). The purpose of these analyses was to learn more about the stability of homogeneous (or uniform) bubble flow and the transition to heterogeneous flow regimes. In most cases, the results of the analyses were again strongly dependent on the exact formulation of the fluid–bubble interaction force, particularly of its lateral component. Several of these authors pinpoint the crucial effects of virtual mass and of (the sign of) the lift force.

Recently, Ghatage et al (2014a, 2014b) reported on linear stability analyses for a three-phase sparged reactor and for solid–liquid fluidized beds, respectively. Ruzicka (2013) used a different technique for investigating the hydrodynamic stability of a bubble column. All these papers illustrate the search for better insight in the mechanisms governing the transition toward heterogeneous flow regimes characterized by the presence of mesoscale structures.

In the meantime, careful experiments by Mudde et al (2009) in a cylindrical bubble column 15 cm in diameter showed that the homogeneous regime was stable up to a gas fraction of 55%, well beyond the predictions on the basis of stability analyses. This remarkable phenomenon was attained by feeding the air via 559 very identical needles arranged in very uniform pattern—such that all bubbles were of the same size and left the needles at exactly the same velocity. Further, great care was taken to ensure that the column was perfectly aligned vertically. This finding has great consequences for computational simulations of bubble columns on the basis of two-fluid models: a uniform supply of the gas phase at the bottom of a bubble column should really result in a homogeneous flow regime. By shutting off certain groups of needles, Mudde et al (2009) observed that immediately mesoscale vortical coherent motions were formed that ruined the homogeneous flow regime. Similarly, only due to a deliberately uneven gas supply should CFD simulations result in a heterogeneous flow field dominated by the occurrence of such chaotic flow structures.

These results are in line with an earlier report by Fligner et al (1994) that the origin of clusters in gas–solid riser flows was seen to be associated with the way the oil is injected at the base of the riser. In addition, it is known in the fluidization literature that too low a pressure drop across the gas distributor may provoke convective instabilities and bubble formation in the bed—an effect that was also found in a linear stability analysis by Medlin and Jackson (1975) for a porous plate distributor.

These findings about the effect of the way the phases are supplied to a column or bed may have enormous implications for our understanding of the occurrence of mesoscale coherent vortical structures in dispersed two-phase flow systems. It would mean that—at least under conditions of industrial relevance—these vortical structures are not the eventual result of infinitesimally small instabilities perturbing the uniform or homogeneous flow regime but are due to large-scale spatial differences in gas or particle fraction induced either intentionally or unintentionally. The resulting density differences may result in spatially varying gravity/buoyancy effect which in their turn may give rise to spatial differences in vertical velocities. This suggests a strong analogy with the way turbulence is created: due to either shear or buoyancy. From this perspective, it is not surprising that also under two-phase flow conditions the slope found in the inertial subrange of the turbulent energy spectrum is (close to) $-5/3$, just like in single-phase turbulent flows (Harteveld et al, 2005; Mudde and Saito, 2001). The same conclusion pertains to aerated stirred vessels where the impeller motion creates the fluid motions and breaks up the bubbles which are then advected by the turbulent flow (in large parts of the vessel).

## 3.3 Running CFD Simulations

In addition to the attempts to better understand the origin of mesoscale structures in dispersed two-phase flows by means of extremum principles and stability analyses, CFD simulations may provide guidance. One may therefore expect that simulations of particle behavior on the basis of a proper description of the fluid–particle interaction force reproduce the formation and dynamics of coherent structures and clusters (see also Capecelatro et al, 2015). After all, CFD simulations allow visualizations of 3D transient flow fields and make it possible to systematically investigate the effect of all types of variables (flow rates, physical properties, geometrical parameters) on flow fields to a degree not feasible experimentally.

Among other investigators, Agrawal et al (2001) report that the clusters and strands observed experimentally can be qualitatively simulated by proper simulations. These authors mention two mechanisms leading to the formation of these clusters and strands: (i) an instability associated with the difference in inertia between the two phases, and (ii) damping of the fluctuating particle motions by the interstitial fluid and by inelastic particle collisions. Several authors report that particles respond to vorticity in the carrier phase and accumulate in low-vorticity regions (see, e.g., Capecelatro et al, 2015; Derksen, 2003; Maxey, 1987; Mei, 1994; see also Fig. 10). Goto and Vassilicos (2008) proposed a sweep-stick mechanism. CFD simulations should be able to reproduce and investigate these mechanisms—reason for a short introduction on two-phase flow CFD.

In general, at least three different types of CFD simulations are available for computationally reproducing dispersed two-phase flow systems, although also the use of mixed versions have been reported:

- Within the Euler–Lagrange framework (see, e.g., Patankar and Joseph, 2001; Sommerfeld and Decker, 2004), first the flow field of the carrier or continuous phase is calculated in an Eulerian framework; then, the flow behavior of the second, dispersed phase is mimicked by tracking the path of individual (usually point) particles in a Lagrangian way. In this procedure, a correct description of the fluid–particle interaction is crucial. This interaction can be treated one way (the particles feel the flow of the carrier phase, while the latter does not feel the particles) or two way and iteratively (rendering the simulation computationally more demanding).

  The flow field of the continuous, or carrier, phase can be simulated by means of various techniques such as the finite-volume technique used in most commercial CFD software, an LB solver (see, e.g., Raabe, 2004), or a pseudo-spectral technique (not unusual in the computational physics community—see, e.g., Eaton, 2009; Homann et al, 2015; Li et al, 2001; Olivieri et al, 2014).

  The effect of treating the particles as point particles (see also Balachandar, 2009) is that the detailed flow between the particles in response to the presence and motion of the particles remains unresolved (see, e.g., Derksen, 2003). As a result, a correlation is required in order to take the fluid–particle interaction into account. The Euler–Lagrange approach is most often used for simulating dilute gas–solid and liquid–solid flows where the particle size is smaller than the smallest turbulent length scale (eddy) considered in the flow simulation of the carrier phase (Balachandar and Eaton, 2010).
- In the Euler–Euler approach of more dense multiphase flows, the dispersed phase may well be conceived as a second continuous phase that

interpenetrates the real continuous or carrier phase; this approach is often referred to as the two-fluid formulation. The inherently "simultaneous" presence of two interpenetrating continua is taken into account by their respective volume fractions. In fact, the existence of a second phase consisting of dispersed, or discrete, particles (or drops or bubbles) not being connected or in permanent contact is overlooked in favor of a more superficial, or continuum, description in terms of volume fractions. Particle size is a nonexisting topic in this conceptual approach.

In the two-fluid formulation, the velocity field of each of these two continuous phases is described by its own continuity equation and Navier–Stokes equation (e.g., Anderson and Jackson, 1967; Rietema and Van den Akker, 1983; Sokolichin et al, 2004; Tabib et al, 2008). Each of these Navier–Stokes equations comprises a mutual phase interaction force. Only in the size, or numerical value, of this phase interaction force, particle size may come out. However, in selecting the computational grid or the numerical technique for solving the flow fields of the two continuous phases, particle size is completely irrelevant—due to the definition of mutually interpenetrating continua.

- In simulating bubbling and circulating fluidized beds, often discrete particle methods (DPMs) or discrete elements methods (DEMs) are used. In such dense systems, the hydrodynamic behavior is assumed to be dominated by particle collisions and the carrier phase hydrodynamics is secondary, particularly when the carrier phase is a gas. The starting point for this type of simulations therefore is the particle phase consisting of discrete particles the motion and collisions of which are tracked—reason why some authors denote this class of simulations also as being of the Euler–Lagrange type. The collisions are described with the help of a restitution coefficient and a friction coefficient which essentially have to account for the energy dissipation in the system.

  The flow field of the carrier phase is not resolved at the scale of the particles, but just in averaged terms, and is strongly determined by the interacting and colliding particle phase. Usually, turbulence in the carrier phase is not included. The fluid–particle interaction is usually described by means of either an empirical correlation—such as the Ergun relation for dense systems or the Wen and Yu relation for more dilute flow regimes—or a similar correlation based on dedicated simulations. In general, one could say that the fluid mechanical contents of DPMs are rather limited and strongly empirical.

The calculation of the particle trajectories and collisions is usually done on the different grid than the CFD of the carrier phase—necessitating specific mapping and filtering techniques as well as closure models and prompting many numerical issues. Recent examples of such filtering studies are due to Radl and Sundaresan (2014) and Peng et al (2014). The study by Kriebitzsch et al (2013) shows that, compared to a DNS which resolves the gas flow field in a fluidized bed completely, an unresolved DEM simulation underpredicts the fluid–particle interaction force substantially.

The class of DPMs and DEMs has been extended to discrete bubble methods (DBMs) for bubbly flows where in addition to bubble coalescence and breakup also the lift force and added mass have to be taken into account.

For more details on this class of DPMs, the reader is referred to, e.g., Hoomans et al (1996), Kuipers and Van Swaaij (1998), Van der Hoef et al (2006), Deen et al (2007), Zhu et al (2007), and Peng et al (2014) as well as to the Chapter Euler-Lagrange modelling of the hydrodynamics of dense multiphase flows by Padding et al.

The attractiveness of DPMs cannot be illustrated better than by Fig. 12—reproduced from Chen et al (2009)—which shows an unprecedented detailed snapshot (though not properly validated yet) of a gas–solid flow obtained by means of a DNS that combines an LB simulation of the gas flow with a discrete element method for the solids phase consisting of 0.1 million solid particles; the code was run a platform comprising 256 graphical processing units (GPUs).

In spite of their claimed successes (see also Zhu et al, 2008), these discrete particle-based approaches are outside the scope of this chapter which strongly focuses on the role of fundamental fluid mechanics in computational simulations and in reproducing the occurrence of coherent structures in multiphase flows.

In both the Euler–Lagrange and the Euler–Euler approach, the interaction force between fluid and particles or between the two interpenetrating phases is a dominating element. As a matter of fact, this is not different from the role of the interaction force in linear stability analyses, as observed by several investigators and touched upon in the preceding section. In selecting proper expressions for this interaction force, all authors recall the interaction force between a fluid and a single particle due to the relative velocity and exploit expressions—usually empirical ones, sometimes based on modeling—from a variety of

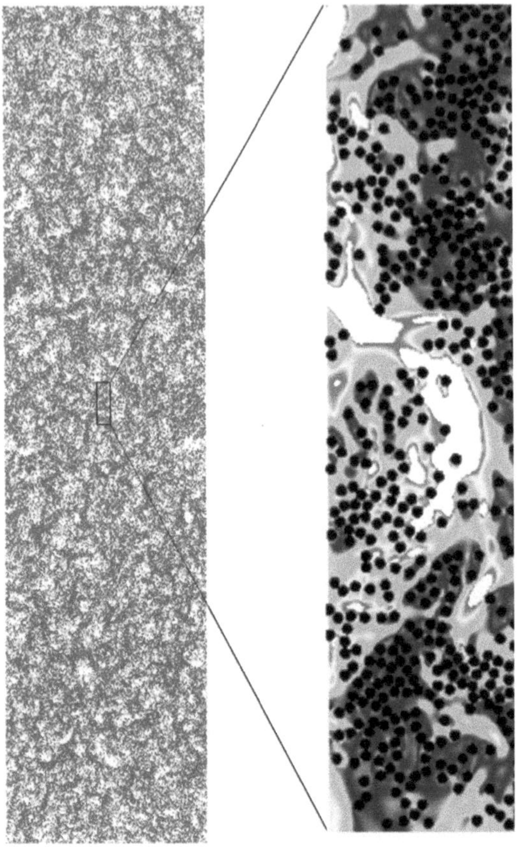

**Figure 12** Snapshots of the solid particle distribution and the local flow structure from a DNS of gas–solid flow carried out by Chen et al (2009). *Reprinted from Chen et al (2009) with permission from Elsevier.*

literature sources. When doing this, the validity range and the conditions for which such expressions have been determined should be adhered strictly.

## 3.4 An Intermediate Conclusion

It is the author's conviction that in many (turbulent) dispersed multiphase flows—except probably in very dense multiphase flow systems—the origin of mesoscale structures is in the fluid–particle interaction, with a secondary role for particle–particle interaction (collisions, coalescence, breakup). Clustering of particles is believed to be intimately connected with the chaotic dynamics of fluid accelerations, as particles converge toward each other where and when the divergence of the acceleration field is positive (Goto and Vassilicos, 2008). These authors proposed a sweep-stick mechanism and investigated this for homogeneous isotropic turbulence. The challenge is to assess the applicability of this (or any other) mechanism for real-life conditions different from homogeneous isotropic turbulence. The best way of doing is by means of computational simulations in which then the fluid–particle interaction should be dealt with correctly. Fluid–particle interaction therefore is the main body of this chapter.

When trying to describe the fluid–particle (in the Euler–Lagrangian approach) or fluid–fluid interaction (in the Euler–Euler approach) in terms of equations and correlations available in the literature, one golden rule has always to be observed: they should not be stretched beyond the original test or model conditions and the validity range of the resulting expressions. The present author observes that this golden rule is not being observed in many papers and simulations. Just two examples:

- It is incorrect to expect that, for a particle immersed in a 3D turbulent fluid in the presence of other particles, the steady-state drag coefficient may be used that has been obtained by dropping a particle (along a 1D path) through a homogeneous stagnant fluid.
- It is incorrect to assemble a fluid–particle interaction force by simply combining via a linear addition a number of forces obtained for various canonical cases. Many authors—see, e.g., Sokolichin and Eigenberger (1994), Lahey (1995), Hill et al (1995), Decker and Sommerfeld (2000), Laín et al (2000), Derksen (2003), Hibiki and Ishii (2007), Niceno et al (2008), Tabib et al (2008), and Dhotre et al (2009)—pursue this "simple" approach. While this procedure may be allowed for mere creeping flow conditions, generally it is not due to the notorious nonlinear character of the Navier–Stokes equation.

Before these individual canonical cases and several cases of misconceptions will be discussed in greater detail, it makes sense to first review the fundamental concept of fluid–particle interaction for both single-particle and multiple particle flows. Note that particle–particle interactions and particle collisions, the concepts of granular temperature and granular flow inclusively, are largely outside the scope of this chapter.

## 4. THE FUNDAMENTALS OF FLUID–PARTICLE INTERACTION

In any undergraduate textbook on fluid mechanics, the concepts of drag force and lift force are explained. Here, the textbook of Munson

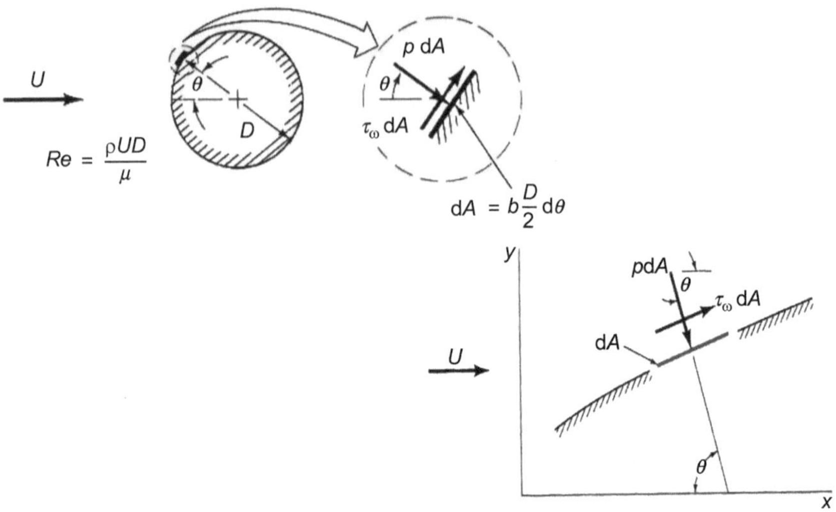


**Figure 13** Illustration of the fluid–particle interaction force on an immersed particle. *Reprinted from Munson et al (2010) with permission from Wiley.*

et al (2010) will be used as a starting point for an elementary refresher. Figure 13, reproduced from this textbook, shows a stationary cylinder (with size $b$ in the direction normal to the plane of the paper) in a uniform steady 1D cross-flow. The presence of this cylinder (or any particle or object) affects the flow as the cylinder forces the flow to deviate from its rectilinear motion, while the fluid velocity has to match the no-slip condition at the surface of the cylinder. This means that the flow field changes and, as a result, also the pressure field changes: pressure is maximum at the "nose" of the cylinder, usually denoted as the stagnation pressure, while—depending on the shape of the body and the inertia of the fluid (the Reynolds number)—more downstream along the cylinder's surface the pressure decreases again, resulting in a pressure drop across the cylinder and, hence, in a force on the cylinder in the direction of the flow: the so-called form drag. The shear rate along the surface results in the so-called friction drag.

The enlarged view in Fig. 13 shows how two forces act on a surface element $\mathrm{d}A$: a force $p\mathrm{d}A$ normal to the surface element due to the pressure $p$, and a shear force $\tau_w\mathrm{d}A$ acting along the surface element due to the local shear rate $\dot{\gamma}$. If the flow direction is denoted as the $x$-direction and $y$ denotes the direction normal to the flow, each of these two forces can be decomposed in an $x$-component and a $y$-component. Together they add up to the $x$- and $y$-components of the total fluid–particle interaction force given by

$$F_{\mathrm{D}} = \int_A \mathrm{d}F_x = \int_A p\cos\theta \mathrm{d}A + \int_A \tau_w \sin\theta \mathrm{d}A \tag{1}$$

and

$$F_{\mathrm{L}} = \int_A \mathrm{d}F_y = -\int_A p\sin\theta \mathrm{d}A + \int_A \tau_w \cos\theta \mathrm{d}A \tag{2}$$

and denoted as the drag force and the lift force, respectively. In the case of the cylinder in the unrestricted uniform steady 1D cross-flow of Fig. 13, the lift force is zero because the contributions from the "upper" and "lower" halves of the flow domain (at either side of the horizontal line of symmetry) cancel. As soon as there is no line or plane of symmetry in the velocity field around an object, the lift force is nonzero.

Drag force and lift force are strongly related and result from one and the same velocity field and one and the same pressure field around an object immersed in a flowing fluid or moving through a fluid. Tabib et al (2008) are certainly incorrect when they state that "the total interfacial force between the two phases may arise from several independent physical effects" such as drag force, lift force, etc. By the way, the term "lift force" may only have a clear meaning in the case of a well-defined (oncoming or main) flow direction. These annotations will be discussed in greater detail in the section on the lift force in a transient nonuniform or turbulent 3D flow field,

One should also realize that the pressure field in the above expressions is just the result of the presence of the particle. This pressure (contribution) is different from, actually is superimposed upon, the background pressure field that drives the oncoming flow and which also exerts a force on the article. This was also appreciated by Tchen (1947). It is also clearly seen in the following 3D derivation due to Rietema and Van den Akker (1983)—see also Van den Akker (1986)—which builds on the paper by Anderson and Jackson (1967) and resembles the Reynolds decomposition technique known from turbulence modeling.

Consider an arbitrarily shaped particle in a flow field. Decompose all local flow variables, denoted by a single prime, into an average value (averaged over a scale large compared to the diameter of the particle considered) and a local perturbation about this average value due to the presence and/or motion of the immersed particle. This perturbation is denoted by a double prime. Hence,

$$p' = p + p''; \quad \tau' = \tau + \tau'' \tag{3}$$

This decomposition may be allowed if the time scales of the average flow and the perturbation are really different (the principle of separation of scales applies)—in the same way as is assumed in linear stability analyses. The local fluid–particle interaction force $\underline{F}'$ exerted by the fluid upon the particle is given by

$$\underline{F}' = -\int_A \underline{n} \cdot \left( p' \underline{\underline{\delta}} + \underline{\underline{\tau}}' \right) \mathrm{d}A \tag{4}$$

in which $\underline{n}$ denotes the outward normal to each surface area element d$A$ of the particle surface $A$ and $\underline{\underline{\delta}}$ stands for the Kronecker delta. Substituting Eq. (3) into Eq. (4) and then applying Gauss' theorem gives

$$\begin{aligned} \underline{F}' &= -\int_A \underline{n} \cdot \left( p\underline{\underline{\delta}} + \underline{\underline{\tau}} \right) \mathrm{d}A - \int_A \underline{n} \cdot \left( p'' \underline{\underline{\delta}} + \underline{\underline{\tau}}'' \right) \mathrm{d}A \\ &= -\left( \nabla p + \nabla \cdot \underline{\underline{\tau}} \right) V_\mathrm{p} - \int_A \underline{n} \cdot \left( p'' \underline{\underline{\delta}} + \underline{\underline{\tau}}'' \right) \mathrm{d}A \end{aligned} \tag{5}$$

where $V_\mathrm{p}$ denotes the particle volume. The second term, with the two perturbation terms, is equivalent to the integrals of Eqs. (1) and (2) but illustrates better the 3D character and the interweaving of pressure field, shear stress field, and interaction force.

There is consensus that in one way or another this second term depends on the relative velocity, or slip velocity, between fluid and particle defined by $v_\mathrm{s} = U_\mathrm{p} - U_\mathrm{f}$. The common interpretation is that the concept of this slip velocity covers both the case of a stationary particle immersed in a flow and that of a particle moving through a stagnant fluid. Generally, the flow field around a particle is characterized with the help of a particle Reynolds number that comprises particle size and this slip velocity. Yet, a warning is appropriate here: whether or not these two cases are identical depends on whether or not the flow and pressure fields in the boundary layer around the particle and in its wake are really the same. Particularly under turbulent flow conditions this is dubious: the turbulent boundary layer around and the wake behind a particle traveling at a high Reynolds number through a stagnant fluid may be different from the turbulent boundary layer and the wake created by a turbulently flowing fluid around a stationary particle, as the dynamic boundary conditions are different.

As far as the first term of the fluid–particle interaction force $\underline{F}'$ of Eq. (5) is concerned: in a uniform homogeneous steady flow of the fluid, it results in a term $-\nabla p$ (Tchen, 1947) or a gravity term (Bird et al, 1960) in the equation of motion for a single particle: see also, e.g., Sokolichin et al (2004). Derksen (2003) includes both the $-\nabla p$ term and the viscous $\mu\nabla^2 U$ term. In a multiple particle system with $N$ particles contained within a volume element of volume $\Delta^3$, the first term of Eq. (5) results in a contribution to the fluid–particle interaction force per unit of volume of the multiple particle system that runs like

$$-\frac{1}{\Delta^3}\sum_N\left(\nabla p+\nabla\cdot\underline{\underline{\tau}}\right)V_\mathrm{p}=-\alpha\left(\nabla p+\nabla\cdot\underline{\underline{\tau}}\right) \tag{6}$$

This result illustrates that both the background pressure field and the background shear stress field (if present) work on a particle or—in a multiple particle system—on all particles.

The great challenge now is how to translate the fundamental expression of the second term of Eq. (5) into an equation—in terms of known variables (particle size, slip velocity, physical properties) and/or the particle Reynolds number (in terms of slip velocity and particle size)—that can be used with confidence in the stability analysis or computational simulation of interest. In computational simulations, spatial and temporal resolution may create additional issues as these scales may interfere with the time and length scales of fluid flow and particle behavior.

Literature on fluid–particle interaction is voluminous and, among other things, contains many reports about the interaction force in very specific canonical cases such as the steady free fall of spheres and particles of different shapes in a stagnant fluid and the lateral motion of particles in a shear flow, and about the effect of acceleration and deceleration, of particle rotation, of free-stream turbulence, and of adjacent particles—essentially the same topics as already discussed in a series of review papers by Torobin and Gauvin (1959a, 1959b, 1959c, 1960a, 1960b, 1961). An elementary introduction on the topic can also be found in Stringham et al (1969). These papers provide a good starting point for this review, particularly with respect to the older literature on fluid–particle interaction, but need an update of course.

## 5. THE STEADY-STATE DRAG FORCE

Torobin and Gauvin (1959a) provide a lot of interesting details about the early experiments on the fluid resistance experienced by single spherical particles during their steady motion through a stagnant fluid under the effect of gravity. The pertinent expression for the drag force $F_D$ in such cases runs as

$$F_D = C_D A \frac{1}{2} \rho_f v_s^2 \tag{7}$$

in which $A$ denotes the (largest) cross-sectional area of the particle (or object) exposed to the oncoming fluid flow, $\rho_f$ stands for the fluid density, and the drag coefficient $C_D$ is a function of the particle Reynolds number defined by $Re = \rho_f |v_s| d / \mu_f$ in which $d$ denotes particle diameter. Eventually, Lapple and Shepherd (1940) collected lots of experimental data into the so-called standard drag curve for a single sphere, being a single plot presenting $C_D$ versus the particle Reynolds number. This plot found its way into all chemical engineering reference books (e.g., Perry's Chemical Engineers Handbook) and—in spite of the way the pertinent experimental data are collected—is widely accepted as being valid also for the case of a stationary particle in a uniform flow.

Several correlations are available for approximating the standard drag curve for a single sphere over a particular range of particle Reynolds numbers. The most widely used correlations are:
Stokes:

$$C_D = \frac{24}{Re} \quad \text{for } Re < 0.1 \text{ (creeping flow)} \tag{8}$$

Schiller and Naumann, see, e.g., Clift et al. (1978):

$$C_D = \frac{24}{Re} \cdot \left(1 + 0.15 Re^{0.687}\right) \quad \text{for } Re < 700 \text{ or } < 1000 \tag{9}$$

Morsi and Alexander (1972):

$$C_D = a_1 + \frac{a_2}{Re} + \frac{a_3}{Re^2} \quad a\text{'s depend on} Re \tag{10}$$

Wallis (1974):

$$C_D = 20.76 Re^{-0.64} \quad \text{for } 1.5 < Re < 100 \tag{11}$$

$$C_D = 6.32 Re^{-0.38} \quad \text{for } 100 < Re < 1080 \tag{12}$$

In the case of (distorted) bubbles, often (see, e.g., Niceno et al, 2008) a very different correlation is used:

$$\text{Ishii and Zuber (1979)}: \quad C_D = 0.667 Eo^{0.5} \tag{13}$$

in which $Eo$ denotes the nondimensional Eötvös number $Eo = \frac{g(\rho_l - \rho_g) d_B^2}{\sigma}$, where $d_B$ stands for the (equivalent) bubble diameter, $\sigma$ for interfacial tension, and the densities relate to liquid (l) and gas (g). It should be kept in mind when using correlations like Eqs. (8)–(13) that the uncertainty (or accuracy) in the $C_D$ values obtained this way is still of the order of 10%. This is illustrated in the systematic CFD study of bubble columns by Tabib et al (2008).

Substituting Eq. (8) into Eq. (7) results in Stokes law for the unidirectional steady-state creeping motion of a sphere in a viscous fluid:

$$F_D = 3\pi\mu v d \quad \text{for } Re < 0.1 \tag{14}$$

It can be derived analytically (see, e.g., Bird et al, 1960) that two-thirds of this Stokes force is due to friction, and—even under this creeping flow conditions—one-thirds results from form drag.

Strictly speaking, the following conditions pertain to both the Stokes derivation and the standard drag curve and the above associated drag coefficient correlations (Tchen, 1947; Torobin and Gauvin, 1959a):

- the particle is spherical, has a smooth surface, and does not rotate;
- it is falling (under the influence of gravity) through a stagnant homogeneous infinite medium (no effects of containing walls) without any disturbance and without any other particle or object in the vicinity;
- it is moving at a constant velocity, the so-called terminal velocity.

The effect of these restrictions is that the boundary layer flow around the particle and—if applicable—the wake behind it as well as the pressure field are well developed, overall steady, and clearly defined.

Of course, using the correlations (8)–(13) under circumstances different from those cited above is not allowed. This means, first of all, that the particle motion through the fluid or the fluid flow around the particle should be steady and strictly 1D over a sufficiently long distance to allow for the development of boundary layer and wake toward a steady state and to permit the use of the standard drag curve. Lateral forces due an asymmetrical flow field should be absent. The development of both the boundary layer around and the wake behind a particle is affected by local accelerations or decelerations of the immersed particle and/or of the embedding fluid, by a steady shear field in the surrounding fluid, by the dynamics of free-stream turbulence, by particle rotations (either externally and deliberately imposed, or as the result of shear flow), by adjacent walls of a container, and by the presence and motion of adjacent particles in a multiple particle system. All such effects may result in different values of the drag force and, in many cases, also in lateral forces absent in the experiments leading to the standard drag curve and the associated drag correlations.

An example illustrating this issue is the derivation (in 1910) by Oseen of the drag coefficient for the case that the particle is stationary in a fluid flow where the inertia can no longer be neglected due to a finite Reynolds number, say in the range $0.1 < Re < 5$. His derivation results in the correlation

$$C_D = \frac{24}{Re}\left(1 + \frac{3}{16}Re + \cdots\right) \quad \text{for } Re < 5 \tag{15}$$

which results in $C_D$—values slightly higher than given by the standard drag curve where Stokes underpredicts. The Oseen solution is discussed in some greater detail by, e.g., Torobin and Gauvin (1959a) and Brenner (1966).

When a particle moves within a containment, the development of a boundary layer and a wake and therefore its motion will be hindered—depending on the ratio of particle diameter to containment diameter. Happel and Brenner (1973) discussed this effect extensively for low particle Reynolds number flows. More recently, Chen and McLaughlin (1995) derived a set of correction factors for the drag force in the vicinity of just a wall.

Adopting the Eqs. (7)–(13) for computationally simulating turbulent multiparticle or multiple bubble systems may ignore the conditions under which the delicate experiments leading to standard drag curve and correlations have been carried out. It is not just a matter of checking the range of particle Reynolds numbers one is dealing with, rather it is about appreciating the complex hydrodynamics of the flow around an immersed particle (development and separation of the boundary layer, wake formation).

## 6. THE EFFECT OF ACCELERATION: ADDED MASS AND BASSET FORCE

According to Torobin and Gauvin (1959c), the study of a particle accelerating while moving through a fluid started with Bessel as early as 1826 who was interested in the motion of a spherical pendulum in both water and air. Usually, however, reference is made to the early investigations (in the era 1888–1927) of Basset, Boussinesq, and Oseen who arrived at the following equation of motion for a particle starting from rest and accelerating unidirectionally in a stagnant viscous fluid while all the time the particle Reynolds remains smaller than unity:

$$F_D = \frac{1}{2} V_p \rho_f \dot{U}_p + 3\pi\mu U_p d + \frac{3}{2}\pi\mu d^2 \left(\frac{1}{\pi\nu}\right)^{1/2} \int_{\theta_o}^{\theta} \left[\dot{U}_p(\theta_1)/(\theta-\theta_1)^{1/2}\right] d\theta_1 \quad \text{for } Re < 0.1 \tag{16}$$

in which $U_p$ and $\dot{U}_p$ denote particle velocity and particle acceleration, respectively. The first term of this so-called BBO equation stands for the added mass of fluid that is accelerated with the particle, the second term is the Stokes' steady-state force, and the third term is a memory integral, or history, contribution usually named after Basset that takes into account that the boundary layer development lags behind during acceleration and which increases the instantaneous drag (see, e.g., Torobin and Gauvin, 1959a; Tchen, 1947).

The following restrictions apply to the use of the original BBO equation:

- the particle is spherical and rigid, has a smooth surface, and does not rotate;
- the particle starts from rest and accelerates unidirectionally from the moment $\theta = \theta_o$ in a homogenous fluid originally at rest.

The Basset force may only be important numerically, when the particle accelerates strongly and/or the particle is large (since the Basset force is proportional to $d^2$, while the Stokes force is proportional to just $d$).

Numerically, the added mass term in the BBO equation, Eq. (16), may be only important in cases where the fluid density is higher than the particle density, i.e., mainly in bubble flows. With creeping flow around spherical particles (including rigid spherical bubbles), the volume of fluid accelerated along with a single particle is found to be equal to half the particle volume. Under different circumstances, the coefficient 1/2 in this term is usually replaced by the added mass coefficient $C_A$ which may be a function of the acceleration number $\mathrm{Ac} = \left(\frac{\dot{U}_p d}{U_p^2}\right)$. For the harmonic oscillation of a sphere, Odar and Hamilton (1964) proposed.

$$C_A = 1.05 - \frac{0.066}{0.12 + \mathrm{Ac}^{-2}} \tag{17}$$

Torobin and Gauvin (1959c) reviewed much of the earlier research on the effect of acceleration on the fluid–particle interaction force; they claim that there are substantial "differences between the wake structures occurring in oscillating and steady motion." Their conclusion is that using a total unsteady flow drag coefficient "has more theoretical justification than the summation of coefficients often employed." Torobin and Gauvin (1959c) report on experiments by Ingebo in the 1950s who found a reduced drag coefficient in the case of a spray accelerating in an air flow and arrived at a modified drag coefficient $C_{DA}$ given by

$$C_{DA} = 27Re^{-0.84} \tag{18}$$

With a view to applying the concept of the BBO equation to turbulent flows, Tchen (1947) extended the BBO equation for cases where both the particle and the fluid undergo acceleration. Corrsin and Lumley (1956) and Maxey and Riley (1982) identified several inconsistencies in Tchen's work. Their resulting equations comprise many more small terms which seem to be important for applications such as the dispersion of very fine particles in a turbulent atmosphere. For a more detailed discussion on these terms and their relevance, the reader is referred to Maxey and Riley (1982) and Olivieri et al (2014). When the density of the particle is sufficiently larger than that of the ambient fluid, the equation of motion may be simplified again (Yamamoto et al, 2001). Armenio and Fiorotto (2001) reported on numerical simulations assessing systematically the importance of added mass and Basset force in turbulent flows.

When the fluctuations felt by a particle immersed in a time varying or turbulent fluid could be characterized by a frequency $f$, the ratio of Basset force to Stokes force is of the order of $0.1\left(\frac{d^2 f}{\nu}\right)^{1/2}$—see also Derksen (2003). At a high frequency, this Basset force—if applicable—would therefore be much more important than the Stokes steady-state drag law. The delicate question is however whether the BBO equation modified in one way or another may really be used when a particle is immersed in a 3D flow field consisting of a dynamic mixture of large and small eddies and therefore continuously accelerates, decelerates, and changes direction as it is subjected to eddies of varying orientations and strength. The boundary layer around such a particle has to adapt continuously to dynamic boundary conditions—which is very different from the gradual development in the original unidirectional BBO setup. The time scales associated with the turbulent flow and with the boundary layer development may be too different.

Another issue needs mentioning: the added mass term comprises a time derivative. Particularly when not only the particle but also the fluid is accelerating, special care must be taken as to which total time derivatives have to be taken: the time derivative of the particle velocity should follow the particle velocity, while the total derivative of the fluid velocity should follow the undisturbed fluid velocity as seen by the particle (see, e.g., Balachandar and Eaton, 2010).

For higher particle Reynolds numbers, when inertial effects in the developing and separating boundary layer and in the wake behind the particle can no longer be ignored, not only is Eq. (16) itself no longer valid—as the derivation falls short outside the viscosity dominated creeping flow regime—but also a linear addition of several terms to arrive at the total interaction is no longer allowed. In a flow where inertial effects may no longer be ignored, decomposing a flow field into several (canonical) contributions is incorrect.

Suppose there are two canonical types of inertial flow around a particle with each its own fluid–particle interaction force, such as a steady-state drag and a history related force. Each of these forces is due to a particular fluid flow field around the particle in question. Unless the principle of separation of scales applies and/or inertial effects can safely be ignored, the linear addition of two flow fields each described by its own Navier–Stokes equation does not yield the total flow field, just because of the convective or inertial terms in the Navier–Stokes equation, e.g.,

$$U_{x1}\frac{\partial U_{y1}}{\partial x} + U_{x2}\frac{\partial U_{y2}}{\partial x} \neq (U_{x1} + U_{x2})\frac{\partial}{\partial x}(U_{y1} + U_{y2}) \tag{19}$$

This observation applies to all such attempts aiming at a correct expression for a total interaction force by combining a number of forces valid for individual canonical cases—at least in cases in which inertial effects cannot be ignored.

## 7. THE LIFT FORCE

As discussed in the context of Eqs. (1) and (2), drag force and lift force are intimately connected as they stem from the same velocity and pressure fields. Lift is only obtained in cases where the flow on either "side" of an object is asymmetrical. In the case of an airfoil, this asymmetry is realized by turning the airfoil such that the air flow approaches it under an angle, the so-called angle of attack, with the view of creating a large pressure difference between "lower" and "upper" side. The lift depends on the shape, particularly the curvature, of the airfoil; a cambered airfoil already generates lift with the angle of attack being zero. On the analogy of the drag force, the lift force is usually (Munson et al, 2010) calculated by means of

$$F_L = C_L A \frac{1}{2} \rho_f v_s^2 \tag{20}$$

in which the lift coefficient is again a function of the Reynolds number related to the particle (or object). The underlying assumption is still that the oncoming free stream is steady and uniform (or homogeneous). The case of the lift force on a (towed) particle in a turbulent flow field has been studied by Homann et al (2015).

Our interest is mostly in (nearly) spherical particles: then lift is less obvious a concept. Let's first consider a situation in which the (main) flow direction of the carrier phase does not align with the particle velocity. An example of this is a particle dropping at its terminal velocity under the action of gravity through a steady and homogeneous horizontal air or water flow. Although the path covered by the particle is (at least) in a 2D plane, the term "lift" is not appropriate here as the particle's motion is not driven by an asymmetric flow field with pertinent large form drag. In such cases, the steady-state fluid–particle interaction force follows from a 3D generalization of Eq. (7) where vectors are denoted by means of an underscore:

$$\underline{F}_D = C_D A \frac{1}{2} \rho_f |\underline{v}_s| \underline{v}_s \quad \text{with } C_D = C_D(Re) \tag{21}$$

in which $\underline{v}_s = \underline{U}_p - \underline{U}_f$ and the particle Reynolds number is given by $Re = \rho_f |\underline{v}_s| d / \mu_f$. This expression for $\underline{F}_D$ along with this definition of $Re$ in terms of the magnitude of the slip velocity vector reflects that particle motion and flow field around the particle are the result of the combined effects of gravity and drag, generally resulting in a curved particle path. As a result, this flow field may be different from that in the rectilinear case for which the common $C_D(Re)$ correlation has been determined. Nevertheless, the common $C_D(Re)$ correlation is being used in this and many other situations in which the particle motion is not aligned with the flow field of the fluid. With the view of tracking the particle path, the force vector $\underline{F}_D$ has to be decomposed in its components, e.g.,

$$F_{D,x} = C_D A \frac{1}{2} \rho_f |\underline{v}_s| v_{s,x} \tag{22}$$

The term "lift" is reserved for cases in which a particle immersed in a flow experiences different velocities, velocity gradients, and pressures on

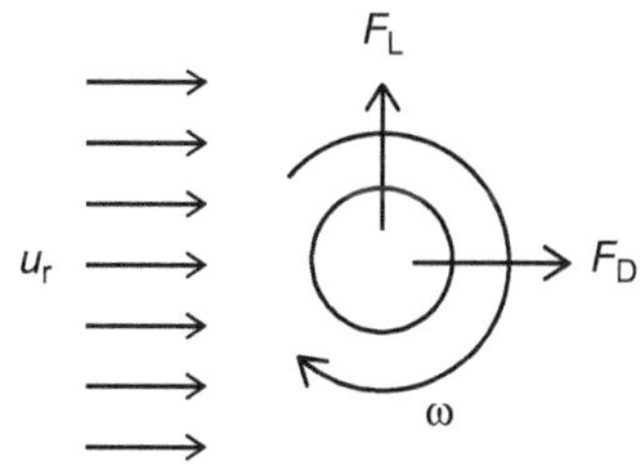


**Figure 14** The Magnus force on a rotating particle in a uniform flow. *Reprinted from Tsuji et al (1985) with permission from the ASME.*

either side. Obvious conditions of interest are then particles moving in a shear flow, such as in a laminar Poiseuille flow, and particles in a turbulent flow. Under such conditions, the particle also starts revolving about its axis. Another example relates to gas–liquid flows in vertical pipes and columns where bubbles close to the wall may experience a lift force due to the near-wall gradient in the vertical liquid velocity.

Nowadays, the lift force in such a case is named after Saffman (1965, 1968) and will be discussed in greater detail below. Another example of lift force is due to a spin deliberately imposed on a particle (like with a tennis ball, a golf ball, or a baseball), named after Magnus who—according to Torobin and Gauvin (1960a)—was one of the first to study this phenomenon as early as 1852. Figure 14 (reproduced from Tsuji et al, 1985) illustrates the Magnus lift effect.

## 7.1 The Saffman Lift Force

Many experimental observations have been reported in the literature about particles which in a shear flow deviate from a rectilinear path. Early examples are those by Tollert (1954), Segré and Silberberg (1962), and Goldsmith and Mason (1962) are in terms of the preferential equilibrium radial or lateral position in the flow field. Torobin and Gauvin (1960a) and Brenner (1966) reviewed these early observations and the first attempts to explain them. The preferential radial or lateral position depends on the density difference between particle and fluid, on the sign of the shear rate, and on whether the particle is lagging behind the fluid flow or just leading.

Nowadays, the generally accepted expression for the lift force on a single small spherical particle in an unbounded linear shear flow is due to Saffman (1965, 1968)—see also Stone (2000):

$$F_{LS} = 1.615\mu d\left(U_f - U_p\right) Re_G^{1/2}\,\mathrm{sgn}(G) \tag{23}$$

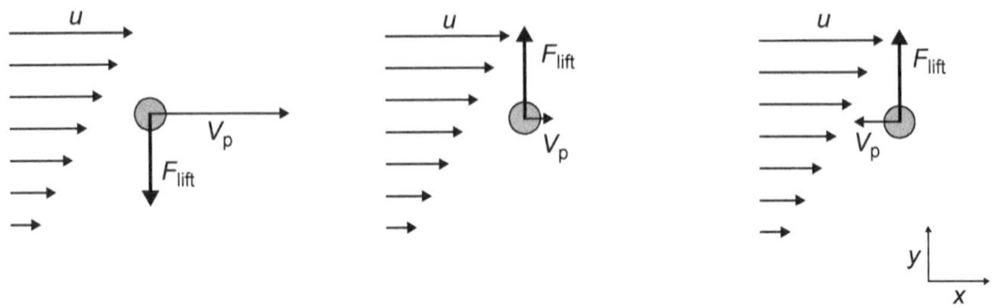


**Figure 15** Direction of the lift force on a particle in a shear flow for varying relative particle velocities. *Reprinted from Van Wageningen (2005).*

in which $G = \mathrm{d}U_\mathrm{f}/\mathrm{d}y$ is the fluid velocity gradient (shear rate) and $Re_\mathrm{G} = |G|\,d^2/\nu$ denotes the Reynolds number related to the (local) shear rate. Figure 15 (from Van Wageningen, 2005) illustrates the direction of the Saffman lift force for a variety of scenarios. The above Saffman expression is valid under a number of restrictions:

- the particle is spherical, rigid, and has a smooth surface;
- the shear flow is steady, unbounded, and linear;
- the particle Reynolds number *Re*—see Eq. (19)—is much smaller than unity;
- the Reynolds number related to the shear rate $Re_\mathrm{G}$ is also much smaller than unity;
- the rotational Reynolds number $Re_\Omega = \Omega d^2/\nu$ is much smaller than unity, where $\Omega$ is the rotational speed of the particle;
- $Re \ll \sqrt{Re_\mathrm{G}}$.

While Hibiki and Ishii (2007) present a table reviewing the essentials of as many as 39 lift force studies and equations, the present chapter highlights just a few. Several authors (Asmolov and McLaughlin, 1999; McLaughlin, 1991; Mei, 1992; Wang et al, 1997) modified the above Saffman expression by adding a coefficient with the view of extending its range of validity. For somewhat higher *Re* numbers, Mei (1992) proposed an expression which—rewritten in terms of the fluid's vorticity $\omega$ (see also Derksen, 2003)—runs as

$$\underline{F}_\mathrm{LS} = 0.1524 \frac{\pi}{4} d^3 \frac{1}{2} \rho_\mathrm{f} \left[ \left( U_\mathrm{f} - U_\mathrm{p} \right) \times \omega \right] \tag{24}$$

Legendre and Magnaudet (1998) developed an empirical correlation on the basis of numerical simulations which can be summarized as

$$\underline{F}_{\mathrm{LS}} = C_{\mathrm{L}\infty}\frac{\pi}{8}d^3\rho_{\mathrm{f}}\left[\left(U_{\mathrm{f}} - U_{\mathrm{p}}\right) \times \omega\right] \tag{25}$$

where $C_{\mathrm{L}\infty}$ in the low to moderate *Re* regime depends on *Re* and the non-dimensional shear rate and in the moderate to high *Re* regime depends on just *Re*.

In the case of (solid) particles toward or in the vicinity of a flat or curved wall, the expression of the lift force needs correction. Several authors considered this case, such as Cox and Hsu, 1977; McLaughlin (1993), Chen and McLaughlin (1995), and Wang et al (1997).

In the case of (larger) bubbles, deformation of the bubbles, slanted wakes behind the bubbles, wobbling motions, and bubble–bubble interaction complicate the issue of lift force in a shear flow substantially. Several authors in the early 1990s (see Sokolichin and Eigenberger, 1994) erroneously denoted the shear-induced lift force on bubbles as the Magnus force (see below) and argued that such a lift force would explain that bubbles in a uniformly aerated bubble column tend to accumulate toward the center of the column. To reproduce this effect in their simulations, those authors did have to change the sign of their lift force: see again Sokolichin and Eigenberger (1994), who further raised the question whether bubbles would really rotate when subjected to a shear field.

Tomiyama et al (2002) carried out measurements on the transverse migration of single bubbles in simple shear flows and distinguished between a shear-induced component of lift force and a lift force component induced by a slanted wake. These authors proposed a complicated empirical correlation for the "net transverse lift coefficient" in which a modified Eötvös number (related to the maximum horizontal dimension of a bubble) occurs predominantly. It is noteworthy that according to Tomiyama's correlation the "net transverse lift coefficient changes sign when deformation becomes substantial"—which would explain the two experimentally observed phenomena (depending on conditions) of "wall peaking" and "core peaking" in the radial void profiles.

This switch in sign of the lift coefficient (or of the lateral migration) was confirmed—at least qualitatively—in numerical simulations of the lateral migration of a single bubble in a shear flow by Sankaranarayanan and Sundaresan (2002) and by Bothe et al (2007) and by experimental data collected and interpreted by Lucas and Tomiyama (2011). Also Mudde et al (2009) observed both wall peaking and core peaking in the radial voidage profiles, referred to Tomiyama's change of sign in the lift but also discussed the role of large coherent structures in switching from wall peaking to core peaking. In the latter experiments, there does not seem to be much room for wakes behind the closely packed bubbles at the higher void fractions—see Fig. 16, reproduced from Mudde et al (2009).

Hibiki and Ishii (2007) derived a model for the effect of bubble deformation that can be decomposed in a shear-induced and a deformation-induced part. Sankaranarayanan and Sundaresan (2002) correlated their lift coefficient at low shear conditions—obtained by means of numerical simulations—with the nondimensional Tadaki number *Ta*, which is a combination of Reynolds and Morton: $Ta = Mo^{0.23}\ Re$; they also found that highly distorted bubbles experience a much larger lift force than circular or spherical bubbles, accompanied by a change in the flow pattern around the bubble and eventually even leading to a negative lift coefficient.

Quite rightly, the lift force on bubbles may still be described as poorly understood, certainly in multiple bubble systems. Bubble deformation plays an important role: more or less spherical bubbles tend to pair next to each other (resulting in slower rise velocities), while deformable bubbles rise faster as they line up vertically. Given the results from DNSs as to clustering of bubbles—see, e.g., Fig. 17 reproduced from Tryggvason et al (2006)—one may even wonder whether lift force studies on an individual bubble are really increasing our understanding of multi-bubble flows. The issue of the direction of the lift force for bubbles, deformable or not, flowing up or down along a wall has also been addressed in DNSs by Tryggvason and Lu (2015).

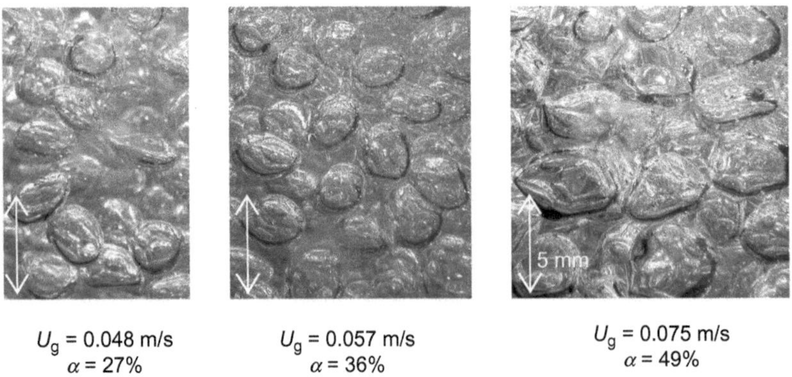


**Figure 16** Snapshots of bubbly flows at high values of gas velocity and bubble holdup, showing the proximity of the bubbles. *Reprinted from Mudde et al (2009) with permission from the ACS.*

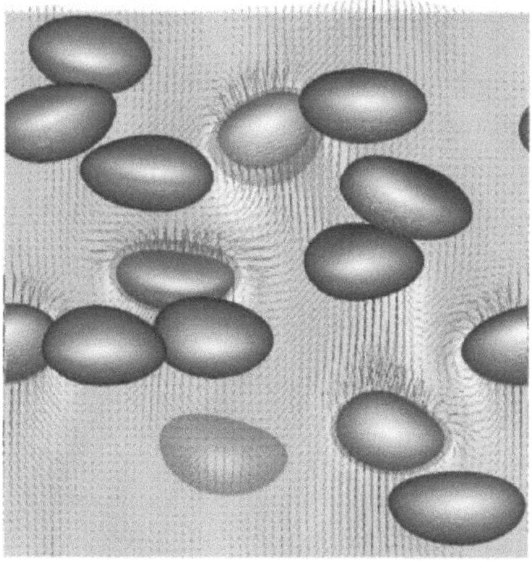

**Figure 17** Snapshot from a simulation of 14 wobbly bubbles in a periodic domain. *Reprinted from Tryggvason et al (2006) with permission from the IoP.*

## 7.2 The Magnus Lift Force

While the examples of spinning tennis balls, golf balls, and baseballs all relate to (very) high Reynolds numbers ($Re > 10^4$), in dispersed two-phase flows a finite range of moderate Reynolds numbers are more interesting. Tsuji et al (1985) carried out experiments in the range $550 < Re < 1600$, while Oesterlé and Bui Dinh (1998) did the same for nondimensional angular velocities $\gamma = \Omega d / (U_p - U_f)$ between 2 and 12 in a uniform flow field for $10 < Re < 140$.

Rubinow and Keller (1961) derived expressions for the steady-state lift force for a spinning sphere in a stagnant viscous fluid for small values of $Re$:

$$\underline{F}_{LM} = \frac{\pi}{4} d^3 \frac{1}{2} \rho_f [\Omega \times U_p] \{1 + O(Re)\} \tag{26}$$

The question is what the expression for the Magnus force is when the ambient medium is not stagnant (initially) but also exhibits vorticity of its own. The idea is that in such a case it is the particle rotational velocity relative to the fluid vorticity, or the relative slip, that drives the Magnus lift force. This idea found its way into the textbook by Crowe et al (1998) and can also be found in, e.g., Yamamoto et al (2001) and Derksen (2003). The most general expression for the Magnus lift force then runs as follows:

$$\underline{F}_{\mathrm{LM}} = \frac{\pi}{4} d^2 \frac{1}{2} \rho_{\mathrm{f}} C_{\mathrm{LM}} |U_{\mathrm{f}} - U_{\mathrm{p}}| \frac{(\omega - 2\Omega) \times (U_{\mathrm{f}} - U_{\mathrm{p}})}{|\omega - 2\Omega|} \tag{27}$$

in which $C_{\mathrm{LM}}$ is a complicated function of the particle Reynolds number $Re$ and of the rotational Reynolds number $Re_{\mathrm{R}} = \left(|(1/2)\omega - \Omega| d^2\right)/\nu$ which is different from $Re_{\Omega}$ referred to earlier in the context of Eq. (23).

## 8. THE IMPACT OF TURBULENCE

The canonical case of a single particle settling at a higher particle $Re$ number (i.e., at $Re \gg 1$) in a stagnant fluid is different from the case of a single stationary particle immersed in a turbulent flow (at a high $Re$ number) and is also different from a single particle moving at a high relative velocity through a turbulent flow. The differences relate to the development of the boundary layer around the particle and the wake behind it, both of which are strongly determined by the turbulent momentum exchange with the bulk of the fluid. The time and length scales of the flow in a laminar and turbulent boundary layer are very different from the spectrum of eddies and the associated time and length scales in a turbulent bulk flow. The Reynolds number of a (turbulent) flow (defined in terms of the absolute linear fluid velocity and a typical macro length scale such as pipe or channel diameter) is rather different from the particle Reynolds number (defined in terms of the relative particle velocity and the particle diameter). These two Reynolds numbers characterize different types of flow and should not be mixed up. The ratio of these two Reynolds numbers is an important parameter describing the impact of free-stream turbulence on the development of the boundary layer induced by the (relative) motion of the (see also Mei, 1994). Also Balachandar (2009) distinguishes between the particle Reynolds number and the bulk flow Reynolds number.

Figure 18 illustrates the effect of free-stream turbulence (vorticity) on the flow around a stationary cylinder: the two pictures show that the development and separation of the boundary layer and the size of the wake in the absence and in the presence of free-stream turbulence are very different, which should be reflected in very different values for the drag coefficient—Torobin and Gauvin (1960b) describe the difference as "startling." These authors discuss various experimental results from the 1950s showing that free-stream turbulence has an effect on the flow field around both stationary and moving particles and on their drag coefficient. All these findings were inconclusive, however, mainly due to the wide

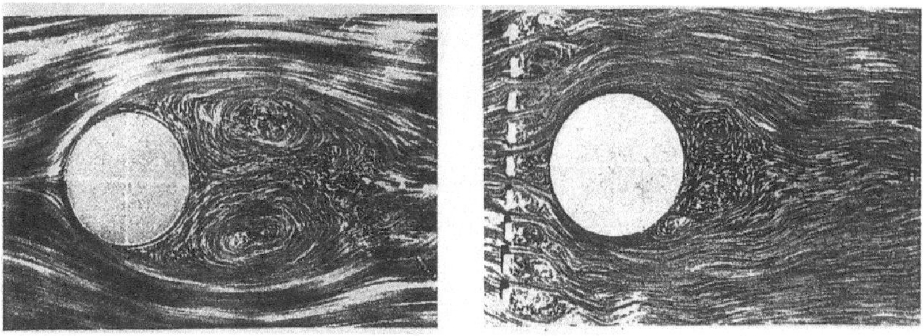

**Figure 18** The flow around a cylinder—left: the incident flow being laminar; right: in a free-stream vorticity. *Pictures taken as early as 1931 by Ahlborn, reprinted from Torobin and Gauvin (1960b) with permission from Wiley.*

variation in experimental conditions investigated and the limited capabilities of experimental techniques in the early days. These authors also reviewed various modeling studies as to the stability of laminar boundary layers and their transition to turbulence is numerous; they concluded that the effect of free-stream turbulence on this transition and, as a result, on the drag coefficient remained unclear.

In 1970, Uhlherr and Sinclair (see Brucato et al, 1998) found an increase of the drag coefficient for spheres as a result of free-stream turbulence, at least at higher values of the turbulence intensity $T = u_{\mathrm{rms}}/v_{\mathrm{s}}$ in which $u_{\mathrm{rms}}$ stands for the root mean square of the turbulent fluctuations in the fluid's axial velocity. Uhlherr and Sinclair considered the pressure difference between nose and wake as well as the position of the boundary layer separation which shifts toward the back of the particle due to the free-stream turbulence. Their correlations are given by

$$C_{\mathrm{D}} = \frac{162T^{1/3}}{Re} \quad \text{for } Re < 50 \text{ and } 0.05 < T < 0.5 \tag{28}$$

$$C_{\mathrm{D}} = 4T + 0.133\left(1 + \frac{150}{Re}\right)^{1.585} \quad \text{for } 50 < Re < 700 \text{ and } 0.07 < T < 0.5 \tag{29}$$

Balachandar (2009) translated turbulence intensity into the ratio of integral length scale to Kolmogorov length scale of the turbulent flow field (see also below) which in fact is a function of the bulk flow Reynolds number. Kim and Balachandar (2012) also emphasized the role of self-induced vortex shedding from the wake and its impact in terms of stochastic components of drag and lift forces at particle Reynolds numbers of 250 and 350.

Mei (1994) recapitulated the understanding obtained in the 1980s and the early 1990s by several experiments and computational simulations concerning the effect of oscillations and turbulence on the settling behavior of a particle in the nonlinear drag range. He particularly distinguished between the so-called trajectory bias which causes particles to collect in regions of high strain rate or low vorticity, and an effect due to the nonlinearity in the drag law. While the trajectory bias has been found to result in higher settling rates, the nonlinearity (i.e., the inertia of the particle) increases the effective particle time scale and as a result reduces the settling velocities as is confirmed by many experimental data among which those measured in stirred vessels by Nienow and Bartlett (1974). On the reverse, several papers report a reduced drag force due to ambient turbulence and even no significant effect (see Balachandar and Eaton, 2010; Torobin and Gauvin, 1960b). Bakker and Van den Akker (1994) used an increased viscosity in the Morsi-Alexander drag coefficient relation, eq. (10), to account for the effect of turbulence. The increase in the drag coefficient was calculated with the help of the turbulent kinetic energy $\kappa$ and its rate of dissipation $\varepsilon$, and results in an increase of the drag coefficient. Brucato et al (1998) considered the net averaged settling rate of a dilute assembly of particles and translated the observed lower settling rate into an increased drag coefficient. Homann et al (2015) found the drag coefficient exerted on a towed sphere increasing with turbulence intensity in the ambient fluid flow and with particle Reynolds number.

The general consensus, however, is that the behavior of a single particle in a turbulent fluid flow is strongly determined by the ratio of various times scales:

- the widely used viscous response (or relaxation) time scale for a particle in the Stokes regime:

$$\tau_{ps} = \frac{\left(\rho_p + (1/2)\rho_f\right) d^2}{18\mu} \tag{30}$$

  where the second term between the brackets stands for the added mass.
- the response time scale for a particle in the inertial subrange as mentioned by, e.g., Mei (1994), Helland et al (2000), and Balachandar (2009):

$$\tau_{pi} = \frac{\left(2\rho_p + \rho_f\right) d^2}{36\mu} \frac{1}{\phi(Re)} \tag{31}$$

in which $\phi(Re)$ is a finite Reynolds number correction to the Stokes drag due to Schiller and Naumann—see Eq. (9).

- the fluid integral time scale $\tau_f$ which in many cases is given by

$$\tau_f = L/U_f, \quad \text{or} \quad \tau_f = L^{2/3}/\varepsilon^{1/3} \tag{32}$$

where $\varepsilon$ stands for the (local) dissipation rate of turbulent kinetic energy and $L$ denotes either the large, or integral, length scale of the flow domain or the size of an $L$-size eddy; in a stirred vessel $\tau_f = 1/N$.

- the typical time scale (eddy lifetime) of the turbulent eddies:

$$\tau_{eddy} = k/\varepsilon \tag{33}$$

in which $k$ denotes the spatially resolved turbulent kinetic energy concentration. Several investigators (e.g., Eaton, 2009; Kulick et al, 1994; Moraga et al, 2003) considered this eddy lifetime rather than the above integral time scale. This eddy lifetime is often used in particle tracking both in a Reynolds-averaged Navier–Stokes (RANS)-based simulation (with the local $k$-value) and in an LES (with $k = k_{SGS}$): after a time interval equal to this eddy lifetime, the particle is subjected to a new eddy as a result of which it will start moving in a different direction.

- the Kolmogorov time scale for a turbulent fluid flow:

$$\tau_k = (\nu/\varepsilon)^{1/2} \tag{34}$$

which is associated with the lifetime of the smallest eddies in a turbulent flow field named after Kolmogorov.

A proper analysis is needed to assess which of the above time scales are relevant in the case of interest. The nondimensional Stokes number is widely used to denote the ratio of the particle relaxation time to a fluid relaxation time (see also Ghatage et al, 2013). Depending on the application and the pertinent fluid–particle interaction model, the fluid relaxation time is either the Kolmogorov time scale (Balachandar, 2009; Collins and Keswani, 2004; Olivieri et al, 2014; Pai and Subramaniam, 2012) or the integral time scale (Derksen, 2003; Lane et al, 2005; Derksen et al, 2008; Van Wageningen et al, 2004) or the eddy lifetime (Eaton, 2009; Moraga et al, 2003).

When a particle is so small that $\tau_{ps}/\tau_k < 1$, it is capable of responding to the dynamic world of the Kolmogorov eddies while obeying Stokes law. A larger particle in the inertial subrange, for which $\tau_k < \tau_p < \tau_f$, is too big to respond to eddies of size smaller than $\tau_p^{3/2}\varepsilon^{1/2}$ and its motion will (mainly) be affected by eddies larger in size than $\tau_p^{3/2}\varepsilon^{1/2}$. When $\tau_p > \tau_f$, a particle is even not capable of adapting to the dynamics of the largest turbulent eddies.

Yet, the dynamic ambient flow around such a large particle—characterized by the abundant presence of eddies of all sizes—may have a definite impact on the development of the flow around it (boundary layer, wake) and therefore on the settling velocity (see again Fig. 18).

Clearly, this simplification in terms of time scales is not the complete story. This can be illustrated by Balachandar's (2009) analysis of particles settling under turbulent flow conditions when a particle continuously experiences accelerations in varying directions as a result of the interaction with turbulent eddies. When $\tau_{ps}/\tau_k < 1$, the gravitational acceleration $g$ competes with the Kolmogorov acceleration $a_k = \varepsilon^{2/3}/\eta^{1/3} = \nu^2/\eta^3$, where $\eta$ denotes the Kolmogorov length scale $(\nu^3/\varepsilon)^{1/4}$. For $(g/a_k) < 1$, gravity dominates the effect of the eddies for particles larger than a critical value which for water at room temperature amounts to 47 μm and in the case of air at room temperature to 284 μm; for particles smaller than such critical values, the settling velocity will be smaller than the relative velocities induced by the turbulence resulting in slow settling. For $(g/a_k) > 1$, gravity dictates the settling. In all these cases, the Reynolds number of the free-stream turbulence does not play a role at all. Balanchandar's (2009) approach in terms of time scales, however, does not consider the details of the dynamic momentum exchange between free-stream turbulence and boundary layer and wake of a particle in the various particle regimes.

Homann et al (2015) concluded the relevant parameter to be the ratio of the viscous boundary layer thickness to the Kolmogorov length scale of the ambient turbulent flow. Other authors—such as Magelli et al (1990), Brucato et al (1998), and Pinelli et al (2004)—considered the ratio of particle size to the Kolmogorov length scale $\eta$ rather than a ratio of time scales and correlate their experimental data with this length scale ratio only. This approach will be discussed in greater detail in the next section.

## 9. MULTIPLE PARTICLE SYSTEMS

In multiple particle systems, particularly when the mutual distances between the particles become smaller, the flow field around a particle in a swarm—i.e., the development and separation of the boundary layer and the formation and dynamic behavior of the wake behind particles—is really affected by the presence and motion of neighboring particles. There are two effects: first of all, there is simply not sufficient space for boundary layer and wake to develop without interference with the adjacent particles and their boundary layers and wakes—and, second, for a given superficial velocity (or flow rate) the presence of multiple particles may affect both the local pressure gradient and the linear velocities at some distance of the particle surfaces, as the fluid not just has to pass around a single particle but must find its way in the tortuous passages between all particles of the swarm. In principle, all these effects are comprised in Eqs. (4)–(6).

Yet, the majority of the literature on fluid–particle interaction starts from the base case of a single particle (or bubble), carries out experiments, derives model equations for the behavior of a single particle under various flow conditions (as discussed), and/or applies single-phase correlations in multiple particle systems. In the pre-CFD era and in the domain of process design and optimization, the usual focus is on the steady-state drag and one takes refuge to simple correlations such as that proposed decades ago by Richardson and Zaki (1954):

$$\frac{v_s}{v_{s\infty}} = (1-\phi)^{n-1} \tag{35}$$

where $v_s$ stands for the slip velocity of a particle in a swarm—see Eq. (21)—and the subscript $\infty$ refers to a single particle, while $\phi$ denotes the volume fraction of the dispersed phase and the empirical exponent $n$ depends on the particle Reynolds number. In Eq. (35), the exponent is $n-1$ rather than just $n$, because the $n$-values reported by Richardson and Zaki (1954) refer to a sedimenting bed surface rather than to the motion of a single particle. The drag force for low particle Reynolds numbers is then given by

$$F_D = 3\pi d\mu v_{s\infty}(1-\phi)^{1-n} \tag{36}$$

The Wen and Yu (1966) correlation

$$F_D = 3\pi d\mu v_{s\infty}\left(1+0.15Re^{0.687}\right)(1-\phi)^{-3.65} \tag{37}$$

is a refinement of the above Richardson and Zaki (1954) relation that introduces the Schiller and Naumann correction factor—see Eq. (9). Both correlations are intended for $\phi$ values in excess of, say, 0.4 and for the range $0.01 < Re < 5000$. Helland et al (2000) used a similar approach when defining a particle relaxation time in a fluidized bed. Kandhai et al (2003) investigated, by means of an LB technique, the applicability of the above Wen an Yu correlation for modeling the interphase drag coefficient in Euler–Euler simulations of fluidized beds, in comparison with the very common Ergun equation (originally derived for packed beds, but also used by, e.g., Tsuo and Gidaspow, 1990).

For more dilute multiphase systems, the correlations developed by Barnea and Mizrahi (1973) are to be preferred because of the underlying physical arguments. Not only did they present a detailed review of the many multiparticle models available at the time, but they also distinguished between three independent effects, viz., (a) a hydrostatic effect, taken into account by using an apparent suspension density rather than the fluid density; (b) a "momentum transfer hindrance" effect due to increased momentum transfer within the fluid due to the presence and motion of adjacent particles taken into account by an apparent viscosity; and (c) a "wall" effect as the adjacent particles have a similar restraining effect on the flow field as the pipe wall for a single particle settling in a pipe.

For the apparent viscosity $\mu_a$, various models starting with Einstein's $\mu_a = \mu(1+2.5\phi)$ are available in the literature; Barnea and Mizrahi (1973) propose

$$\mu_a = \mu \exp\left(\frac{K_1\phi}{1-K_2\phi}\right) \quad \text{with } K_1 = \frac{5}{3} \text{ and } K_2 = 1 \tag{38}$$

The "wall" effect is taken into account by means of a term $\left[1+K_3\phi^{1/3}\right]$ with $K_3 = 1$; Van den Akker and Rietema (1979b) suggest $K_3$ to be a function of $Re$, decreasing from $K_3 = 1$ at $Re = 0.03$ to $K_3 = 0.5$ at $Re = 0.5$. When the dispersed particles are drops, the apparent viscosity takes a different form that also contains the viscosity of the internal liquid (Barnea and Mizrahi, 1975).

According to Barnea and Mizrahi (1973), the drag force in the creeping flow regime then obeys to

$$F_D = 3\pi d\mu v_s\left(1+K_3\phi^{1/3}\right)\exp\left(\frac{K_1\phi}{1-K_2\phi}\right) \tag{39}$$

while the combined gravity/buoyancy force is given by

$$F_g = V_p\left(\rho_p - \rho_f\right)g(1-\phi) \tag{40}$$

When inertia can no longer be ignored, the drag force runs as

$$F_D = C_D A \frac{1}{2}\rho_f v_s^2\left(1+K_3\phi^{1/3}\right) \tag{41}$$

in which $C_D = m\,Re^n$, just like in Eqs. (11) and (12), with $m$ and $n$ depending on the $Re$ range of interest as before, but now with $Re = \rho_f v_s d/\mu_a$. Rietema (1982) illustrated that the Barnea and Mizrahi model predicts that upon increasing $\phi$ from 0 to 0.2, the ratio $v_s/v_{s\infty}$ drops must faster than according to the Richardson and Zaki (1954) correlation.

The topic of the drag coefficient in multiple particle systems has regained interest because of the CFD simulations of *turbulent* multiple particle systems such as agitated solid suspensions, aerated stirred vessels, solid–liquid fluidized beds, and gas–solids cyclones. A recent review on this topic can be found in Ghatage et al (2013) who also presented new experimental data on the slip velocity of a foreign particle in solid–liquid fluidized beds.

Following in the footsteps of Magelli et al (1990), Brucato et al (1998) collected settling data in a Couette-type device by means of a residence time technique which yields the net settling rate of the particles. Their data show that the turbulence in the vessel reduces the net, or evident, settling rate substantially; these authors translate this into a higher value of the drag coefficient and adopt the ratio of particle size to Kolmogorov length scale $\eta$ as the only correlating parameter according to

$$\frac{C_D}{C_{D0}} = 1 + 8.76 \times 10^{-4} \left(\frac{d}{\eta}\right)^3 \tag{42}$$

in which $C_{D0}$ denotes the drag coefficient for a particle moving through a stagnant fluid. Brucato et al (1998) themselves report that particularly the data for the smallest glass beads deviate from the above correlation. Several researchers pursued their line of approach in which $d/\eta$ is taken as the controlling parameter and suggested slightly different correlations: see, e.g., Khopkar et al (2006) and the review by Sardeshpande and Ranade (2012).

With respect to the Brucato et al (1998) approach, one may really wonder however whether the idea of translating lower net averaged settling rates into a higher drag coefficient for a single particle is appropriate, as the path followed by the individual particles may not be rectilinear at all: the approach completely overlooks and ignores the response of the particles to all Taylor vortices and eddies in the annular Couette space in which these authors made their observations, as they focus on the overall, or evident, settling rate. One may expect that the various types of vortex flow described in the appendix of their paper would advect the particles, resulting in convoluted paths covered by the particles. There is no mentioning in their paper about the effect of these turbulent vortex flows on the exact settling behavior of the particles. Adopting the above correlation in the context of CFD simulations for calculating the detailed dynamic response of individual particles to varying local turbulence levels and tracking their paths therefore does not seem appropriate and is certainly not recommended.

In addition, one may raise doubts as to the role of the ratio $d/\eta$ as correlating parameter: while for all experimental conditions this ratio is smaller than unity indeed and therefore suggests all data are in the same regime, in fact the Stokes number—defined as $\tau_{ps}/\tau_k$—varies between O(0.1) and O (10), depending on impeller speed and particle size. This latter variation suggests a strongly varying impact of carrier phase turbulence, as according to Balachandar (2009) particles are capable of responding to the dynamics of a turbulent flow field in cases where $\tau_{ps}/\tau_k < 1$, while in cases where $\tau_{ps}/\tau_k > 1$ they are not, or at least not rigorously. The smallest glass beads data for which $\tau_{ps}/\tau_k < 1$ do seem to be in agreement with Balachandar's analysis that gravity dictates the settling when the ratio $g/a_k$ is much greater than unity—as is the case in all tests. Ghatage et al (2013) present a plot of a normalized slip velocity versus a Stokes number; this plot for a very wide range of Stokes exhibits a wide spread in experimental results from different authors—leaving the topic of the effect of turbulence on drag and slip velocity essentially unresolved.

# 10. CONVENTIONAL CFD SIMULATIONS

## 10.1 The Euler–Lagrange Approach

The above extensive explanation about fluid–particle interaction demonstrates how complex a topic this is when the approach is to draw up generally applicable empirical correlations with the view of using them in the common CFD simulations by means of the (commercial) CFD software. Except for the Stokes regime of particle motions when the particle Reynolds number is smaller than 0.1 (or, less strictly, smaller than unity), simply adding in any type of CFD simulation a number of expressions derived for specific canonical cases to get the correct total interaction force conflicts with the nonlinear character of fluid flow and the Navier–Stokes equation. The common drag force correlations have been obtained under specific conditions: steady-state, no turbulence in the ambient stagnant fluid, no lift force (neither a result of shear flow in the fluid, nor due to particle rotation), and no adjacent particles or containment walls. Using the steady-state drag coefficient correlations for conditions in which a particle continuously experiences varying hydrodynamic conditions ignores the time scales needed for boundary layers and wakes to adapt to such variations.

Obviously, inertial particles are not capable of exactly following the fluid velocity fluctuations (see also Dehbi, 2008); a drift correction velocity, or a drifting velocity, and an extra time scale pertinent to the turbulence as seen viewed by the particles may be used to take this effect into account (Mudde and Simonin, 1999; Viollet and Simonin, 1994). The result of this drifting velocity is that the particles get dispersed. While usually proper attention is being paid to the Reynolds number restrictions for the use of empirical correlations, these more generic considerations with respect to the details of the fluid mechanics hardly receive attention.

In the Euler–Lagrangian approach of two-phase flow (see, e.g., Crowe et al, 1996), the particles are treated as point particles: the finite volume of the particles is not considered and the flow around the particles is not resolved. The motion of the particle is simulated by means of Newton's second law and that is why the fluid–particle interaction force is needed and the empirical correlations enter. Although the flow around the particles is not resolved, any empirical correlation does reflect the hydrodynamics of the canonical case involved. The use of the Euler–Lagrange approach, or point-particle method, is usually restricted to the more dilute gas–solid and liquid–solid systems. Ignoring the mutual interaction of particles is therefore not too serious a simplification. The fluid–particle interaction can be treated in the simpler one-way mode or according to the more complicated two-way coupling mode in which the particles also affect the carrier phase flow field (Decker and Sommerfeld, 2000; Derksen, 2003; Derksen et al, 2008).

In addition, we should distinguish between particle tracking in a flow field obtained by means of the very common RANS equations and particle tracking in a flow field simulated by the more advanced LES technique. The difference between RANS and LES has been described elsewhere, e.g., Van den Akker (2006). Now, only the aspects relevant for a proper application of the fluid–particle interaction force in these types of simulation will be highlighted.

In RANS-based simulations, the focus is on the average fluid flow as the complete spectrum of turbulent eddies is modeled and remains unresolved. When nevertheless the turbulent motion of the particles is of interest, this can only be estimated by invoking a stochastic tracking method mimicking the instantaneous turbulent velocity fluctuations. Various particle dispersion models are available, such as discrete random walk models (among which the eddy lifetime or eddy interaction model) and continuous random walk models usually based on the Langevin equation (see, e.g., Decker and

Sommerfeld, 2000; Dehbi, 2008). Given the various and numerous assumptions of any turbulence model used for the RANS simulation and the degree of sophistication of the random walk model applied, a very accurate representation of the fluid–particle interaction force may not be very essential. Hence, very realistic information on particle trajectories cannot be expected on the basis of RANS-based flow fields.

In LESs (see, e.g., Crowe et al, 1996; Van den Akker, 2006), most of the turbulent motions of the flow field of the carrier phase are resolved. A particle being tracked in an LES flow field feels all the associated velocity fluctuations. Now, just the subgrid scale (SGS) fluctuations are not resolved while they still may have an effect on the particle motions; modeling their effect is usually done by means of a Gaussian stochastic procedure (Derksen et al, 2008), on the basis of the idea that these SGS velocity fluctuations (in the absence of particles) are more or less isotropic. It may be anticipated, however, that the resolved velocity fluctuations are much stronger than the SGS fluctuations. When, in addition, the particle relaxation time is larger than the time step applied in the LES, Derksen (2003) argued that particle motion is hardly influenced by the SGS velocities.

## 10.2 The Euler–Euler Approach

For computationally simulating more dense two-phase flow systems, the Euler–Euler approach, based on the two-fluid model, is widely exploited. In this approach, the two phases are conceived as two interpenetrating continua and the fact that the dispersed phase in real life is made up by individual particles, drops, or particles is overlooked. The only term where particle size pops up is in the phase interaction force—see Eqs. (5) and (6)—for which empirical expressions derived from single-particle behavior are used which contain particle size in various ways. All above generic remarks about the combined use of the various correlations again apply however.

Most of the Euler–Euler two-phase flow simulations are of the RANS-type, with the turbulence model being adapted to a varying extent with the view of taking the effect of the second phase into account. As a result, this type of simulations is based on a huge number of modeling assumptions which may lead to many (partially compensating) approximations. Due to the large number of particle differential equations involved, the huge computational burden results in a need to restrict the number of computational cells and in computational stability issues and a slow convergence rate of the simulation. Overall, the resolution of the simulation results is limited and the data produced should be viewed with caution.

In a series of papers on dispersed bubbly flows—Deen et al (2001), Dhotre et al (2008, 2009), and Niceno et al (2008, 2009)—a so-called Euler–Euler LES (EELES) concept is being explored. A typical feature of this EELES concept is the Milelli criterion that requires the grid cell size to be larger than the bubble size which in most of the above five papers was taken to be 4 mm. The reasoning behind this criterion is that the cell size should be large enough to contain all the bubble interface details, probably including the flow field in the immediate vicinity of a bubble. Apart from the fact that no mentioning is made of the effect of the large wakes behind such 4 mm bubbles and their mutual interaction, this criterion misinterprets the two-fluid concept of two interpenetrating continua. According to a proper understanding of this concept, the only role of particle or bubble size is in estimating (alternatively via a separate routine) the phase interaction force(s); there is no reason why this force should not be allowed to have the same value in adjacent cells.

Furthermore, the consequence of the Milelli criterion is that a substantial proportion of the turbulent spectrum of bubbly flows, the so-called bubble-induced turbulence, has to be incorporated as a (major) contribution to the SGS turbulence—as done in most of the above papers. Adhering to the Milelli criterion therefore contrasts with the fundamental idea of an LES through which most of the energetic eddies of a turbulent flow field are resolved thanks to the application of a very fine grid. The present author agrees with Tryggvason and Lu (2015) that refraining from resolving a (relatively) large range of scales leads to "LES-like" simulations at most—for the sake of which separate SGS models have to be derived which should take this bubble-induced turbulence into account.

Several papers on Euler–Euler simulations (both RANS based and of the above EELES type) report about the need to include a separate "turbulent dispersion" term (or force) for reproducing a correct spatial distribution of suspended particle clouds or bubble swarms. Various models for this turbulent dispersion have been suggested, all of them modeling this effect in terms of the gradient of the particle or bubble volume fraction. Sometimes, this additional term is in the continuity equations (Bakker and Van den Akker, 1994; Sardeshpande and Ranade, 2012; Tamburini et al, 2014) with the justification that is due to fluctuations in the volume fractions; however, as the volume fraction which anyhow is an averaged variable does not vary on the scale of the particles or bubbles—see the discussion on Eq. (3)—this justification does not have a sound basis (see also Li et al, 2015).

Other authors (e.g., Dhotre et al, 2008, 2009; Hill et al, 1995; Laín et al, 2000; Masood and Delgado, 2014; Moraga et al, 2003; Mudde and Simonin, 1999; Niceno et al, 2008) have incorporated such a term into the phase interaction force in the momentum balances, its basis being in the above concept of a drifting velocity (Dehbi, 2008; Viollet and Simonin, 1994). In all cases, assumptions have to be made about the (isotropic) turbulent diffusivity or the turbulent diffusivity tensor in this term. Also this topic is covered by the recent review by Pourtousi et al (2014).

## 10.3 Computational Results from the Literature

Numerous papers have been published on particularly the topic of this chapter: the role of the fluid–particle interaction force in *RANS-based* simulations of either the Euler–Lagrange or the Euler–Euler type. The reason for this massive effort is in the common perception that the current understanding about the fluid–particle interaction is insufficient and responsible for the generally modest predictive performance of the current RANS-based two-phase flow simulations. This in itself may already be indicative of a serious flaw in the way the fundamental knowledge on fluid–particle interaction in the fluid mechanics literature has been condensed in the conventional CFD software. An exhaustive review of all RANS-based two-phase simulations reported in the literature is virtually impossible and also not needed, as various reviews are available, sometimes restricted to a particular canonical interaction force and/or a particular class of two-phase flows (Balachandar and Eaton, 2010; Hibiki and Ishii, 2007; Kuipers and Van Swaaij, 1998; Li et al, 2015; Ochieng and Onyango, 2010; Pourtousi et al, 2014; Sokolichin et al, 2004; Tabib et al, 2008; Visuri et al, 2011; Shah et al, 2015).

All RANS-based two-phase flow CFD studies eventually use the detested technique of adding several canonical correlations—or only use the steady-state drag force which by the way always is taken into account. Many papers report on the effect of using alternative correlations for the drag coefficient and of adding or dropping added mass, Basset term, lift force, turbulence correction, and turbulent dispersion (see, e.g., Shah et al, 2015). Most papers, e.g., Becker et al (1994), Hill et al (1995), Lahey (1995), Hoekstra et al (1999), Laín et al (2000), Oey et al (2003), Lucas et al (2007), Elsayed and Lacor (2011), Lucas and Tomiyama (2011), Gupta and Roy (2013), and Masood and Delgado (2014), compare simulation results with usually a limited set of experimental data. The comparison is usually restricted to radial profiles of variables such as particle volume fraction, averaged axial velocities, and root mean square (RMS) turbulent velocities. The overall conclusion from all these studies is that generally neither point-particle methods based on RANS-based simulations nor Euler–Euler simulations are capable of reproducing the basic average characteristics of a wide variety of two-phase flows, let alone the coherent vortical structures which dominate so many two-phase flows.

LESs have been proven to yield better representations of the turbulence characteristics of single-phase flows (Hartmann et al, 2004; Roux et al, 2005; Shah et al, 2015), since rather than exploiting approximate turbulence models they calculate most of the turbulent eddies thanks to the use of much finer computational grids. Hence, the inherently transient LESs provide a better basis for Euler–Lagrange, or point-particle tracking, simulations, as the spatial and temporal resolutions of a great part of the turbulence spectrum are promising a much more realistic representation of particle trajectories or dynamic volume fraction distributions. Again, the focus of most research is on the fluid–particle interaction force because this is by far the greatest source of uncertainty in these simulations.

In the literature, the number of LES studies of two-phase flows is rather limited, as LESs are computationally very demanding (compared to RANS-based simulations) and in the case of Euler–Lagrange simulations particle tracking should preferably done for large numbers of particles. The latter drawback can be cured by switching to the DSMC method—see, e.g., Tsuji et al (1998).

Because of a vivid interest in spray combustion, quite a few papers deal with the effect of finely dispersed particles or droplets on the turbulence characteristics of jets and sprays. The interaction of particle dynamics and turbulence in these flows has prompted very fundamental stochastic approaches involving filtering and averaging techniques, probability density functions, and quadrature-based moment methods (Fevrier et al, 2005; Fox, 2012; Labourasse et al, 2007) which are beyond the scope of this chapter. Riber et al (2009) and Senoner et al (2009) obtained LES results for recirculating and evaporating two-phase flows, respectively, by both Euler–Lagrange and Euler–Euler methods and compared them mutually and with experimental data.

Derksen (2003), Shalaby et al (2008), and De Souza et al (2012) tracked point particles in a gas flow field obtained by means of LESs with the view of studying separation performance of gas–solid cyclones, while Derksen et al

(2008) did the same with two-way coupling between gas flow field and solid particles. Derksen (2003), Hartmann et al (2006), and Ayranci et al (2013) studied solid suspension in agitated vessels by means of Euler (LES)–Lagrange simulations, while Wardle (2011) did the same for a centrifugal liquid–liquid contactor. In most of these studies, the LESs were carried out with the help of LB techniques rather than the common finite-volume method, as LB techniques are computationally less demanding—see, e.g., Nourgaliev et al (2003), Raabe (2004), and Van den Akker (2006). Finally, the EELES approach discussed above has been used for simulating some specific bubble columns for which experimental data are available for validation.

While assessment of any simulation results should preferably be carried out with the help of quantitative data, such data are often restricted to radial profiles of averaged variables such as bubble fraction, liquid velocity profiles, and sometimes RMS data for turbulent velocities. Most often, no reference is made is to the coherent mesoscale structures of interest in this chapter. The inherently transient LESs do comprise the presence and dynamics of such structures, as is illustrated in the appealing snapshots of Fig. 9. Unfortunately, no quantitative experimental data are available for assessing the correctness of these self-organizing structures shown in Fig. 9. As a result, there is no way of verifying the effect of applying the various assumptions and correlations with respect to the fluid–particle interaction. This is partly due to the inaccessibility of not too dilute dispersed two-phase flows for nonintrusive optical measuring techniques while inserting probes is believed to modify the structures of interest. Radiation-based experimentation such as computer-automated radioactive particle tracking (CARPT) and tomographical techniques provide options for assessing the existence and exact shape and dynamics of these mesoscale structures.

Whatever experimental techniques are used, enormous amounts of data have to be collected, mined, processed, and interpreted before meaningful and significant quantitative information can be presented. The same applies to all data obtained via the inherently transient LESs. These considerations have led many researchers to take refuge to statistical methods such as the use of probability density functions. This approach contrasts with the more mechanistic approach of identifying individual mesoscale structures and labeling them in terms of, e.g., shape, size, mobility, deformation rate, characteristic lifetime, and even chaos parameters. Researchers in the field of single-phase (highly) turbulent flows, e.g., Gaither et al (2012), may provide guidance with respect to the necessarily advanced data mining techniques.

Van den Akker (1998) and Mudde et al (1999) present several fingerprints selected from gamma-ray signals in turbulent fluidized beds. Helland et al (2000) reproduce time series of particle clusters in a circulating fluidized bed obtained by means of an Euler (LES)–Lagrange simulation. Tanaka et al (1995, 1996) and Tsuji et al (1998) present cluster patterns in vertical gas–solid flows resulting from DSMC simulations. Comparing experimental and computational information on the structure and dynamics of particle clusters may be an interesting way of assessing the usefulness of LESs for reproducing and interpreting the formation and dynamics of mesoscale coherent structures.

## 11. DIRECT NUMERICAL SIMULATIONS

Both in RANS-based CFD simulations and in LESs, decisions are needed as to how to model the fluid–particle interaction force due to the unresolved aspects of the flow and pressure fields and as to which empirical correlations would be applicable. The same issue pops up when in DEMs or DBMs the Milelli criterion is obeyed. The obvious way of avoiding the use of such models and the combined or selective use of correlations derived for the various canonical cases of particle motions is by just resolving the entire fluid flow around a particle and by carrying out the integration of Eq. (4). In DNSs, a flow field is completely resolved by using just the Navier–Stokes equation without any further model.

While in principle this is possible in the cases of both laminar and turbulent flows, a time-dependent DNS of turbulent flow requires a (very) fine computational grid, down to the scale of the so-called Kolmogorov eddies, and correspondingly (very) small time steps, depending on the Reynolds number of the flow. In the case of (very) turbulent flows, the size of the computational domain is therefore usually subject to serious restrictions, although with the steep increase in computational power these restrictions quickly relax. An example are the DNSs of a marginally turbulent stirred vessel (liquid-only) carried out by Gillissen and Van den Akker (2012) on 128 CPUs with the help of a grid of some 4 billion cells. Donzis et al (2008) and Gaither et al (2012) even carried out DNSs for homogeneous isotropic single-phase turbulence in a cubic domain on $O(10^4)$ processors. In bubbly flows, DNSs for swarms comprising hundreds of bubbles have become feasible these days (Tryggvason and Lu, 2015).

In the context of the current issues of fluid–particle interaction and the formation and dynamics of particle swarms and clusters, DNSs also seem to be perfectly suited for completely resolving the fluid flow around a single particle or around multiple particles. An early example of such a DNS is the study by Legendre and Magnaudet (1998) about the lift force felt by a spherical bubble in a viscous linear shear flow. Other examples are the LB simulations of a single sphere settling in a tank under gravity by Ten Cate et al (2002) and Rohde et al (2008). Zhang and Prosperetti (2005) reported on a DNS study concerning the sedimentation of 1024 solid particles at a single-particle Reynolds number of about 10.

In turbulent flows, two important requirements have to be satisfied: (a) the particle diameter should be larger than the grid spacing by an order of magnitude in order to reproduce the flow around the particle to a reasonable degree; and (b) the grid spacing should be maximum twice the Kolmogorov length scale with the view of resolving the turbulent spectrum of the fluid phase in which the particles are embedded. These requirements may restrict the size of the domain for which simulations can be run unless massive computer power is available.

Balachandar and Eaton (2010) mentioned a small number of DNS studies for a single particle in a turbulent flow as well as for swarms of maximum O (1000) particles in turbulent channel flow. Derksen (2012) reported on a DNS for liquid–solid flow in an agitated tank under "mildly turbulent conditions" (impeller-based Reynolds number of 2000): the edge of the rectangular "tank" was just 3.0 cm and the impeller diameter 1.2 cm only, while the 3600 uniformly sized spherical particles were 1.0 mm in diameter. Tryggvason and Lu (2015) review earlier DNS results from the Tryggvason group for upward and downward flows of hundreds of bubbles in vertical channels. Thanks to a unique computer platform at the Beijing Institute of Process Engineering of the Chinese Academy of Sciences, Chen et al (2009) performed a DNS of a turbulent gas–solid flow comprising 0.1 million solid particles on a platform comprising 256 GPUs.

A more affordable option for a DNS implementation is to refrain from attempting to simulate turbulent multiphase flow in the whole domain of any process device of industrial scale and to focus on a "periodic box" in the bulk of a two-phase flow—which term is short for "box with periodic walls." The use of periodic walls in a simulation protects the inside of the box against the restricting effects of solid boundaries (such as the common no-slip condition), suggesting that the fluid flow extends across the boundaries of the box. The homogeneous flow conditions in such a periodic-box DNS are imposed via a spectral or linear forcing method (Alvelius, 1999; Carroll and Blanquart, 2013; Lundgren, 2003; Rosales and Meneveau, 2005).

Not every periodic-box DNS including particles resolves the flow field around and between the particles: Gualteri et al (2013) and Olivieri et al (2014) studied clustering and turbulence modulation in particle-laden flows by considering point-like particles obeying Stokes law and moving through a carrier fluid flow simulated by a pseudo-spectral type of DNS. In this chapter, however, the focus is on DNSs of the fluid–particle system resolving all fluid motions in the presence of the moving particles, by virtue of the usual no-slip boundary condition at the particle boundaries. In this way, particles are treated as extended objects, with a finite volume, subject to fluid forces calculated from first principles (the Navier–Stokes equation) rather than parameterized, while reversibly exerting forces on the fluid.

An illustrative example is a snapshot in Fig. 19 from a periodic-box DNS of a liquid–solid suspension with monodisperse spherical particles colliding as a result of the turbulence level imposed via a random forcing technique; it was performed by Ten Cate et al (2004) by means of an LB technique. The snapshot shows both the instantaneous liquid velocity field (denoted by the arrows) and the spatial distribution of the rate of energy dissipation in the liquid (denoted by the color code) along with the instantaneous position of the particles in some cross-sectional area through the periodic box. In addition, Derksen and Van den Akker (2007) demonstrated the usefulness of the periodic-box DNS implementation for investigating the effect of

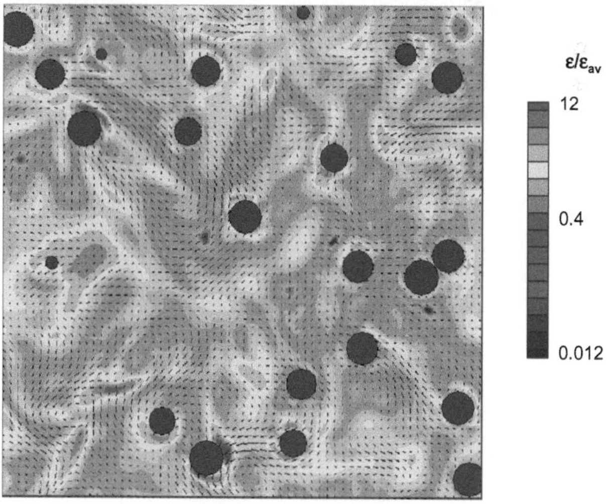


**Figure 19** Snapshot from a DNS simulation of colliding mono-sized spherical particles in homogeneous isotropic turbulence in a periodic box as carried out by Ten Cate et al (2004). *Reprinted from Van den Akker (2010) with permission from the ACS.*

varying turbulence levels on the dynamic processes of coalescence and redispersion of droplets in the flow of an immiscible second liquid.

In fact, this Fig. 19 is the computational equivalent of multiphase flow fields obtained by means of PIV under carefully chosen experimental conditions (such as refractive index matching between particles, fluid, and containing walls): Fig. 20, from Ghatage et al (2013), presents details of the flow field in between the solid particles of a solid–liquid fluidized bed. Of course, it is hard to compare such computational and experimental flow fields quantitatively, due to the inherent dynamic chaotic particle patterns. The validity of the computational techniques used may better be assessed in well-controlled flows—as was done by Ten Cate et al (2002) and Rohde et al (2008) for a single sphere settling under gravity and approaching and hitting the bottom of the containing vessel.

Periodic-box DNSs for multiphase flows can be used for a variety of purposes. Ten Cate et al (2001) report how a few periodic box DNSs could elucidate details of particle behavior including particle collisions in a crystallizer. Van den Akker (2010) builds on this and suggests a truly multiscale simulations by concurrently combining a LES for a two-phase flow process with tens of periodic box DNSs. Tryggvason et al (2006) reviewed a number of numerical methods—among which the volume-

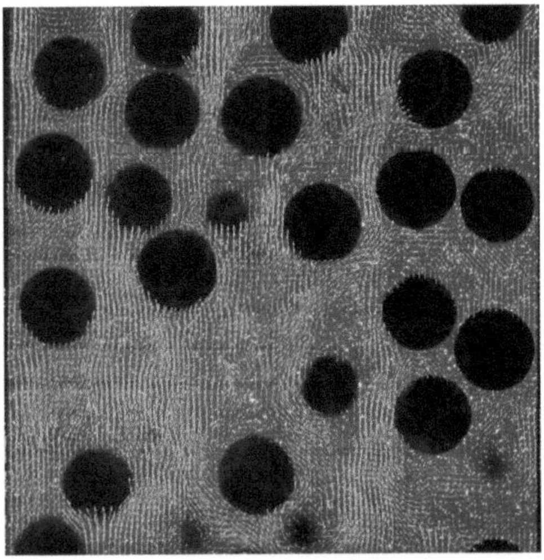

**Figure 20** Snapshot from a processed PIV data set showing the liquid velocities between the solid particles in a solid–liquid fluidized bed. *Reprinted from Ghatage et al (2013) with permission from Elsevier.*

of-fluid and front tracking techniques—suitable for such DNSs for dispersed bubbly flows. Sankaranarayanan and Sundaresan (2002) performed periodic-box DNSs for deriving closure relations for the lift force in ordered arrays of bubbles in a shear flow at various gas fractions. Kandhai et al (2003) did the same for gas flow through disordered fixed arrays of mono-sized spherical solid particles in a periodic box and investigated for which solids fractions the Wen and Yu and the Ergun interphase drag correlations can best be used. Holloway et al (2010) used a similar approach in poly-disperse suspensions with the view of assessing various gas–particle drag correlations. Statistical methods may be needed for such a comparison between experimental and computational data and for deriving closure relations which can be used in, e.g., LESs (see also Tryggvason and Lu, 2015).

The above papers illustrate the potential and promise of DNSs for the purpose of computationally simulating multiphase flows in industrial-scale process equipment. As such, DNSs are a major step forward in comparison with RANS-based simulations and LESs which—though to a different degree—rely on models. In addition, DNSs are useful for assessing the validity and applicability of experimental correlations for fluid–particle interaction under specific conditions—both for a single particle and for multiple particles. Examples of DNS studies in randomly or orderly arranged particle or bubble assemblies carried out to this purpose are Hill et al (2001), Kandhai et al (2003), Beetstra et al (2007), Holloway et al (2010), Tenneti et al (2011), Gillissen et al (2011), Tenneti and Subramaniam (2014), and Tang et al (2015). Most of these studies focus on the effect of voidage on drag and slip velocity.

So far, however, and to the best of my knowledge, the general inadequacy of the linear addition of the various very specific interaction forces in more dilute particle systems has never been questioned—let alone investigated by means of such periodic-box DNSs.

One way of assessing the current practice of adding the various "partial" forces would be to carry out such a periodic-box DNS of a turbulent fluid carrying finite-size particles, to perform the summation of the fluid–particle interaction forces $\underline{F}'$—see Eq. (5)—felt by all particles in this periodic box on the basis of the periodic-box DNS results, and to insert this total interaction force into a coarse-grained simulation of either the carrier phase only or a two-fluid model. Such a procedure would completely circumvent the use of empirical correlations for dealing with the second term of Eq. (5) and would fit into the multiscale approach proposed by Van den Akker (2010).

## 12. CONCLUSIONS

Many industrial processes are still designed on the basis of the assumptions of plug flow and steady-state uniform two-phase flow. For this chapter, much evidence has been collected with respect to the abundant occurrence of transient mesoscale coherent structures, strands, or clusters in various turbulent multiphase flows, at least at scales and under conditions relevant to industrial processes. This evidence is from a variety of sources: experimental observations in academic laboratories, results from hydrodynamic stability analyses, and computational simulation studies (both of the LES and the DNS type). Unfortunately, this evidence comprises many indecisive and even contradictory reports about the drivers behind these structures, clusters, and strands, and about their dependence on density ratio, particle size, volume fractions, operating conditions, and so on.

On the basis of all information gathered, it is fair to conclude that one of the major drivers behind the occurrence, the shape, and the dynamics of these mesoscale structures is the fluid–particle interaction force that plays a dominant role, both in stability analyses and in CFD simulations of any type. This role is related to the difference in inertia of the two phases and, as a result, to the temporally and spatially varying difference in velocities of dispersed phase (particles) and carrier (or continuous) phase. Cluster and strand formation seem to be closely related to the continuous chaotic accelerations in a turbulent carrier fluid (in the Euler–Lagrange approach) or a turbulent continuous phase (in the Euler–Euler or two-fluid approach). An interesting explanation for cluster formation is the sweep-stick mechanism proposed by Goto and Vasillicos (2008).

For his reason, the main body of this chapter deals with fluid–particle interaction and addresses such questions as: what is the fluid mechanical basis of it, for which canonical cases do we have reliable data and correlations, how does the most general correlation of the general case look like, what are the effects of turbulence in the carrier phase and of neighboring particles in a swarm, how is the fluid–particle interaction treated in the various CFD approaches (point-particle tracking, two-fluid models), and what is the effect of all the variations of the interaction force applied by so many investigators on the computational results, particularly with respect to the mesoscale structures of interest? These issues have been addressed in great detail in the context of both Euler–Lagrange and Euler–Euler simulations. (Other important effect, such as particle–particle collisions, coalescence and redispersion of bubbles and drops, and modulation of the carrier phase turbulence due to the presence and motion of the dispersed-phase particles, have not been considered.)

The review of this chapter resulted in a number of significant conclusions with respect to the fluid–particle interaction force:

- Apart from the Stokes regime of (relative) particle motion, the linear addition of several canonical expressions to find the actual interaction force is not allowed, as superimposing flow fields is generally not allowed due to the nonlinear characteristic of fluid flow and of the Navier–Stokes equation.
- It is incorrect to expect that for the purpose of describing the motion or path of a particle immersed in a 3D turbulent fluid in the presence of other particles, the steady-state drag coefficient may be used with confidence, as the latter has been obtained by dropping a single particle (along a 1D path) through a stagnant fluid.
- Adopting the Brucato correlation Eq. (42)—or similar correlations involving the ratio of particle size to Kolmogorov length scale—for the purpose of calculating the detailed dynamic response of individual particles to varying local turbulence levels and the paths of individual particles in the context of CFD simulations is not a valid approach, as it is based on the translation of a lower overall settling rate of a (dilute) swarm into an increased drag coefficient. Rather, the analysis should be in terms of time scales.
- The Milelli criterion that prescribes that a bubble should be smaller than a grid cell size in an Euler–Euler simulation is based on an incorrect interpretation of the concept of two interpenetrating continua and unnecessarily introduces another averaging procedure. Furthermore, the Milelli criterion bars the option of a two-phase LES that aims at resolving the energetic eddies of a turbulent flow field which requires the application of a fine grid.
- Combining existing correlations for the purpose of approximating the fluid–particle interaction force pertinent to a particular case of interest may not result in a realistic, truth-like representation of particle trajectories or dispersed-phase volume fractions. As a result, the details of the formation of clusters, strands, or coherent structures should be viewed with great caution.

Given the above five major conclusions and many more interaction force issues discussed in great detail in this chapter, it is recommended to perform DNSs, preferably in a periodic box anywhere in a two-phase flow field, with the help of the local (shear or turbulent) flow conditions. As in a DNS, the interstitial flow field is completely resolved, the next step then is applying Eq. (4) that expresses the fluid–particle interaction force in terms of the flow and pressure fields around a particle. This approach has already been pursued by several investigators in order to propose new correlations, or assess existing correlations, for the interaction force in specific canonical cases. A more generally valid option would be to integrate all fluid–particle interaction forces in a periodic box and feed this total force as a local force in a coarse-grained simulation of the two-phase flow of interest.

## REFERENCES

Agrawal K, Loezos PN, Syamlal M, Sundaresan S: The role of meso-scale structures in rapid gas-solid flows, *J Fluid Mech* 445:151–185, 2001. http://dx.doi.org/10.1017/S0022112001005663.

Alvelius K: Random forcing of three-dimensional homogeneous turbulence, *Phys Fluids* 11:1880–1889, 1999. http://dx.doi.org/10.1063/1.870050.

Anderson TB, Jackson R: A fluid mechanical description of fluidized beds, *Ind Eng Chem Fundam* 6:527–539, 1967. http://dx.doi.org/10.1021/i160024a007.

Anderson TB, Jackson R: Fluid mechanical description of fluidized beds: stability of state of uniform fluidization, *Ind Eng Chem Fundam* 7:12–21, 1968. http://dx.doi.org/10.1021/i160025a003.

Armenio V, Fiorotto V: The importance of the forces acting on particles in turbulent flows, *Phys Fluids* 13:2437–2440, 2001. http://dx.doi.org/10.1063/1.1385390.

Asmolov ES, McLaughlin JB: The inertial lift on an oscillating sphere in a linear shear flow, *Int J Multiphase Flow* 25:739–751, 1999. http://dx.doi.org/10.1016/S0301-9322(98)00063-9.

Ayranci I, Kresta SM, Derksen JJ: Experiments and simulations on bidisperse solids suspension in a mixing tank, *Chem Eng Technol* 36:1957–1967, 2013. http://dx.doi.org/10.1002/ceat.201300409.

Bakker A, Van den Akker HEA: A computational model for the gas-liquid flow in stirred reactors, *Chem Eng Res Des* 72:594–606, 1994.

Balachandar S: A scaling analysis for point-particle approaches to turbulent multiphase flows, *Int J Multiphase Flow* 35:801–810, 2009. http://dx.doi.org/10.1016/j.ijmultiphaseflow.2009.02.013.

Balachandar S, Eaton JK: Turbulent dispersed multiphase flow, *Annu Rev Fluid Mech* 42:111–133, 2010. http://dx.doi.org/10.1146/annurev.fluid.010908.165243.

Barnea E, Mizrahi J: A generalized approach to the fluid dynamics of particulate systems: part 1. General correlation for fluidization and sedimentation in solid multiparticle systems, *Chem Eng J* 5:171–189, 1973. http://dx.doi.org/10.1016/0300-9467(73)80008-5.

Barnea E, Mizrahi J: A generalized approach to the fluid dynamics of particulate systems: part 2. Sedimentation and fluidization of clouds of spherical liquid droplets, *Can J Chem Eng* 53:461–468, 1975. http://dx.doi.org/10.1002/cjce.5450530501.

Bauer M, Eigenberger G: A concept for multi-scale modeling of bubble columns and loop reactors, *Chem Eng Sci* 54:5109–5117, 1999. http://dx.doi.org/10.1016/S0009-2509(99)00264-X.

Becker S, Sokolichin A, Eigenberger G: Gas-liquid flow in bubble columns and loop reactors: part II. Comparison of detailed experiments and flow simulations, *Chem Eng Sci* 49:5747–5762, 1994. http://dx.doi.org/10.1016/0009-2509(94)00290-8.

Beek WJ: Oscillations and vortices in a batch of liquid sustained by a gas flow. In *Proc symp two-phase flow, Exeter, UK*, 1965, pp F401–F416.

Beetstra R, Van der Hoef MA, Kuipers JAM: Drag force of intermediate Reynolds number flow past mono- and bidisperse arrays of spheres, *AIChE J* 53:489–501, 2007. http://dx.doi.org/10.1002/aic.11065.

Bird RB, Stewart WE, Lightfoot EN: *Transport phenomena*, New York, NY, 1960, Wiley.

Bombardelli FA, Buscaglia GC, Garcia MH, Dari EA: Simulation of wandering phenomena in bubble plumes, *Mec Comput* 23:1969–1994, 2004.

Bothe D, Schmidtke M, Warnecke HJ: Direct numerical computation of the lift force on single bubbles. In *Proceedings of the 5th joint ASME/JSME fluids engineering summer conference*, vol. 2, 2007, pp 309–310.

Brenner H: Hydrodynamic resistance of small particles at small Reynolds numbers, *Adv Chem Eng* 6:287–438, 1966.

Brucato A, Grisafi F, Montante G: Particle drag coefficient in turbulent fluids, *Chem Eng Sci* 53:3295–3684, 1998. http://dx.doi.org/10.1016/S0009-2509(98)00149-3.

Capecelatro J, Desjardins O, Fox RO: On fluid-particle dynamics in fully developed cluster-induced turbulence, *J Fluid Mech* 780:578–635, 2015. http://dx.doi.org/10.1017/jfm.2015.459.

Carroll P, Blanquart G: A proposed modification to Lundgren's physical space velocity forcing method for isotropic turbulence, *Phys Fluids* 25:105114, 2013. http://dx.doi.org/10.1063/1.4826315.

Cassanello M, Larachi F, Kemoun A, Al-Dahhan MH, Dudoković MP: Inferring liquid chaotic dynamics in bubble columns using CARPT, *Chem Eng Sci* 56:6125–6134, 2001. http://dx.doi.org/10.1016/S0009-2509(01)00218-4.

Chen RC, Reese J, Fan LS: Flow structure in a three-dimensional bubble column and three-phase fluidized bed, *AIChE J* 49:1093–1104, 1994. http://dx.doi.org/10.1002/aic.690400702.

Chen MW, McLaughlin JB: A new correlation for the aerosol deposition rate in vertical ducts, *J Colloid Interface Sci* 169:437–455, 1995. http://dx.doi.org/10.1006/jcis.1995.1054.

Chen FG, Ge W, He XF, et al: Multi-scale HPC system for multi-scale discrete simulation—development and application of a supercomputer with 1 Peta-flops peak performance in single precision, *Particuology* 7:332–335, 2009. http://dx.doi.org/10.1016/j.partic.2009.06.002.

Clift R, Grace JR, Weber ME: *Bubbles, drops, and particles*, New York, 1978, Academic Press.

Collins LR, Keswani A: Reynolds number scaling of particle clustering in turbulent aerosols, *New J Phys* 6:119, 2004. http://dx.doi.org/10.1088/1367-2630/6/1/119.

Corrsin S, Lumley J: On the equation of motion for a particle in turbulent fluid, *Appl Sci Res* 6:114–116, 1956.

Cox RG, Hsu SK: The lateral migration of solid particles in a laminar flow near a plane, *Int J Multiphase Flow* 3:201–222, 1977. http://dx.doi.org/10.1016/0301-9322(77)90001-5.

Crowe CT, Troutt TR, Chung JN: Numerical models for two-phase turbulent flows, *Annu Rev Fluid Mech* 28:11–43, 1996. http://dx.doi.org/10.1146/annurev.fl.28.010196.000303.

Crowe CT, Sommerfeld M, Tsuji Y: *Multiphase flows with drops and particles*, Boca Raton, 1998, CRC Press.

Davidson JF, Harrison D: *Fluidised particles*, London, UK, 1963, Cambridge University Press.

De Nevers N: Bubble driven fluid circulations, *AIChE J* 14:222–226, 1968. http://dx.doi.org/10.1002/aic.690140205.

De Souza FJ, Salvo RD, Martins DAD: Large eddy simulation of the gas–particle flow in cyclone separators, *Sep Purif Technol* 94:61–70, 2012. http://dx.doi.org/10.1016/j.seppur.2012.04.006.

Decker S, Sommerfeld M: Numerical calculations of two-phase flows in agitated vessels using the Euler/Lagrange approach. In *Proceedings of ASME 2000 fluid engineering division summer meeting, Boston, MA*, FEDSM'00 11154, 2000, pp 1–8.

Deen NG, Solberg T, Hjertager BH: Large eddy simulation of the gas-liquid flow in a square cross-sectioned bubble column, *Chem Eng Sci* 56:6341–6349, 2001. http://dx.doi.org/10.1016/S0009-2509(01)00249-4.

Deen NG, Van Sint Annaland M, Van der Hoef MA, Kuipers JAM: Review of discrete particle modeling of fluidized beds, *Chem Eng Sci* 62:28–44, 2007. http://dx.doi.org/10.1016/j.ces.2006.08.014.

Degaleesan S, Duduković MP, Pan Y: Experimental study of gas-induced liquid-flow structures in bubble columns, *AIChE J* 47:1913–1931, 2001. http://dx.doi.org/10.1002/aic.690470904.

Dehbi A: A CFD model for particle dispersion in turbulent boundary layer flows, *Int J Multiphase Flow* 34:819–828, 2008. http://dx.doi.org/10.1016/j.ijmultiphaseflow.2008.03.001.

Derksen JJ: Numerical simulation of solids suspension in a stirred tank, *AIChE J* 49:2700–2714, 2003. http://dx.doi.org/10.1002/aic.690491104.

Derksen JJ, Van den Akker HEA: Multi-scale simulations of stirred liquid-liquid dispersions, *Chem Eng Res Des* 85:697–702, 2007. http://dx.doi.org/10.1205/cherd06161.

Derksen JJ, Van den Akker HEA, Sundaresan S: Two-way coupled large-eddy simulations of the gas-solid flow in cyclone separators, *AIChE J* 54:872–885, 2008. http://dx.doi.org/10.1002/aic.11418.

Derksen JJ: Highly resolved simulations of solids suspension in a small mixing tank, *AIChE J* 58:3266–3278, 2012. http://dx.doi.org/10.1002/aic.13889.

Devanathan N, Duduković MP, Lapin A, Lübbert A: Chaotic flow in bubble columns, *Chem Eng Sci* 50:2661–2667, 1995. http://dx.doi.org/10.1016/0009-2509(95)00070-L.

Dhotre MT, Niceno B, Smith BL: Large eddy simulation of a bubble column using dynamic sub-grid scale model, *Chem Eng J* 136:337–348, 2008. http://dx.doi.org/10.1016/j.cej.2007.04.016.

Dhotre MT, Niceno B, Smith BL, Simiano M: Large-eddy simulations (LES) of the large scale bubble plumes, *Chem Eng Sci* 64:2692–2704, 2009. http://dx.doi.org/10.1016/j.ces.2009.02.040.

Donzis DA, Yeung PK, Pekurovsky D: Turbulence simulations on $O(10^4)$ processors, 2008: In *Proceedings of the Tera-Grid 2008 conference, Las Vegas, NV*, 2008. http://p3dfft.googlecode.com/svn-history/r488/docs/TG08_DNS.pdf.

Eaton JK: Two-way coupled turbulence simulations of gas-particle flows using point-particle tracking, *Int J Multiphase Flow* 35:792–800, 2009. http://dx.doi.org/10.1016/j.ijmultiphaseflow.2009.02.009.

Elder JW: Laminar free convection in a vertical slot, *J Fluid Mech* 23:77–98, 1965. http://dx.doi.org/10.1017/S0022112065001246.

Elsayed K, Lacor C: The effect of cyclone inlet dimensions on the flow pattern and performance, *Appl Math Model* 35:1952–1968, 2011. http://dx.doi.org/10.1016/j.apm.2010.11.007.

Fevrier P, Simonin O, Squires KD: Partitioning of particle velocities in gas-solid turbulent flows into a continuous field and a spatially uncorrelated random distribution: theoretical formalism and numerical study, *J Fluid Mech* 533:146, 2005. http://dx.doi.org/10.1017/S0022112005004088.

Fligner M, Schipper PH, Sapre AV, Krambeck FJ: Two phase cluster model in riser factors: impact and radial density distribution on yields, *Chem Eng Sci* 49:5813–5818, 1994. http://dx.doi.org/10.1016/0009-2509(94)00377-7.

Fox RO: Large-eddy-simulation tools for multiphase flows, *Annu Rev Fluid Mech* 44:47–76, 2012. http://dx.doi.org/10.1146/annurev-fluid-120710-101118.

Freedman W, Davidson JF: Hold-up and liquid circulation in bubble columns, *Trans Inst Chem Eng* 47:T251–T262, 1969.

Gaither KP, Childs H, Schulz KW, et al: Visual analytics for finding critical structures in massive time-varying turbulent-flow simulations, *IEEE Comput Graph Appl* 32(4):34–45, 2012. http://dx.doi.org/10.1109/MCG.2012.63.

Garg SK, Pritchett JW: Dynamics of gas-fluidized beds, *J Appl Phys* 46:4493–4500, 1975. http://dx.doi.org/10.1063/1.321421.

Geurst JA: Virtual mass in two-phase bubbly flow, *Phys A* 129:233–261, 1985. http://dx.doi.org/10.1016/0378-4371(85)90168-2.

Geurst JA: Drift mass, multifluid modeling of two-phase bubbly flow and superfluid hydrodynamics, *Phys A* 152:1–28, 1988. http://dx.doi.org/10.1016/0378-4371(88)90063-5.

Ghatage SV, Sathe MJ, Doroodchi E, Joshi JB, Evans GM: Effect of turbulence on particle and bubble slip velocity, *Chem Eng Sci* 100:120–136, 2013. http://dx.doi.org/10.1016/j.ces.2013.03.031.

Ghatage SV, Bhole MR, Padhiyar N, Joshi JB, Evans GM: Prediction of regime transition in three-phase sparged reactors using linear stability analysis, *Chem Eng J* 235:307–330, 2014a. http://dx.doi.org/10.1016/j.cej.2013.08.095.

Ghatage SV, Peng ZB, Sathe MJ, et al: Stability analysis in solid-liquid fluidized beds: experimental and computational, *Chem Eng J* 256:169–186, 2014b. http://dx.doi.org/10.1016/j.cej.2014.06.026.

Gillissen JJJ, Sundaresan S, Van den Akker HEA: A lattice Boltzmann study on the drag force in bubble swarms, *J Fluid Mech* 679:101–121, 2011. http://dx.doi.org/10.1017/jfm.2011.125.

Gillissen JJJ, Van den Akker HEA: Direct numerical simulation of the turbulent flow in a baffled tank driven by a Rushton turbine, *AIChE J* 58:3878–3890, 2012. http://dx.doi.org/10.1002/aic.13762.

Goldsmith HL, Mason SG: Flow of suspensions through tubes .1. Single spheres, rods, and discs, *J Colloid Sci* 17:448–476, 1962. http://dx.doi.org/10.1016/0095-8522(62)90056-9.

Goto S, Vassilicos JC: Sweep-stick mechanism of heavy particle clustering in fluid turbulence, *Phys Rev Lett* 100:054503, 2008. http://dx.doi.org/10.1103/PhysRevLett.100.054503.

Grbavcic ZB, Garic RV, Hadzismajlovic DE, et al: Variational model for prediction of the fluid particle interphase drag coefficient and particulate expansion of fluidized and sedimenting beds, *Powder Technol* 68:199–211, 1991. http://dx.doi.org/10.1016/0032-5910(91)80046-L.

Groen JS, Mudde RF, Van den Akker HEA: Time dependent behaviour of the flow in a bubble column, *Trans Inst Chem Eng* 74:615–621, 1995.

Groen JS, Oldeman RSGC, Mudde RF, Van den Akker HEA: Coherent structures and axial dispersion in bubble column reactors, *Chem Eng Sci* 51:2511–2520, 1996. http://dx.doi.org/10.1016/0009-2509(96)00110-8.

Groen JS: *Scales and structures in bubbly flows,* PhD thesis, Delft, The Netherlands, 2004, Delft University of Technology.

Gualteri P, Picano F, Sardina G, Casciola CM: Clustering and turbulence modulation in particle laden shear flows, *J Fluid Mech* 715:134–162, 2013. http://dx.doi.org/10.1017/jfm.2012.503.

Gupta A, Roy S: Euler-Euler simulation of bubbly flow in a rectangular bubble column: experimental validation with radioactive particle tracking, *Chem Eng J* 225:818–836, 2013. http://dx.doi.org/10.1016/j.cej.2012.11.012.

Happel J, Brenner H: *Low Reynolds number hydrodynamics,* ed 2, Englewood Cliffs, NJ, 1973, Prentice Hall.

Harris AT, Davidson JF, Thorpe RB: The prediction of particle cluster properties in the near wall region of a vertical riser (200157), *Powder Technol* 127:128–143, 2002. http://dx.doi.org/10.1016/S0032-5910(02)00114-6.

Harteveld WK, Mudde RF, Van den Akker HEA: Dynamics of a bubble column: influence of gas distribution on coherent structures, *Can J Chem Eng* 81:389–394, 2003. http://dx.doi.org/10.1002/cjce.5450810308.

Harteveld WK, Mudde RF, Van den Akker HEA: Estimation of turbulence power spectra for bubbly flows from laser Doppler anemometry signals, *Chem Eng Sci* 60:6160–6168, 2005. http://dx.doi.org/10.1016/j.ces.2005.03.037.

Hartmann H, Derksen JJ, Montavon C, Pearson J, Hamill IS, Van den Akker HEA: Assessment of large eddy and RANS stirred tank simulations by means of LDA, *Chem Eng Sci* 59:2419–2432, 2004. http://dx.doi.org/10.1016/j.ces.2004.01.065.

Hartmann H, Derksen JJ, Van den Akker HEA: Numerical solution of a dissolution process in a stirred tank reactor, *Chem Eng Sci* 61:3025–3032, 2006. http://dx.doi.org/10.1016/j.ces.2005.10.058.

Helland E, Occelli R, Tadrist L: Numerical study of cluster formation in a gas-particle circulating fluidized bed, *Powder Technol* 110:210–221, 2000. http://dx.doi.org/10.1016/S0032-5910(99)00260-0.

Hibiki T, Ishii M: Lift force in bubbly flow systems, *Chem Eng Sci* 62:6457–6474, 2007. http://dx.doi.org/10.1016/j.ces.2007.07.034.

Hill DP, Wang DM, Gosman AD, Issa RI: Numerical prediction of two-phase bubbly flow in a pipe. In *Proceedings of the 2nd international conference on multiphase flow'95, Kyoto, Japan, MO3*, 1995, pp 1–6.

Hill RJ, Koch DL, Ladd AJC: Moderate-Reynolds-number flows in ordered and random arrays of spheres, *J Fluid Mech* 448:243–278, 2001. http://dx.doi.org/10.1017/S0022112001005936.

Hills JH: Radial non-uniformity of velocity and voidage in a bubble column, *Trans Inst Chem Eng* 52:1–9, 1974.

Hoekstra AJ, Derksen JJ, Van den Akker HEA: An experimental and numerical study of turbulent swirling flow in gas cyclones, *Chem Eng Sci* 54:2055–2065, 1999. http://dx.doi.org/10.1016/S0009-2509(98)00373-X.

Holloway W, Yin XL, Sundaresan S: Fluid-particle drag in inertial polydisperse gas-solid suspensions, *AIChE J* 56:1995–2004, 2010. http://dx.doi.org/10.1002/aic.12127.

Homann H, Bec J, Grauer R: Effect of turbulent fluctuations on the drag and lift forces on a towed sphere and its boundary layer, *J Fluid Mech* 721:155–179, 2015. http://dx.doi.org/10.1017/jfm.2013.66.

Hoomans PBP, Kuipers JAM, Briels WJ, Van Swaaij WPM: Discrete particle simulation of bubble and slug formation in a two-dimensional gas-fluidised bed: a hard-sphere approach, *Chem Eng Sci* 51:99–118, 1996. http://dx.doi.org/10.1016/0009-2509(95)00271-5.

Horio M, Kuroki H: Three-dimensional flow visualization of dilutely dispersed solids in bubbling and circulating fluidized beds, *Chem Eng Sci* 49:2413–2421, 1994. http://dx.doi.org/10.1016/0009-2509(94)E0071-W.

Hyre M, Glicksman L: Brief communication on the assumption of minimum energy dissipation in circulating fluidized beds, *Chem Eng Sci* 52:2435–2438, 1997. http://dx.doi.org/10.1016/S0009-2509(96)00508-8.

Ishii M, Zuber N: Drag coefficient and relative velocity in bubbly, droplet or particulate flows, *AIChE J* 25:843–855, 1979. http://dx.doi.org/10.1002/aic.690250513.

Jackson R: The mechanics of fluidized beds: the motion of fully developed bubbles, *Trans Inst Chem Eng* 41:22, 1963.

Joshi JB, Sharma MM: A circulation cell model for bubble columns, *Trans Inst Chem Eng* 57:244–251, 1979.

Joshi JB, Sharma MM: Correspondence, *Trans Inst Chem Eng* 60:256, 1982.

Joshi JB: Computational flow modelling and design of bubble column reactors, *Chem Eng Sci* 56:5893–5933, 2001. http://dx.doi.org/10.1016/S0009-2509(01)00273-1.

Kandhai D, Derksen JJ, Van den Akker HEA: Interphase drag coefficients in gas-solid flows, *AIChE J* 49:1060–1065, 2003. http://dx.doi.org/10.1002/aic.690490423.

Kaufmann A, Moreau M, Simonin O, Helie J: Comparison between Lagrangian and mesoscopic Eulerian modelling approaches for inertial particles suspended in decaying isotropic turbulence, *J Comput Phys* 227:6448–6472, 2008. http://dx.doi.org/10.1016/j.jcp.2008.03.004.

Khopkar AR, Kasat GR, Pandit AB, Ranade VV: Computational fluid dynamics simulation of the solid suspension in a stirred slurry reactor, *Ind Eng Chem Res* 45:4416–4428, 2006. http://dx.doi.org/10.1021/ie050941q.

Kim J, Balachandar S: Mean and fluctuating components of drag and lift forces on an isolated finite-sized particle in turbulence, *Theor Comput Fluid Dyn* 26:185–204, 2012. http://dx.doi.org/10.1007/s00162-010-0219-1.

Kriebitzsch SHL, Van der Hoef MA, Kuipers JAM: Fully resolved simulation of a gas-fluidized bed: a critical test of DEM models, *Chem Eng Sci* 91:1–4, 2013. http://dx.doi.org/10.1016/j.ces.2012.12.038.

Kuipers JAM, Van Swaaij WPM: Computational fluid dynamics applied to chemical reaction engineering, *Adv Chem Eng* 24:227–328, 1998. http://dx.doi.org/10.1016/S0065-2377(08)60094-0.

Kulick JD, Fessler JR, Eaton JK: Particle response and turbulence modification in fully-developed channel flow, *J Fluid Mech* 277:109–134, 1994. http://dx.doi.org/10.1017/S0022112094002703.

Labourasse E, Lacanette D, Toutant A, et al: Towards large eddy simulation of isothermal two-phase flows: Governing equations and a priori tests, *Int J Multiphase Flow* 33:1–39, 2007. http://dx.doi.org/10.1016/j.ijmultiphaseflow.2006.05.010.

Lahey RT: The CFD analysis of multidimensional phenomena in multiphase flow. In *Proceedings of the 2nd international conference on multiphase flow'95, Kyoto (Japan), MO2*, 1995, pp 1–9.

Laín S, Bröder D, Sommerfeld M: Numerical modelling of the hydrodynamics in a bubble column using the Euler-Lagrange approach. In *Proceedings of the MFTP-2000—international symposium on multiphase flow and transport phenomena, Antalya (Turkey)*, 2000.

Lane GL, Schwarz MP, Evans GM: Numerical modelling of gas-liquid flow in stirred tanks, *Chem Eng Sci* 60:2203–2214, 2005. http://dx.doi.org/10.1016/j.ces.2004.11.046.

Lapin A, Lübbert A: Numerical simulation of the dynamics of two-phase gas-liquid flows in bubble columns, *Chem Eng Sci* 49:3661–3674, 1994. http://dx.doi.org/10.1016/0009-2509(94)E0121-6.

Lapple CE, Shepherd CB: Calculation of particle trajectories, *Ind Eng Chem* 32:605–617, 1940. http://dx.doi.org/10.1021/ie50365a007.

Lübbert A, Lippert J, Korte T, Schügerl K: Strukturuntersuchungen im Zweiphasensystem Blasensäule, *Chem Ing Tech* 54:1193, 1982.

Lee Y, Korpela SA: Multicellular natural-convection in a vertical slot, *J Fluid Mech* 126:91–121, 1983. http://dx.doi.org/10.1017/S0022112083000063.

Lee D, Huh JY, Jeong D, Shin J, Yun A, Kim J: Physical, mathematical, and numerical derivations of the Cahn–Hilliard equation, *Comput Mater Sci* 81:216–225, 2014. http://dx.doi.org/10.1016/j.commatsci.2013.08.027.

Legendre D, Magnaudet J: The lift force on a spherical bubble in a viscous linear shear flow, *J Fluid Mech* 368:81–126, 1998. http://dx.doi.org/10.1017/S0022112098001621.

León-Becerril E, Liné A: Stability analysis of a bubble column, *Chem Eng Sci* 56:6135–6141, 2001. http://dx.doi.org/10.1016/S0009-2509(01)00254-8.

Li JH, Kwauk MS: Multiscale nature of complex fluid-particle systems, *Ind Eng Chem Res* 40:4227–4237, 2001. http://dx.doi.org/10.1021/ie0011021.

Li Y, McLaughlin JB, Kontomaris K, Portela L: Numerical simulation of particle-laden turbulent channel flow, *Phys Fluids* 13:2957–2967, 2001. http://dx.doi.org/10.1063/1.1396846.

Li ZP, Derksen JJ, Gao ZM: Models and applications for simulating turbulent solid-liquid suspensions in stirred tanks, *J Chem Eng Jpn* 48:329–336, 2015. http://dx.doi.org/10.1252/jcej.14we056.

Lin T, Reese J, Hong T, Fan LS: Quantitative analysis and computation of two-dimensional bubble columns, *AIChE J* 42:301–318, 1996. http://dx.doi.org/10.1002/aic.690420202.

Liu X, Glasser BJ, Howley MA: Instability of bounded gas-particle fluidized beds, *AIChE J* 53:811–824, 2007. http://dx.doi.org/10.1002/aic.11134.

Liu X, Howley MA, Johri J, Glasser BJ: The hydrodynamic stability of a fluid-particle flow: instabilities in gas-fluidized beds, *Chem Eng Educ* 42(4):179–184, 2008.

Lucas D, Prasser HM, Manera A: Influence of the lift force on the stability of a bubble column, *Chem Eng Sci* 60:3609–3619, 2005. http://dx.doi.org/10.1016/j.ces.2005.02.032.

Lucas D, Krepper E, Prasser HM, Manera A: Investigations on the stability of the flow characteristics in a bubble column, *Chem Eng Technol* 29:1066–1072, 2006. http://dx.doi.org/10.1002/ceat.200600150.

Lucas D, Krepper E, Prasser HM: Use of models for lift, wall and turbulent dispersion forces acting on bubbles for poly-disperse flows, *Chem Eng Sci* 62:4146–4157, 2007. http://dx.doi.org/10.1016/j.ces.2007.04.035.

Lucas D, Tomiyama A: On the role of the lateral lift force in poly-dispersed bubbly flows, *Int J Multiphase Flow* 37:1178–1190, 2011. http://dx.doi.org/10.1016/j.ijmultiphaseflow.2011.05.009.

Lundgren TS: Linearly forced isotropic turbulence, 2003: In *Annual Research Briefs*, Center for Turbulence Research, Stanford, CA, 2003, pp 461–473. http://www.dtic.mil/cgi-bin/GetTRDoc?AD=ADA420749#page=454.

Magelli F, Fajner D, Nocentini M, Pasquali G: Solids distribution in vessels stirred with multiple impellers, *Chem Eng Sci* 45:615–625, 1990. http://dx.doi.org/10.1016/0009-2509(90)87005-D.

Martyushev LM, Seleznev VD: Maximum entropy production principle in physics, chemistry and biology, *Phys Rep* 426:1–45, 2006. http://dx.doi.org/10.1016/j.physrep.2005.12.001.

Masood RMA, Delgado A: Numerical investigation of the interphase forces and turbulence closure in 3D square bubble columns, *Chem Eng Sci* 108:154–168, 2014. http://dx.doi.org/10.1016/j.ces.2014.01.004.

Maxey MR, Riley JJ: Equation of motion for a small rigid sphere in a non-uniform flow, *Phys Fluids* 26:883–889, 1982. http://dx.doi.org/10.1063/1.864230.

Maxey MR: The gravitational settling of aerosol particles in homogeneous turbulence and random flow fields, *J Fluid Mech* 174:441–465, 1987. http://dx.doi.org/10.1017/S0022112087000193.

McLaughlin JB: Inertial migration of a small sphere in linear flows, *J Fluid Mech* 224:261–274, 1991. http://dx.doi.org/10.1017/S0022112091001751.

McLaughlin JB: The lift on a small sphere in wall-bounded linear shear flows, *J Fluid Mech* 246:249–265, 1993. http://dx.doi.org/10.1017/S0022112093000114.

Medlin J, Jackson R: Fluid mechanical description of fluidized beds—effect of distributor thickness on convective instabilities, *Ind Eng Chem Fundam* 14:315–321, 1975. http://dx.doi.org/10.1021/i160051a017.

Mei R: An approximate expression for the shear lift force on a spherical-particle at finite Reynolds-number, *Int J Multiphase Flow* 18:145–147, 1992. http://dx.doi.org/10.1016/0301-9322(92)90012-6.

Mei R: Effect of turbulence on the particle settling velocity in the nonlinear drag range, *Int J Multiphase Flow* 20:273–284, 1994. http://dx.doi.org/10.1016/0301-9322(94)90082-5.

Minev PD, Lange U, Nandakumar K: A comparative study of two-phase flow models relevant to bubble column dynamics, *J Fluid Mech* 394:73–96, 1999. http://dx.doi.org/10.1017/S0022112099005546.

Monahan SM, Fox RO: Linear stability analysis of a two-fluid model for air-water bubble columns, *Chem Eng Sci* 62:3159–3177, 2007. http://dx.doi.org?10.1016/j.ces.2007.03.021.

Moraga FJ, Larreteguy AE, Drew DA, Lahey RT: Assessment of turbulent dispersion models for bubbly flows in the low Stokes number limit, *Int J Multiphase Flow* 29:655–673, 2003. http://dx.doi.org/10.1016/S0301-9322(03)00018-1.

Morsi SA, Alexander AJ: Investigation of particle trajectories in 2-phase flow systems, *J Fluid Mech* 55:193–208, 1972. http://dx.doi.org/10.1017/S0022112072001806.

Mudde RF, Lee DJ, Reese J, Fan LS: The role of coherent structures on Reynolds stresses in a 2-D bubble column, *AIChE J* 43:913–926, 1997a. http://dx.doi.org/10.1002/aic.690430407.

Mudde RF, Groen JS, Van den Akker HEA: Liquid velocity field in a bubble column: LDA experiments, *Chem Eng Sci* 52:4217–4224, 1997b. http://dx.doi.org/10.1016/S0009-2509(97)88935-X.

Mudde RF, Harteveld WK, Van den Akker HEA, Van der Hagen THJJ, Van Dam H: Gamma radiation densitometry for studying the dynamics of fluidized beds, *Chem Eng Sci* 54:2047–2054, 1999. http://dx.doi.org/10.1016/S0009-2509(98)00372-8.

Mudde RF, Simonin O: Two- and three-dimensional simulations of a bubble plume using a two-fluid model, *Chem Eng Sci* 54:5061–5069, 1999. http://dx.doi.org/10.1016/S0009-2509(99)00234-1.

Mudde RF, Saito T: Hydrodynamical similarities between bubble column and bubbly pipe flow, *J Fluid Mech* 437:203–228, 2001. http://dx.doi.org/10.1017/S0022112001004335.

Mudde RF: Gravity-driven bubbly flows, *Annu Rev Fluid Mech* 37:393–423, 2005. http://dx.doi.org/10.1146/annurev.fluid.37.061903.175803.

Mudde RF, Harteveld WK, Van den Akker HEA: Uniform flow in bubble columns, *Ind Eng Chem Res* 48:148–158, 2009. http://dx.doi.org/10.1021/ie8000748.

Munson BR, Young DF, Okiishi TH, Huebsch WW: *Fundamentals of fluid mechanics,* New York, 2010, Wiley.

Mutsers SMP, Rietema K: Effect of interparticle forces on expansion of a homogenous gas-fluidized bed, *Powder Technol* 18:239–248, 1977. http://dx.doi.org/10.1016/0032-5910(77)80014-4.

Niceno B, Dhotre MT, Deen NG: One-equation sub-grid scale (SGS) modelling for Euler-Euler large eddy simulation (EELES) of dispersed bubbly flow, *Chem Eng Sci* 63:3923–3931, 2008. http://dx.doi.org/10.1016/j.ces.2008.04.050.

Niceno B, Boucker M, Smith BL: Euler-Euler large-eddy simulation of a square cross-sectional bubble column using the Neptune CFD code, *Sci Technol Nucl Ins,* Article ID 410272, 8 pp, 2009. http://dx.doi.org/10.1155/2009/410272.

Nienow AW, Bartlett R: The measurement and prediction of particle-slip velocities in agitated vessels. In *Proceedings of the 1st European conference on mixing & centrifugal separation, Cambridge, UK, B1,* 1974, pp 1–15.

Nourgaliev R, Dinh T, Theofanous T, Joseph D: The lattice Boltzmann equation method: theoretical interpretation, numerics and implications, *Int J Multiphase Flow* 29:117–169, 2003. http://dx.doi.org/10.1016/S0301-9322(02)00108-8.

Ochieng O, Onyango MS: CFD simulation of solids suspension in stirred tanks: review, *Hem Ind* 64(5):365–374, 2010. http://dx.doi.org/10.2298/HEMIND100714051O.

Odar F, Hamilton WS: Forces on a sphere accelerating in a viscous fluid, *J Fluid Mech* 18:302–314, 1964. http://dx.doi.org/10.1017/S0022112064000210.

Oesterlé B, Bui Dinh T: Experiments on the lift of a spinning sphere in a range of intermediate Reynolds numbers, *Exp Fluids* 25:16–22, 1998. http://dx.doi.org/10.1007/s003480050203.

Oey RS, Mudde RF, Van den Akker HEA: Sensitivity study on interfacial closure laws in two-fluid bubbly flow simulations, *AIChE J* 49:1621–1636, 2003. http://dx.doi.org/10.1002/aic.690490703.

Olivieri S, Picano F, Sardina G, Iudicone D, Brandt L: The effect of the Basset history force on particle clustering in homogeneous and isotropic turbulence, *Phys Fluids* 26:041704, 2014. http://dx.doi.org/10.1063/1.4871480.

Pai MG, Subramaniam S: Two-way coupled stochastic model for dispersion of inertial particles in turbulence, *J Fluid Mech* 700:29–62, 2012. http://dx.doi.org/10.1017/jfm.2012.89.

Patankar NA, Joseph DD: Modeling and numerical simulation of particulate flows by the Eulerian-Lagrangian approach, *Int J Multiphase Flow* 27:1659–1684, 2001. http://dx.doi.org/10.1016/S0301-9322(01)00021-0.

Pellew A, Southwell RV: On maintained convective motion in a fluid heated from below, *Proc R Soc Lond A* 176:0312–0343, 1940. http://dx.doi.org/10.1098/rspa.1940.0092.

Peng ZB, Doroodchi E, Luo C, Moghtaderi B: Influence of void fraction calculation on fidelity of CFD-DEM simulation of gas-solid bubbling fluidized beds, *AIChE J* 60:2000–2018, 2014. http://dx.doi.org/10.1002/aic.14421.

Pinelli D, Montante G, Magelli F: Dispersion coefficients and settling velocities of solids in slurry vessels stirred with different types of multiple impellers, *Chem Eng Sci* 59:3081–3089, 2004. http://dx.doi.org/10.1016/j.cplett.2004.04.033.

Pourtousi M, Sahu JN, Ganesan P: Effect of interfacial forces and turbulence models on predicting flow pattern inside the bubble column, *Chem Eng Process* 75:38–47, 2014. http://dx.doi.org/10.1016/j.cep.2013.11.001.

Raabe D: Overview of the LB method for nano- and microscale fluid dynamics in materials science and engineering, *Model Simul Mater Sci Eng* 12:R13–R46, 2004. http://dx.doi.org/10.1088/0965-0393/12/6/R01.

Radl S, Sundaresan S: A drag model for filtered Euler-Lagrange simulations of clustered gas-particle suspensions, *Chem Eng Sci* 117:416–425, 2014. http://dx.doi.org/10.1016/j.ces.2014.07.011.

Reese J, Chen RC, Tzeng JW, Fan LS: Characterization of the macroscopic flow structure in gas-liquid and gas-liquid-solid fluidization systems using particle image velocimetry, *Int Video J Eng Res* 3:17, 1993.

Reese J, Fan LS: Transient flow structure in the entrance region of a bubble column using particle image velocimetry, *Chem Eng Sci* 49:5623–5636, 1994. http://dx.doi.org/10.1016/0009-2509(94)00269-X.

Riber E, Moureau V, Garcia M, Poinsot T, SImonin O: Evaluation of numerical strategies for large eddy simulation of particulate two-phase recirculating flows, *J Comput Phys* 228:539–564, 2009. http://dx.doi.org/10.1016/j.jcp.2008.10.001.

Richardson JF, Zaki WN: The sedimentation of a suspension of uniform spheres under conditions of viscous flow, *Chem Eng Sci* 3:65–73, 1954. http://dx.doi.org/10.1016/0009-2509(54)85015-9.

Rietema K, Ottengraf SPP: Laminar liquid circulation and bubble street formation in a gas-liquid system, *Trans Inst Chem Eng* 48:T54–T62, 1970.

Rietema K: Science and technology of dispersed 2-phase systems—parts I and II, *Chem Eng Sci* 37:1125–1150, 1982. http://dx.doi.org/10.1016/0009-2509(82)85058-6.

Rietema K, Van den Akker HEA: On the momentum equations in dispersed 2-phase systems, *Int J Multiphase Flow* 9:21–36, 1983. http://dx.doi.org/10.1016/0301-9322(83)90004-6.

Rohde M, Derksen JJ, Van den Akker HEA: An applicability study of advanced lattice-Boltzmann techniques for moving, no-slip boundaries and local grid refinement, *Comput Fluids* 37:1238–1252, 2008. http://dx.doi.org/10.1016/j.compfluid.2007.10.012.

Rosales C, Meneveau C: Linear forcing in numerical simulations of isotropic turbulence: physical space implementations and convergence properties, *Phys Fluids* 17:095106, 2005. http://dx.doi.org/10.1063/1.2047568.

Ross J, Corian AD, Müller SC: Proposed principles of maximum local entropy production, *J Phys Chem B* 116:7858–7865, 2012. http://dx.doi.org/10.1021/jp302088y.

Roux S, Lartigue G, Poinsot T, Meier U, Bérat C: Studies of mean and unsteady flow in a swirled combustor using experiments, acoustic analysis and large eddy simulations, *Combust Flame* 141:40–54, 2005. http://dx.doi.org/10.1016/j.combustflame.2004.12.007.

Rubinow SI, Keller JB: The transverse force on a spinning sphere moving in a viscous fluid, *J Fluid Mech* 11:447–459, 1961. http://dx.doi.org/10.1017/S0022112061000640.

Ruzicka MC: On stability of a bubble column, *Chem Eng Res Des* 91:191–203, 2013. http://dx.doi.org/10.1016/j.cherd.2012.07.011.

Saffman PG: Lift on a small sphere in a slow shear flow, *J Fluid Mech* 22:385–400, 1965. http://dx.doi.org/10.1017/S0022112065000824.

Saffman PG: Corrigendum to 'Lift on a small sphere in a slow shear flow', *J Fluid Mech* 31:624, 1968.

Sankaranarayanan K, Sundaresan S: Lift force in bubbly suspensions, *Chem Eng Sci* 57:3521–3542, 2002. http://dx.doi.org/10.1016/S0009-2509(02)00269-5.

Sardeshpande MV, Ranade VV: Computational fluid dynamics modelling of solid suspension in stirred tanks, *Curr Sci* 102:1539–1551, 2012.

Segré G, Silberberg A: Behaviour of macroscopic rigid particles in Poiseuille flow - part II. Experimental results and interpretation, *J Fluid Mech* 14:136–157, 1962.

Senoner JM, Sanjosé M, Lederlin T, et al: Eulerian and Lagrangian large-eddy simulations of an evaporating two-phase flow, *C R Méc* 337:458–468, 2009. http://dx.doi.org/10.1016/j.crme.2009.06.002.

Shah RSSRE, Sajjadi B, Raman AAA, Ibrahim S: Solid-liquid mixing analysis in stirred vessels, *Rev Chem Eng* 31:119–147, 2015. http://dx.doi.org/10.1515/revce-2014-0028.

Shalaby H, Wozniak K, Wozniak G: Numerical calculation of particle-laden cyclone separator flow using LES, *Eng Appl Comput Fluid Mech* 2:382–392, 2008. http://dx.doi.org/10.1080/19942060.2008.11015238.

Shnip AI, Kolhatkar RV, Swamy D, Joshi JB: Criteria for the transition from the homogeneous to the heterogeneous regime in two-dimensional bubble column reactors, *Int J Multiphase Flow* 18:705–726, 1992. http://dx.doi.org/10.1016/0301-9322(92)90040-N.

Sokolichin A, Eigenberger G: Gas-liquid flow in bubble columns and loop reactors: part I. Detailed modelling and numerical simulation, *Chem Eng Sci* 49:5735–5746, 1994. http://dx.doi.org/10.1016/0009-2509(94)00289-4.

Sokolichin A, Eigenberger G, Lapin A, Lübbert A: Dynamic numerical simulation of gas-liquid two-phase flows—Euler/Euler versus Euler/Lagrange, *Chem Eng Sci* 52:611–626, 1997. http://dx.doi.org/10.1016/S0009-2509(96)00425-3.

Sokolichin A, Eigenberger G, Lapin A: Simulation of buoyancy driven bubble flow: established simplifications and open questions, *AIChE J* 50:24–45, 2004. http://dx.doi.org/10.1002/aic.10003.

Sommerfeld M, Decker S: State of the art and future trends in CFD simulation of stirred vessel hydrodynamics, *Chem Eng Technol* 27:215–224, 2004. http://dx.doi.org/10.1002/ceat.200402007.

Stone HA: Philip Saffman and viscous flow theory, *J Fluid Mech* 409:165–183, 2000. http://dx.doi.org/10.1017/S0022112099007697.

Stringham GE, Simons DB, Guy HP: *The behavior of large particles in quiescent liquids, 1969:* Washington, 1969, US Government Printing Office. Geological Survey Professional Paper 562-C, http://pubs.usgs.gov/pp/0562c/report.pdf.

Tabib MV, Roy SA, Joshi JB: CFD simulation of bubble column: an analysis of interphase forces and turbulence models, *Chem Eng J* 139:589–614, 2008. http://dx.doi.org/10.1016/j.cej.2007.09.015.

Tamburini A, Cipollina A, Micale G, Brucato A, Ciofalo M: Influence of drag and turbulence modelling on CFD predictions of solid liquid suspensions in stirred vessels, *Chem Eng Res Des* 92:1045–1063, 2014. http://dx.doi.org/10.1016/j.cherd.2013.10.020.

Tanaka T, Yonemura S, Tsuji Y: Effects of particle properties on the structure of clusters. In *Gas-particle flows, ASME Fluids Engineering Division*, vol. 228, 1995, pp 297–302.

Tanaka T, Yonemura S, Kiribayashi K, Tsuji Y: Cluster formation and particle-induced instability in gas-solid flows predicted by the DSMC method, *JSME Int J Series B* 39:239–245, 1996.

Tang Y, Peters EAJF, Kuipers JAM, Kriebitzsch SHL, Van der Hoef MA: A new drag correlation from fully resolved simulations of flow past monodisperse static arrays of spheres, *AIChE J* 61:688–698, 2015. http://dx.doi.org/10.1002/aic.14645.

Tchen CM: *Mean value and correlation problems connected with the motion of small particles suspended in a turbulent fluid,* The Netherlands, 1947, Delft University of Technology. PhD Thesis.

Ten Cate A, Derksen JJ, Kramer HJM, Van Rosmalen GM, Van den Akker HEA: The microscopic modelling of hydrodynamics in industrial crystallisers, *Chem Eng Sci* 56:2495–2509, 2001. http://dx.doi.org/10.1016/S0009-2509(00)00433-4.

Ten Cate A, Nieuwstad CH, Derksen JJ, Van den Akker HEA: Particle imaging velocimetry experiments and lattice-Boltzmann simulations on a single sphere settling under gravity, *Phys Fluids* 14:4012–4025, 2002. http://dx.doi.org/10.1063/1.1512918.

Ten Cate A, Derksen JJ, Portela LM, Van den Akker HEA: Fully resolved simulations of colliding monodisperse spheres in forced isotropic turbulence, *J Fluid Mech* 519:233–271, 2004. http://dx.doi.org/10.1017/s0022112004001326.

Tenneti S, Garg R, Subramaniam S: Drag law for monodisperse gas-solid systems using particle-resolved direct numerical simulation of flow past fixed assemblies of spheres, *Int J Multiphase Flow* 37:1072–1092, 2011. http://dx.doi.org/10.1016/j.ijmultiphaseflow.2011.05.010.

Tenneti S, Subramaniam S: Particle-resolved direct numerical simulation for gas-solid flow model development, *Annu Rev Fluid Mech* 46:199–230, 2014. http://dx.doi.org/10.1146/annurev-fluid-010313-141344.

Tollert H: Die Wirkung der Magnus-kraft in laminaren Strömungen, *Chem Ing Technik* 26:270–277, 1954.

Tomiyama A, Tamai H, Zun I, Hosokawa S: Transverse migration of single bubbles in simple shear flows, *Chem Eng Sci* 57:1849–1858, 2002. http://dx.doi.org/10.1016/S0009-2509(02)00085-4.

Torobin LB, Gauvin WH: Fundamental aspects of solids-gas flow—part I: introductory concepts and idealized sphere motion in viscous regime, *Can J Chem Eng* 38:129–141, 1959a.

Torobin LB, Gauvin WH: Fundamental aspects of solids-gas flow—part II: the sphere wake in steady laminar fluids, *Can J Chem Eng* 38:167–176, 1959b.

Torobin LB, Gauvin WH: Fundamental aspects of solids-gas flow—part III: accelerated motion of a particle in a fluid, *Can J Chem Eng* 38:224–236, 1959c.

Torobin LB, Gauvin WH: Fundamental aspects of solids-gas flow—part IV: the effects of particle rotation, roughness and shape, *Can J Chem Eng* 39:142–153, 1960a.

Torobin LB, Gauvin WH: Fundamental aspects of solids-gas flow—part V: the effects of fluid turbulence on the particle drag coefficient, *Can J Chem Eng* 39:189–200, 1960b.

Torobin LB, Gauvin WH: Fundamental aspects of solids-gas flow—part VI: multiparticle behavior in turbulent fluids, *Can J Chem Eng* 40:113–120, 1961.

Tryggvason G, Esmaeeli A, Lu JC, Biswas S: Direct numerical simulations of gas/liquid multiphase flows, *Fluid Dyn Res* 38:660–681, 2006. http://dx.doi.org/10.1016/j.fluiddyn.2005.08.006.

Tryggvason G, Lu J: Direct numerical simulations of bubble flows, *Bull JSME Mech Eng Rev* 2: paper no. 15-00220, 2015. http://dx.doi.org/10.1299/mer.15-00220.

Tsuji Y, Morikawa Y, Mizuno O: Experimental measurement of the Magnus force on a rotating sphere at low Reynolds numbers, *J Fluids Eng* 107:484–488, 1985. http://dx.doi.org/10.1115/1.3242517.

Tsuji Y, Tanaka T, Yonemura S: Cluster patterns in circulating fluidized beds predicted by numerical simulation: discrete particle model versus two-fluid model, *Powder Technol* 95:254–264, 1998. http://dx.doi.org/10.1016/S0032-5910(97)03349-4.

Tsuo YP, Gidaspow D: Computation of flow patterns in circulating fluidized beds, *AIChE J* 36:885–896, 1990. http://dx.doi.org/10.1002/aic.690360610.

Van den Akker HEA: *Liquid-liquid spray columns: hydrodynamic stability and reduction of axial mixing,* PhD thesis, The Netherlands, 1978, Eindhoven University of Technology.

Van den Akker HEA, Rietema K: Flow patterns and axial mixing in liquid-liquid spray columns—part 3. The reduction of axial mixing by applying a positive temperature gradient, *Trans Inst Chem Eng* 57:84–93, 1979a.

Van den Akker HEA, Rietema K: Flow patterns and axial mixing in liquid-liquid spray columns—part 4. Detection of a multiple circulation pattern generated by the application of a positive temperature gradient, *Trans Inst Chem Eng* 57:147–155, 1979b.

Van den Akker HEA, Rietema K: Correspondence, *Trans Inst Chem Eng* 60:255–256, 1982.

Van den Akker HEA: Momentum equations in dispersed two-phase flows. In Cherimisinoff NP, editor: The Encyclopedia of Fluid Mechanics, Houston, Tx, 1986, *Gulf Publishing Company*. vol 3, pp 371–400.

Van den Akker HEA: Coherent structures in multiphase flows, *Powder Technol* 100:123–136, 1998. http://dx.doi.org/10.1016/S0032-5910(98)00133-8.

Van den Akker HEA: The details of turbulent mixing processes and their simulation, *Adv Chem Eng* 31:151–229, 2006. http://dx.doi.org/10.1016/S0065-2377(06)31003-4.

Van den Akker HEA: Toward a truly multi-scale computational strategy for simulating turbulent two-phase flow processes, *Ind Eng Chem Res* 49:10780–10797, 2010. http://dx.doi.org/10.1021/ie1006382.

Van der Hoef MA, Ye M, Van Sint Annaland M, Andrews AT, Sundaresan S, Kuipers JAM: Multiscale modeling of gas-fluidized beds, *Adv Chem Eng* 31:65–149, 2006. http://dx.doi.org/10.1016/S0065-2377(06)31002-2.

Van Wageningen WFC, Mudde RF, Van den Akker HEA: Numerical simulation of growing Cu particles in a Kenics static mixer reactor in which Cu2+ is reduced by carbohydrates, *Chem Eng Sci* 59:5193–5200, 2004. http://dx.doi.org/10.1016/j.ces.2004.07.102.

Van Wageningen WFC: *Design of a static mixer reactor for copper recovery from waste streams,* PhD thesis, Delft, The Netherlands, 2005, Delft University of Technology.

Verloop WC, Boersma D, Van den Akker HEA, Hein KRG: The fluid dynamics of particles in the freeboard of a pressurized fluidized bed combustor. In *Proceedings of the 12th international conference on fluidized bed combustion, San Diego, CA.* 1993.

Verloop WC: *The fluid dynamics of the freeboard in a fluidized bed reactor,* PhD thesis, The Netherlands, 1994, Delft University of Technology.

Verloop WC, Van den Akker HEA, Boersma D, Hein KRG: Transient modelling of the combined gas-particle flow in the freeboard of a fluidized bed reactor. In *Proceedings of the 13th international conference on fluidized bed combustion, Orlando, FL.* 1995.

Viollet PL, Simonin O: Modelling dispersed two-phase flows: closure, validation and software development, *Appl Mech Rev* 47(6):S80–S84, 1994. Part 2, ASME.

Visuri O, Liiri M, Alopaeus V: Comparison and validation of CFD models in liquid-solid suspensions, *Can J Chem Eng* 89:699–706, 2011. http://dx.doi.org/10.1002/cjce.20458.

Wallis GB: *One-dimensional two-phase flow,* New York, NY, 1969, McGraw-Hill.
Wallis GB: The terminal speed of single drops or bubbles in an infinite medium, *Int J Multiphase Flow* 1:491–511, 1974. http://dx.doi.org/10.1016/0301-9322(74)90003-2.
Wang Q, Squires KD, Chen M, McLaughlin JB: On the role of the lift force in turbulence simulations of particle deposition, *Int J Multiphase Flow* 23:749–763, 1997. http://dx.doi.org/10.1016/S0301-9322(97)00014-1.
Wang J: A review of Eulerian simulation of Geldart A particles in gas-fluidized beds, *Ind Eng Chem Res* 48:5567–5577, 2009. http://dx.doi.org/10.1021/ie900247t.
Wang W, Lu B, Zhang N, Shi ZS, Li JH: A review of multiscale CFD for gas-solid CFB modeling, *Int J Multiphase Flow* 36:109–118, 2010. http://dx.doi.org/10.1016/j.ijmultiphaseflow.2009.01.008.
Wardle KE: Open-source CFD simulations of liquid-liquid flow in the annular centrifugal contactor, *Sep Sci Technol* 46:2409–2417, 2011. http://dx.doi.org/10.1080/01496395.2011.600748.
Wen CY, Yu YH: Mechanics of fluidization, *CEP Symp Series* 62:100–111, 1966.
Wesselingh JA: The velocity of particles, drops and bubbles, *Chem Eng Sci* 21:9–14, 1987. http://dx.doi.org/10.1016/0255-2701(87)80002-8.
Westerterp KR, Van Dierendonck LL, De Kraa JA: Interfacial areas in agitated gas-liquid contactors, *Chem Eng Sci* 18:157–176, 1963. http://dx.doi.org/10.1016/0009-2509(63)85002-2.
Whalley PB, Davidson JF: Liquid circulation in bubble columns. In *IChemE symposium series no. 38: paper J5,* 1974. 27 pp.
Wijffels JB, Rietema K: Flow patterns and axial mixing in liquid-liquid spray columns—part 1. Theory, *Trans Inst Chem Eng* 50:224–233, 1972a.
Wijffels JB, Rietema K: Flow patterns and axial mixing in liquid-liquid spray columns—part 2. Experiments, *Trans Inst Chem Eng* 50:233–239, 1972b.
Yamamoto Y, Potthoff M, Tanaka T, Kajishima T, Tsuji Y: Large-eddy simulation of turbulent gas-particle flow in a vertical channel: effect of considering inter-particle collisions, *J Fluid Mech* 442:303–334, 2001. http://dx.doi.org/10.1017/s0022112001005092.
Yan XK, Holloway W, Sundaresan S: Periodic flow structures in vertical gas-particle flows, *Powder Technol* 241:174–180, 2013. http://dx.doi.org/10.1016/j.powtec.2013.03.024.
Yonemura S, Tanaka T, Tsuji Y: Cluster formation in dispersed gas-solid flow. In *Proceedings of the 2nd international conference on multi-phase flows, Kyoto, Japan.* 1995.
Zehner P: Impuls-, Stoff- und Wärmetransport in Blasensäulen—Teil 1. Strömungsmodell der Blasensäule und Flüssigkeitsgeschwindigkeiten, *Verfahrenstechnik* 16:347–351, 1982.
Zehner P: Momentum, mass and heat transfer in bubble columns—part 1. Flow model of the bubble column and liquid velocities, *Int Chem Eng* 16:22–28, 1986.
Zhang ZZ, Prosperetti A: Sedimentation of 1,024 particles. In *Proceedings of the ASME FED summer conference, Houston, TX, vol. 1,* vol. 1, 2005, pp 919–928.
Zhang D, Deen NG, Kuipers JAM: Numerical simulation of the dynamic flow behavior in a bubble column: a study of closures for turbulence and interface forces, *Chem Eng Sci* 61:7593–7608, 2006. http://dx.doi.org/10.1016/j.ces.2006.08.053.
Zhu HP, Zhou ZY, Yang RY, Yu AB: Discrete particle simulation of particulate systems: theoretical developments, *Chem Eng Sci* 62:3378–3396, 2007. http://dx.doi.org/10.1016/j.ces.2006.12.089.
Zhu HP, Zhou ZY, Yang RY, Yu AB: Discrete particle simulation of particulate systems: a review of major applications and findings, *Chem Eng Sci* 63:5728–5770, 2008. http://dx.doi.org/10.1016/j.ces.2008.08.006.
Zuber N, Findlay JA: Average volumetric concentration in two-phase flow systems, *ASME J Heat Transf* 87(4):453–468, 1965. http://dx.doi.org/10.1115/1.3689137.
Zwietering TN: Suspending in solid particles in liquid by agitators, *Chem Eng Sci* 8:244–258, 1958. http://dx.doi.org/10.1016/0009-2509(58)85031-9.

# INDEX

Note: Page numbers followed by "*f*" indicate figures and "*t*" indicate tables.


## A

Additive manufacturing
  materials, selection of, 50–51
  selective laser melting, 50–51, 51*f*
  shapes, variability of, 50–51
  tensile stress, 51–52, 52*f*
Adhesion forces, 38–39, 51–52
Agglomeration, 10
All-at-once approach, optimal control setting, 59
Analytical (ultra-)centrifugation (AUC), 15, 44

## B

Balance equations, 141–145
Batchelor's equation, 208–209
BBO equation, 312
Bernoulli equation, 146–147
β-lactoglobulin (BLG), 30
Bifurcation
  gas–liquid systems, 263, 265*f*
  gas–solid systems, 263, 264*f*
  stability condition, 263
Binary droplet–droplet interactions
  droplet collision, 173, 174*f*
  satellite droplet, 175, 175*f*
  spray processes, 172–173
  water droplets, 176, 176*f*
Bottom-up methods
  chemical thermodyamics, 24
  DFT, 22–23
  mass transfer, 23
  nucleation burst, 23–24
  particle formation dynamics, 22–23
  particle synthesis, 22–23, 22*f*
  PBE, 25
  PSD, 23
  quasi-monodisperse, 23–24
  shape control, 24
  supersaturation, 22–23
  transport controlled systems, 23
Brownian forces, 139–140
Bubble column reactors
  CFD simulation, 270–276
  chemisorption, 170–172, 173*f*
  EMMS drag, 272–273*f*
  fluid flow, mass transfer, and chemical reaction, 170–172, 171*f*
  gas holdup, 274*f*
  liquid phase dispersion, 170, 171*f*
  multiphase flow, 268–269
  wire meshes, 170–172, 172*f*
Bubble interactions
  contaminated liquids, 167–168
  drag force, 168
  Weber criterion, 169
Bubbles coalescence, 251–254
  bubble column reactors, 270–276
  liquid viscosity and surface tension, 260
  stability condition, 254
Buoyancy force, 146–147

## C

Cahn–Hilliard concept, 295–296
Cahn–Hilliard free energy approach, 295–296
Camsizer® XT device, 103
CFD. *See* Computational fluid dynamics (CFD)
Chemical turbulence, 248–249
Circulation cell models, parade of, 285, 286*f*
Classical fluid mechanics (hydrodynamics), 282
Cluster formation, 286–287
Coating process, fluidized bed, 85
Collisional heat transfer, 203, 205–206
Collision dynamics
  binary droplet–droplet interactions, 172–176
  numerical techniques, 153–156
  particle–particle interaction, 165

Computational fluid dynamics (CFD), 25, 197–198, 211, 284, 286, 288–289
closure law, 268–270
grid-averaged equations, 180
multiphase flow, 268–269
numerical techniques, 153–156
simulations, 87, 298–302, 302*f*
stability-constrained multifluid, 269
Computer-automated radioactive particle tracking (CARPT), 284, 286
Conduction heat transfer, 208, 217–219, 225*f*
Continuous fluid phase
balance equations, 141
CFD, 180
Eulerian grid to Lagrangian positions, 148–151
pseudo-turbulent, 144
smoothed macro-quantities, 142
turbulence theory, 144
Continuum approach, 196–197, 238
Continuum fluid field, 201
Correction coefficient, 206, 215–217, 216*f*
Coupling schemes
discrete particles, 211
translational and rotational motions, 211
Crystallization process, 25, 53, 67

## D

DBS model. *See* Dual-bubble-size (DBS) model
DEM. *See* Discrete element methods (DEMs)
Density functional theory (DFT), 22–23, 46, 48
Di Felice correlation, 199–200
Direct numerical simulations (DNSs), 289, 335–339, 337–338*f*
direct numerical simulation
droplet–droplet collisions, 186
Euler–Lagrange model, 186
kinematic viscosities, 187
volume-of-fluid (VOF), 186–187
particle–fluid interactions, 159–165
Direct simulation Monte Carlo (DSMC) model, 289
binary droplet–droplet interactions
droplet collision, 173, 174*f*
satellite droplet, 175, 175*f*
spray processes, 172–173
water droplets, 176, 176*f*
droplet sprays for spray drying
collision frequencies, 178, 179*f*
droplet size distribution, 178, 179*f*
liquid spray system, 176, 177*f*
Sauter mean diameter, 177–178, 178*f*
numerical techniques, 153–156
Discrete bubble method (DBM)
bubble column reactors
chemisorption, 170–172, 173*f*
fluid flow, mass transfer, and chemical reaction, 170–172, 171*f*
liquid phase dispersion, 170, 171*f*
wire meshes, 170–172, 172*f*
bubble interactions
contaminated liquids, 167–168
drag force, 168
Weber criterion, 169
Discrete element methods (DEMs), 17, 87, 92–95, 93*f*, 197–198, 206, 211, 300–301
Discrete particle methods (DPMs), 300, 87. *See also* Discrete element methods (DEMs)
gas–solid fluidized beds, 167, 167*f*
particle–fluid interactions
DNS, 160–161
drag force/correlations, 160, 161–162*f*
slip velocities, 163, 163*f*
particle–particle interaction
restitution coefficient, 166
soft-sphere model, 165
wetted plate analysis, 166, 166*f*
Discrete particle simulation, 236–238
Discrete phases
balance equations, 141–145
Eulerian grid to Lagrangian positions, 148–151
interphase momentum transfer, 145–148
motion equations, 145
numerical techniques, 151–156
Double taper cone model, 209–210, 209*f*
DPM. *See* Discrete particle methods (DPMs)
Drag force/correlations, 160, 161–162*f*
Droplet sprays for spray drying
collision frequencies, 178, 179*f*
droplet size distribution, 178, 179*f*

liquid spray system, 176, 177*f*
Sauter mean diameter, 177–178, 178*f*
Dual-bubble-size (DBS) model
distribution, 259*f*
energy dissipation, 258
Dual gamma-ray densitometry technique, 287–288, 288*f*

## E

Effective thermal conductivity (ETC)
calculated and measured, 213–214, 214*f*
comparison of, 211–212, 212*f*
effects of, 219–220, 219*f*
variation of, 220–221, 220*f*
Elastic contact deformation with Hertz theory, 104–105
Elastic deformation, 203, 208
Elasticity, modulus of, 105
Elastic stiffness, 105
Electroacoustic techniques, 44–45
Electron microscopy techniques, 17
EMMS model. *See* Energy-minimization multiscale (EMMS) model
Energy dissipation
DBS model, 258–260
EMMS model, 253–254
gas–liquid systems, 277
Energy-minimization multiscale (EMMS) model, 251–255, 295
coalescence rate, 276*f*
conservation equations, 252–253
drag coefficients, 253
energy dissipation, 275–276
gas–liquid systems., 251–252, 252*f*
stability condition, 254
Equation-free modeling, 46
Ergun equation, 199–200
ETC. *See* Effective thermal conductivity (ETC)
Euler–Euler approach, 154, 299, 330–332
Eulerian CFD fashion, 140
Eulerian–Eulerian model, 269
Eulerian grid to Lagrangian positions
mapping function, 148–151, 151*f*
polynomial function, 150
Euler–Lagrange models, 299, 328–330
bubbles, DBMs
column reactors, 170–172
interactions, 167–170
continuous and discrete phases
balance equations, 141–145
Eulerian grid to Lagrangian positions, 148–151
interphase momentum transfer, 145–148
motion equations, 145
numerical techniques, 151–156
definition, 140–141
discrete particle models
gas–solid fluidized beds, 167, 167*f*
particle–fluid interactions, 159–165
particle–particle interaction, 165–166, 166*f*
droplet collision outcomes, DNS, 186–188
droplets, DSMC
binary droplet–droplet interactions, 172–176
droplet sprays for spray drying, 176–180
filtered fluid equation, 182–186
multiphase flows
Archimedes number, 159
capillary number, 159
Eötvös number, 158
Morton number, 159
Ohnesorge number, 159
Reynolds number, 157
viscosity ratio, 157

## F

Field effect transistors (FETs), 9
Filtered fluid equation
filter function, 184
microscopic equations, 182–183
Newtonian stress, 185
smoothed momentum, 182–183
smoothed velocity field, 185–186
Finite element method (FEM), 197, 204–205
Fischer–Tropsch, 266–267
Flowability, powder properties, 50–51
Fluid–bubble interaction force, 297
Fluid–fluid interaction, 303
Fluidization behavior, disruption of, 86

Fluidized beds (FB), 85
coarse particles
CFD–DEM model, 224
conduction heat transfer, 225*f*
density distributions, 224–226
gas fluidization, 221
gas superficial velocity, 221–222, 223*f*
heating process, 222*f*
hot spheres, 221–224
HTC–Ug relationship, 222–224, 223*f*
particle–fluid–particle conduction, 224
time-averaged heat transfer coefficients, 223*f*
conductive heat transfer, 203–209
convective heat transfer, 202–203
fine particles
CFD–DEM approach, 227
flow regimes, heating process, 227, 228*f*
Hamaker constant, 227
heat fluxes, convective and conductive, 227, 228*f*
macroscopic heating rate, 227, 229*f*
particle size, 227
powders, 227
van der Waals force, 227
fluidization behavior, disruption of, 86
formation of stagnation zones, 86
heat transfer
characteristics, 235, 237*f*
complicated gas–solid flow, 235, 236*f*
convective and conductive heat fluxes, 231*f*
Fluent commercial software, 232–234, 234*f*
gas velocity, 231
high tube temperatures, 232–234, 234*f*
$\kappa_s$, effect of, 231*f*, 232
particle residence time, 229–230
particle velocity, 235
porosity, 229–230
positron emission particle tracking method, 229–230
solid flow pattern, 230*f*
temperature fields, 235
three-dimensional bed, 230
tube position, 231
large overspray ratio, 86
macroscale behavior, 255–256
multiphase flow, 268–269
multiphase reactors, 249, 249*f*
radiative, 202–203
undesired particle agglomeration, 86
Wurster coater, 88–89, 89*f*
Fluid–particle interaction, 303–307, 304*f*
Fluid–particle interaction force, 307
Fluid–phase equations, 143
Force–displacement relationship, 105–107
Fourier-transformed infrared spectroscopy (FT-IR), 17–18, 28–29
Free energy functional, 295–296

## G

Gas fluidization, 217–219, 221, 230
Gas kinematic viscosity, 94
Gas–liquid systems
bifurcation and stability condition, 265*f*
DBS model–intrinsic model, 258–260
EMMS, 252*f*, 258–268
energy dissipation, 277
gas–liquid and gas–solid systems, 261–263
*vs.* gas–solid systems, 251–252
liquid viscosity and surface tension, 260–261
stability condition, 254
Gas–particle flow, 289, 289*f*
Gas phase synthesis, 19, 24–25, 60–61
Gas–solid cyclones, 289, 290*f*
Gas–solid fluidized beds, 167, 167*f*
Gas–solid systems
bifurcation/stability condition, 264*f*
EMMS models, 251, 261–263
Gas–solid twophase system, 289, 290*f*
Graphical processing units (GPUs), 301
Gravitational force, 199

## H

Hamaker constant, 227
Heat transfer coefficient (HTC), 196, 218*f*, 223*f*
Heat transfer, in packed and fluidized beds
CFD–DEM approach, 197–198
conclusions and future work, 235–238
contact conditions, 206–207, 207*f*
continuum approach, 196–197
fluid bed reactors, 196

HTC, 196
mathematical modeling, 196–197
mechanisms, 202, 202*f*
model application
fluidized beds, 221–235
packed beds, 211–221
model description
coupling schemes, 211
governing equations, 198–201, 200*t*
heat transfer, 202–211
Navier–Stokes equations, 197–198, 201
particle flow, 196–197
particle–fluid systems, 196–197
thermal behavior, 196–197
Hertz-Mindlin discrete element model, 102
Hertz theory, 104–105
Heterogeneous multiscale methods, 46
Hierarchical products
agglomeration, 10
Aspen software package, 10
building blocks, 6–7
charge transfer, 10–11
continuous fluid phase, 7
correlation function, 8
deficiency reasons, 10
electronic defect states, 9
FETs, 9
flow-sheeting software, 10
fundamental hierarchy, 7, 7*f*
macroscopic scale, 6–7
mass transport, 10–11
material design, 6–7
mesocrystals, 11, 12*f*
microscopic tailoring, 8–9
molecular and nanoscale design, 10–11
molecular interactions, 8
opto-electronic devices, 9
particulate interactions, 7
particulate products, interconnections for, 6–7, 6*f*
process parameters, 9–10
process–structure functions, 8–9, 8*f*
property functions, 9
roll-to-roll processes, 9
statistical thermodynamics, 8
structure–property functions, 8–9, 8*f*
thin films, 9, 13, 39–41
Hot sphere approach, 224–226
Hydrodynamic flow. *See* Euler–Lagrange models
Hydrodynamic instability, 248–249
*Hydrodynamics* (Lamb, Sir Horace), 282–283

## I

Interactions, particulate products
β-lactoglobulin (BLG), 30, 31*f*
chemical functionalization, 27–28
chemical surface effects, 30
classical electrostatics, 27–28
classical force laws, 27–28
colloids, 26
DLVO theory, 26
equilibrium systems, 26
fluid interfaces, 30
hard mechanical forces, 25–26
macroscopic properties, 27–28
molecular interactions, 26
nanoparticle systems, 26
particle interactions, 26, 27*f*
protein foams, 30, 31*f*
repulsive interactions, 30
soft microscopic interactions, 25–26
van der Waals adhesion forces, 26
Interfacial turbulence, 248–249
Internal-loop airlift reactor, 273–275
Interphase momentum transfer
bernoulli equation, 146–147
buoyancy force, 146
numerical techniques, 151–153
Intrinsic model
energy dissipation, 258
gas–liquid *vs.* gas–solid systems, 261–263
Ishii–Zuber. *See* Schiller–Naumann
Isotropic turbulence, 254–255, 275

## K

Kernel function, 270, 275
Kolmogorov-scale turbulence, 289–291
Kolmogorov turbulent, 140

## L

Lagrangian point, 148–151, 151*f*
Large-eddy simulations (LESs), 289
Laser Doppler anemometry (LDA), 284, 286
Lattice Boltzmann (LB) methods, 199–200, 295–296

Lift force
  Magnus lift force, 315, 315*f*, 319–320
  Saffman lift force, 315–318, 316*f*, 318–319*f*
Linear hydrodynamic stability theory, 296
Linear stability analyses, 296–298
Liquid–liquid spray column, 284, 285*f*
Liquid–solid suspension, 289–291, 291–292*f*
Liquid viscosity, 260–261, 260–261*f*
Local porosity, 208, 231–232, 232*f*

## M

Magnus lift force, 315, 315*f*, 319–320
Mapping function, 148–149, 151*f*
Mass balance accounting, 64
Mass transfer and reaction
  gas holdup/experiments, 266–267, 267*f*
  reactor model, 264–265, 266*f*
  syngas conversion, 266–267, 267*f*
Matched asymptotics, 46
Material function (MF), 3, 46–48
MD. *See* Molecular dynamics (MD)
Mechanical particle properties (MPPs), 17, 25–26
Mechanistic approach, 197
Mesocrystals, ZnO quantum dots, 11
Mesoscale flow structures
  acceleration effect, 310–313
  conventional CFD simulations
    Euler–Euler approach, 330–332
    Euler–Lagrange approach, 328–330
    literature computational results, 332–335
  direct numerical simulations, 335–339, 337–338*f*
  evidence of
    cluster formation, 286–287
    computational fluid dynamics (CFD), 284, 286, 288–289
    computer-automated radioactive particle tracking (CARPT), 284, 286
    direct numerical simulations (DNSs), 289
    direct simulation Monte Carlo (DSMC) method, 289
    dual gamma-ray densitometry technique, 287–288, 288*f*
    gas–particle flow, 289, 289*f*
    gas–solid cyclones, 289, 290*f*
    gas–solid twophase system, 289, 290*f*
    large-eddy simulations (LESs), 289
    laser Doppler anemometry (LDA), 284, 286
    liquid–liquid spray column, 284, 285*f*
    liquid–solid suspension, 289–291, 291–292*f*
    parade of circulation cell models, 285, 286*f*
    particle image velocimetry (PIV), 284, 338
    steady-state model, 284
  fluid–particle interaction, 303–307, 304*f*
  lift force
    Magnus lift force, 315, 315*f*
    Saffman lift force, 315–318, 316*f*, 318–319*f*
  multiple particle systems, 324–328
  origin
    CFD simulations, 298–302, 302*f*
    extremum principles, 294–296
    fluid–fluid interaction, 303
    linear stability analyses, 296–298
  steady-state drag force, 308–310
  turbulence, 320–324, 321*f*
Mesoscale transport phenomena and mechanisms
  dissipation, 253–254
  EMMS model, 251–255, 252*f*
    DBS model–intrinsic model, 258–260
    gas–liquid *vs.* gas–solid systems, 261–263
    liquid viscosity, 260–261, 260–261*f*
    mass transfer and reaction, coupling, 264–267
    surface tension, 261, 262*f*
  macroscale behavior, perspective, 255–257, 257*f*
  multifluid CFD Model
    bubble column reactors, 270–276
    closure law, 268–270
  multiphase reactors, 248–249, 248*f*
MF. *See* Material function (MF)
Micro quantities, 142
Model-reduction techniques, 55–56
Molecular dynamics (MD), 22–23, 27–28, 48

Molecular interactions, 8
Momentum equations
bubble column reactors, 170–172
DNC, 186–187
numerical techniques, 151–153
MSS. *See* Multiscale modeling and simulation (MSS)
Multiphase computational fluid dynamics (CFD) simulations, 87
Multiphase flows, 138, 139*f*
Archimedes number, 159
capillary number, 159
Eötvös number, 158
Morton number, 159
Ohnesorge number, 159
Reynolds number, 157
viscosity ratio, 157
Multiphase flows, mesoscale structures, 293–294
Multiphase reactors, mesoscales, 248*f*
Multiphase thermodynamics, 8
Multiple-bubble-size (MBS) models
annealing method, 258–260
distribution, 259*f*
Multiple circulatory flow modes, 284, 285*f*
Multiple particle systems, 324–328
Multiscale, 42–43, 45–46
Multiscale modeling and simulation (MSS)
automatic upscaling and adaptive downscaling, 45–46
breakage function, 46–48
breakage probability, 46–48
bridging time scales, 45–46
CFD–DEM, 48–49
continuum fracture mechanics, 48
DFTQM, 45–46
external parameters, 46–48
intrinsic material properties, 48–49
machine function, 46–48
material function, 46, 47*f*
material properties, 46–48
milling, hierarchy of, 48, 48*f*
operational parameters, 46–48
process optimization, 46
Multiscale simulation framework
macroscale, 90–92, 91*f*
mesoscale
calculation algorithm of, 93–94, 93*f*
discrete elements, 92–95, 95*f*
uniform wetting zone, 95–97
Wurster coater, 97, 97*f*
microscale
discrete particle model (DPM), 98–100
scaled discrete particle model, 100–101
modeling application, 88, 89*f*
submicroscale, 101–103

## N

Navier–Stokes equations, 197–198, 201, 300
Navier–Stokes momentum equations, 138
Newtonian fluid phase, 138, 144
Nuclear magnetic resonance (NMR) spectroscopy, 17–18, 28–29
Numerical simulations, 198–199
Numerical techniques
deterministic approach, 151–153
stochastic approach, 153–156
Nusselt number (Nu), 202–203

## O

ODE. *See* Ordinary differential equations (ODEs)
1D drift flux model, 283
*One-Dimensional Two-Phase Flow* (Wallis), 283
Optimization
cloaking problem, 70*f*
continuum mechanical applications, 67
empiric parameters, 61
experiments, design of, 60
shape dependency, 67–68
shape-differential calculus, 69–70
shape gradient, 68
shape optimization, 67*f*
shape optimization, particle technology, 66–70, 69–70*f*
Si nanoparticles, gas phase, 60–63, 62*f*
volume representation, 68
ZnO quantum dots, liquid phase, 64–66, 65*f*
Ordinary differential equations (ODEs), 53–54, 56–59, 94–95

## P


Packed beds
  predictions *vs.* measurements
    CFD–DEM approach, 215
    continuum models, 211–212
    convective and conductive heat transfer coefficient, 217–219, 218*f*
    correction coefficient *vs.* Young's modulus, 215–217, 216*f*
    dimensionless variables, 211–212, 212*f*
    dynamic fluid, 211–212
    ETC, comparison of, 211–212, 212*f*
    microscopic models, 211–212
    packing structure, 211–212, 215
    radiation heat transfer, 213–214
    radiative conductivity, 213–214
    relative contributions, 214, 215*f*
    soft-sphere approach, 215–217
    solid–solid heat conduction, 213, 213*f*
    stagnant fluid, 213*f*, 214
    structure-based approach, 211–212
    thermal behavior, 211–212
  some variables, effects of
    aspect ratio, 220–221
    bed ETC, 219–221, 219–220*f*
    particle shape, 220
    thermal conductivities, 219–220
Packing density, 50–51
Parade of circulation cell models, 285, 286*f*
Parameter-dependent initial value problems
  adjoint equation, 58
  adjoint method, 57
  derivative-based methods, 56–57
  final value backward problem, 58
  linear sensitivity model, 57
  parameter change, 56–57
  sensitivity, 57
Partial differential equation (PDE), 53–54, 59, 68
Particle flow, 196–197
Particle–fluid convection, 198, 224–226
Particle–fluid flow system, 199
Particle–fluid interaction force, 199–200
Particle–fluid interactions
  DNS, 160–161
  drag force/correlations, 160, 161–162*f*
  slip velocities, 163, 163*f*
Particle formation
  bottom-up methods
    chemical thermodyamics, 24
    DFT, 22–23
    mass transfer, 23
    nucleation burst, 23–24
    particle formation dynamics, 22–23
    particle synthesis, 22–23, 22*f*
    PBE, 25
    PSD, 23
    quasi-monodisperse, 23–24
    reaction controlled processes, 23–24
    shape control, 24
    supersaturation, 22–23
    transport controlled systems, 23
  top-down approaches
    fracture, 19–21
    impaction, 19–21
    local stress energy density, 21
    scaling parameter, value of, 21*t*
    size reduction, 19–21
    spatial distribution, 21
Particle image velocimetry (PIV), 284, 338
Particle–particle collision, 208
Particle–particle conduction, 198, 224
Particle–particle interaction, 199–200
  restitution coefficient, 166
  soft-sphere model, 165
  wetted plate analysis, 166, 166*f*
Particle–particle interactions, 303
Particle–particle static contact, 208
Particles, fluidized bed (FB) process
  collision tests of, 111–113, 112*f*
  compression test of
    elastic contact deformation with Hertz theory, 104–105
    experimental method, 104, 104*f*
    force-displacement curves, 106–107, 106–107*f*, 109, 110*t*
    mechanical material parameters, 108, 108*t*
    results, 106–109, 106*f*
    Young's modulus, 108, 109*f*
  instantaneous particle positions and particle velocities, 126*f*
  shear test of, 109–110, 110–111*f*, 111*t*
  time-averaged distribution, 121*f*

time-averaged particle-particle collision frequency, 122*f*
time-averaged particle-particle collision velocity distribution, 121*f*
time-averaged radial distribution, 120*f*, 125*f*
Wurster tube, 122*t*
Particle size distribution (PSD), 14–15, 21, 23, 86
Particulate products, unified design strategies for
disperse systems, 2–3
driving forces, 4*f*
hierarchical organization, 6–12
processes and properties
model-based optimal control setting, 58–59
optimization, examples of, 60–70
parameter-dependent initial value problems, 56–58
product design
characterization, 44–45
interactions, 25–30
multiscale modeling and simulation, 45–49
particle formation, 19–25
particles into devices, integration of, 49–52
structure formation, 30–43
product properties, 12–19
unit operations, 2–3
Particulate thin films
charge carrier transport, 49–50
energy conversion, 49
particle formation, 49–50
particulate buildings blocks, 49
printable opto-electronics, 49, 50*f*
roll-to-roll production, 49–50
PBE. *See* Population balance equation (PBE)
PBM. *See* Population balance modeling (PBM)
PDE. *See* Partial differential equation (PDE)
PEPT. *See* Positron emission particle tracking (PEPT)
Pertinent parameters, 197
Photoluminescence, 28–29
Photon correlation spectroscopy (PCS), 44–45
Poisson ratio, 105
Population balance equation (PBE), 25, 62*f*, 63
Population balance modeling (PBM), 87, 90–91
Positron emission particle tracking (PEPT) method, 86–87, 229–230
Powder bed, 50–51
Prandtl number, 202–203
Pressure gradient force, 199–200
Process chain
off-line, 44
high-resolution TEM, 44
particulate system, 44
online and inline measurements, 44–45
size characterization, 44
size distributions, 44–45
structure characterization, 45
Process design
disperse properties optimization, 53
local and global optima, 55*f*
measured signal, 53
meta-heuristics, 54–55
model-based determination, 53
particle technology, 53
performance measure, 54, 54*f*
process–structure functions, 53
structure–property functions, 53
trial and error, 54–55
Processes and properties, design
general model-based optimal control setting, 58–59
optimization examples, 60–70
parameter-dependent initial value problems, 56–58
Product design
interactions, 25–30
multiscale modeling and simulation, 45–49
particle formation
bottom-up methods, 21–25
top-down approaches, 19–21
particles into devices, integration of, 49–52
process chain characterization, 44–45
structure formation, 30–43

Product properties
  absorption properties, 15–17, 16*f*
  agglomerates, 13
  complexity generation, particles, 13*f*
  DEMs, 17
  dispersity, 12–13
  filter cakes, 13
  gas phase synthesis, 19
  morphology, 17
  MPPs, 17
  particulate structure, 13
  particulate surface properties, 18–19
  powder technologies, 17
  property function, 12
  PSD, 12
  quantum dots, bandgap of, 14*f*
  quantum size effect, 15
  scanning electron microscopy (SEM), 15
  size-dependent bandgap, 14–15
  sphero-cylinders, frictional ratio of, 15–17, 16*f*
  surface plasmon resonances, 15
  transmission electron microscopy (TEM), 15
  Young's modulus, 18*f*, 19
PSD. *See* Particle size distribution (PSD)
Pseudo-turbulent, 144

## Q

Quantum mechanics (QM), 27–28
Quasi-monodisperse colloids, 60–61
Quasi-steady state solution, 203

## R

Raman spectroscopy, 17–18, 28–29
Reaction controlled processes, 23–24
Real-life two-phase flow reactors, 284
Reynolds numbers, 94, 202–203, 313, 320
Reynolds stress, 144, 146
  multiphase flows, 157
Rolling friction torque, 199
Roll-to-roll processes, 9
Rosin–Rammler distribution, 177

## S

Saffman lift force, 315–318, 316*f*, 318–319*f*
Scaled discrete particle model, 100–101
Scanning electron microscopy (SEM), 15
Schiller–Naumann, 271–273
Second law of thermodynamics, 294
Semi-infinite-media assumption, 204
Sherwood number, 96
Si nanoparticles, 60–61
Single-bubble-size (SBS) model
  DBS model, 258–260
Small-angle neutron scattering (SANS), 44–45
Small-angle X-ray scattering (SAXS), 28, 44–45
Smoothed equations, 141, 148
Soft-sphere approach, 215–217
Soft-sphere model, 165–166
Solid–solid heat conduction, 213, 213*f*
Space discretization, 205*f*
Stability condition
  bifurcation, 264–265*f*
  CFD models, 268–270
  EMMS model, 254–255
  energy dissipate, 277
  gas–liquid and gas–solid systems, 263
Steady-state creeping motion, 309
Steady-state drag force, 308–310
Steady-state model, 284
Stochastic simulations, 87
Structure formation
  building blocks, 32
  equilibrium structures
    biomineralization, 36
    bonding sites, 33–34
    colloidal crystals, 32–33
    crystallization, 32–33
    Janus particles, 33–34
    mesocrystals, 36
    nanorods, 37–38
    nonspherical particles, 37–38
    oriented aggregation, 35–36, 36*f*
    oriented attachment, 35–36
    polydimethylsiloxane stamps, 34–35
    polyhedra, different categories of, 33–34, 34*f*
    quantum dots self-assembly, 33–34, 34*f*
    roll-to-roll processes, 34–35
    3D arrangement, nanoparticles, 36, 36*f*
    topographic patterns, 34–35
    zirconia-based inverse opal pigments, 35, 35*f*

field forces, 32
interactions, 30–32
molecules organization, 30–32
nonequilibirium structures
aggregation kinetics, 41–42
assuming, 38
coagulation processes, 41–42
coarse-grained quasi particles, 42–43
crack-free assemblies, 41–42
DEM simulation, 38
discrete particle models, 42–43
drying particulate (thin)films, 39–41, 41*f*
FCC catalyst, particle clusters for, 43*f*
fractal dimensions, gas phase, 39, 40*f*
fractal-like structures, 38–39
liquid crystalline structures, 41–42
packings, 38
solid–liquid separation, 39–41
spray drying, 42
transient dynamic particulate, 42
particle synthesis, 32
principles of, 33*f*
statistical thermodynamics, 32
transport processes, 32
Sum frequency generation (SFG), 28–29
Superficial gas velocity, 255–256, 271–273
Superficial velocity, 146
Surface force microscopy, 45
Surface tension, 261, 262*f*
Symmetric coordinate system, 204–205, 205*f*

## T

Taylor dispersion concept, 283–284
Texture Analyser, 104
Thermal behavior, 196–197, 211–212, 227
Thermodynamics, second law of, 294
Thermogravimetry, 28–29
3D Printing. *See* Additive manufacturing
Top-down approaches
fracture, 19–21
impaction, 19–21
local stress energy density, 21
scaling parameter, value of, 21*t*
size reduction, 19–21
spatial distribution, 21
Transmission electron microscopy (TEM), 15
Triple-bubble-size (TBS)
annealing method, 258–260
bubble size distribution, 259*f*
Turbulence, 248–249, 320–324, 321*f*
Turbulent single-phase flow, 293
Two bubble classes (TBCs)
EMMS model, 252–253
energy dissipation, 258
Two-fluid model, 198–199
Two-phase flow, 282

## U

Ultrasound spectroscopy, 44–45
Unifying principles, 5, 19, 49
UV–vis absorption spectroscopy, 28–29, 44

## V

van der Waals force, 201
Viscous damping force, 199
Volume-based particle size distribution, 103, 103*f*
Voronoi element network, 209–210
Voronoi polyhedron, 206–208
Voronoi tessellation, 163–164, 206

## W

Wet cold grinding process, 51–52
Wurster coater apparatus. *See also* Wurster fluidized bed apparatus
cross-cut view of, 114*f*
particle dynamics, 119
segmented gas distributor plate of, 114*f*
simulated velocity distribution, 118*f*
steady-state flow, 118*f*
tetrahedral mesh of, 115*f*
Wurster fluidized bed apparatus
material parameters
collision tests of particles, 111–113, 112*f*
compression test of particles, 104–109, 104*f*, 106–107*f*, 108*t*, 109*f*, 110*t*
shear test of particles, 109–110, 110–111*f*, 111*t*
studied materials, 103, 103*f*
multiscale simulation framework
macroscale, 90–92, 91*f*
mesoscale, 92–97, 93*f*, 95*f*, 97*f*

Wurster fluidized bed apparatus (*Continued*)
microscale, 98–101
modeling application, 88, 89*f*
submicroscale, 101–103
process behavior analysis
apparatus geometry, 113–115, 114–115*f*
macroscale calculations, 130, 130*t*, 131*f*
and material parameters, 115–116, 116–117*t*, 118*f*
mesoscale model, 127–130, 127*t*, 128–129*f*
microscale simulation results, 117–127, 118–126*f*, 122–123*t*

## X

X-ray tomography, 45

## Y

Young's modulus, 108, 109*f*, 206, 215–217, 216*f*

## Z

Zeta-potential measurement
aqueous alumina suspensions of, 29*f*
particles, charge of, 29
suspension rheology, 29
ZnO nucleation, 64

# CONTENTS OF VOLUMES IN THIS SERIAL

**Volume 1 (1956)**
J. W. Westwater, *Boiling of Liquids*
A. B. Metzner, *Non-Newtonian Technology: Fluid Mechanics, Mixing, and Heat Transfer*
R. Byron Bird, *Theory of Diffusion*
J. B. Opfell and B. H. Sage, *Turbulence in Thermal and Material Transport*
Robert E. Treybal, *Mechanically Aided Liquid Extraction*
Robert W. Schrage, *The Automatic Computer in the Control and Planning of Manufacturing Operations*
Ernest J. Henley and Nathaniel F. Barr, *Ionizing Radiation Applied to Chemical Processes and to Food and Drug Processing*

**Volume 2 (1958)**
J. W. Westwater, *Boiling of Liquids*
Ernest F. Johnson, *Automatic Process Control*
Bernard Manowitz, *Treatment and Disposal of Wastes in Nuclear Chemical Technology*
George A. Sofer and Harold C. Weingartner, *High Vacuum Technology*
Theodore Vermeulen, *Separation by Adsorption Methods*
Sherman S. Weidenbaum, *Mixing of Solids*

**Volume 3 (1962)**
C. S. Grove, Jr., Robert V. Jelinek, and Herbert M. Schoen, *Crystallization from Solution*
F. Alan Ferguson and Russell C. Phillips, *High Temperature Technology*
Daniel Hyman, *Mixing and Agitation*
John Beck, *Design of Packed Catalytic Reactors*
Douglass J. Wilde, *Optimization Methods*

**Volume 4 (1964)**
J. T. Davies, *Mass-Transfer and Inierfacial Phenomena*
R. C. Kintner, *Drop Phenomena Affecting Liquid Extraction*
Octave Levenspiel and Kenneth B. Bischoff, *Patterns of Flow in Chemical Process Vessels*
Donald S. Scott, *Properties of Concurrent Gas–Liquid Flow*
D. N. Hanson and G. F. Somerville, *A General Program for Computing Multistage Vapor–Liquid Processes*

**Volume 5 (1964)**
J. F. Wehner, *Flame Processes—Theoretical and Experimental*
J. H. Sinfelt, *Bifunctional Catalysts*
S. G. Bankoff, *Heat Conduction or Diffusion with Change of Phase*
George D. Fulford, *The Flow of Lktuids in Thin Films*
K. Rietema, *Segregation in Liquid–Liquid Dispersions and its Effects on Chemical Reactions*

**Volume 6 (1966)**
S. G. Bankoff, *Diffusion-Controlled Bubble Growth*
John C. Berg, Andreas Acrivos, and Michel Boudart, *Evaporation Convection*
H. M. Tsuchiya, A. G. Fredrickson, and R. Aris, *Dynamics of Microbial Cell Populations*
Samuel Sideman, *Direct Contact Heat Transfer between Immiscible Liquids*
Howard Brenner, *Hydrodynamic Resistance of Particles at Small Reynolds Numbers*

**Volume 7 (1968)**

Robert S. Brown, Ralph Anderson, and Larry J. Shannon, *Ignition and Combustion of Solid Rocket Propellants*

Knud Østergaard, *Gas–Liquid–Particle Operations in Chemical Reaction Engineering*

J. M. Prausnilz, *Thermodynamics of Fluid–Phase Equilibria at High Pressures*

Robert V. Macbeth, *The Burn-Out Phenomenon in Forced-Convection Boiling*

William Resnick and Benjamin Gal-Or, *Gas–Liquid Dispersions*

**Volume 8 (1970)**

C. E. Lapple, *Electrostatic Phenomena with Particulates*

J. R. Kittrell, *Mathematical Modeling of Chemical Reactions*

W. P. Ledet and D. M. Himmelblau, *Decomposition Procedures foe the Solving of Large Scale Systems*

R. Kumar and N. R. Kuloor, *The Formation of Bubbles and Drops*

**Volume 9 (1974)**

Renato G. Bautista, *Hydrometallurgy*

Kishan B. Mathur and Norman Epstein, *Dynamics of Spouted Beds*

W. C. Reynolds, *Recent Advances in the Computation of Turbulent Flows*

R. E. Peck and D. T. Wasan, *Drying of Solid Particles and Sheets*

**Volume 10 (1978)**

G. E. O'Connor and T. W. F. Russell, *Heat Transfer in Tubular Fluid–Fluid Systems*

P. C. Kapur, *Balling and Granulation*

Richard S. H. Mah and Mordechai Shacham, *Pipeline Network Design and Synthesis*

J. Robert Selman and Charles W. Tobias, *Mass-Transfer Measurements by the Limiting-Current Technique*

**Volume 11 (1981)**

Jean-Claude Charpentier, *Mass-Transfer Rates in Gas–Liquid Absorbers and Reactors*

Dee H. Barker and C. R. Mitra, *The Indian Chemical Industry—Its Development and Needs*

Lawrence L. Tavlarides and Michael Stamatoudis, *The Analysis of Interphase Reactions and Mass Transfer in Liquid–Liquid Dispersions*

Terukatsu Miyauchi, Shintaro Furusaki, Shigeharu Morooka, and Yoneichi Ikeda, *Transport Phenomena and Reaction in Fluidized Catalyst Beds*

**Volume 12 (1983)**

C. D. Prater, J, Wei, V. W. Weekman, Jr., and B. Gross, *A Reaction Engineering Case History: Coke Burning in Thermofor Catalytic Cracking Regenerators*

Costel D. Denson, *Stripping Operations in Polymer Processing*

Robert C. Reid, *Rapid Phase Transitions from Liquid to Vapor*

John H. Seinfeld, *Atmospheric Diffusion Theory*

**Volume 13 (1987)**

Edward G. Jefferson, *Future Opportunities in Chemical Engineering*

Eli Ruckenstein, *Analysis of Transport Phenomena Using Scaling and Physical Models*

Rohit Khanna and John H. Seinfeld, *Mathematical Modeling of Packed Bed Reactors: Numerical Solutions and Control Model Development*

Michael P. Ramage, Kenneth R. Graziano, Paul H. Schipper, Frederick J. Krambeck, and Byung C. Choi, *KINPTR (Mobil's Kinetic Reforming Model): A Review of Mobil's Industrial Process Modeling Philosophy*

**Volume 14 (1988)**

Richard D. Colberg and Manfred Morari, *Analysis and Synthesis of Resilient Heat Exchange Networks*
Richard J. Quann, Robert A. Ware, Chi-Wen Hung, and James Wei, *Catalytic Hydrometallation of Petroleum*
Kent David, *The Safety Matrix: People Applying Technology to Yield Safe Chemical Plants and Products*

**Volume 15 (1990)**

Pierre M. Adler, Ali Nadim, and Howard Brenner, *Rheological Models of Suspenions*
Stanley M. Englund, *Opportunities in the Design of Inherently Safer Chemical Plants*
H. J. Ploehn and W. B. Russel, *Interations between Colloidal Particles and Soluble Polymers*

**Volume 16 (1991)**

*Perspectives in Chemical Engineering: Research and Education*
Clark K. Colton, *Editor*
Historical Perspective and Overview
L. E. Scriven, *On the Emergence and Evolution of Chemical Engineering*
Ralph Landau, *Academic—industrial Interaction in the Early Development of Chemical Engineering*
James Wei, *Future Directions of Chemical Engineering*
Fluid Mechanics and Transport
L. G. Leal, *Challenges and Opportunities in Fluid Mechanics and Transport Phenomena*
William B. Russel, *Fluid Mechanics and Transport Research in Chemical Engineering*
J. R. A. Pearson, *Fluid Mechanics and Transport Phenomena*
Thermodynamics
Keith E. Gubbins, *Thermodynamics*
J. M. Prausnitz, *Chemical Engineering Thermodynamics: Continuity and Expanding Frontiers*
H. Ted Davis, *Future Opportunities in Thermodynamics*
Kinetics, Catalysis, and Reactor Engineering
Alexis T. Bell, *Reflections on the Current Status and Future Directions of Chemical Reaction Engineering*
James R. Katzer and S. S. Wong, *Frontiers in Chemical Reaction Engineering*
L. Louis Hegedus, *Catalyst Design*
Environmental Protection and Energy
John H. Seinfeld, *Environmental Chemical Engineering*
T. W. F. Russell, *Energy and Environmental Concerns*
Janos M. Beer, Jack B. Howard, John P. Longwell, and Adel F. Sarofim, *The Role of Chemical Engineering in Fuel Manufacture and Use of Fuels*
Polymers
Matthew Tirrell, *Polymer Science in Chemical Engineering*
Richard A. Register and Stuart L. Cooper, *Chemical Engineers in Polymer Science: The Need for an Interdisciplinary Approach*
Microelectronic and Optical Material
Larry F. Thompson, *Chemical Engineering Research Opportunities in Electronic and Optical Materials Research*
Klavs F. Jensen, *Chemical Engineering in the Processing of Electronic and Optical Materials: A Discussion*
Bioengineering
James E. Bailey, *Bioprocess Engineering*
Arthur E. Humphrey, *Some Unsolved Problems of Biotechnology*
Channing Robertson, *Chemical Engineering: Its Role in the Medical and Health Sciences*
Process Engineering
Arthur W. Westerberg, *Process Engineering*
Manfred Morari, *Process Control Theory: Reflections on the Past Decade and Goals for the Next*
James M. Douglas, *The Paradigm After Next*

George Stephanopoulos, *Symbolic Computing and Artificial Intelligence in Chemical Engineering: A New Challenge*
The Identity of Our Profession
Morton M. Denn, *The Identity of Our Profession*

**Volume 17 (1991)**
Y. T. Shah, *Design Parameters for Mechanically Agitated Reactors*
Mooson Kwauk, *Particulate Fluidization: An Overview*

**Volume 18 (1992)**
E. James Davis, *Microchemical Engineering: The Physics and Chemistry of the Microparticle*
Selim M. Senkan, *Detailed Chemical Kinetic Modeling: Chemical Reaction Engineering of the Future*
Lorenz T. Biegler, *Optimization Strategies for Complex Process Models*

**Volume 19 (1994)**
Robert Langer, *Polymer Systems for Controlled Release of Macromolecules, Immobilized Enzyme Medical Bioreactors, and Tissue Engineering*
J. J. Linderman, P. A. Mahama, K. E. Forsten, and D. A. Lauffenburger, *Diffusion and Probability in Receptor Binding and Signaling*
Rakesh K. Jain, *Transport Phenomena in Tumors*
R. Krishna, *A Systems Approach to Multiphase Reactor Selection*
David T. Allen, *Pollution Prevention: Engineering Design at Macro-, Meso-, and Microscales*
John H. Seinfeld, Jean M. Andino, Frank M. Bowman, Hali J. L. Forstner, and Spyros Pandis, *Tropospheric Chemistry*

**Volume 20 (1994)**
Arthur M. Squires, *Origins of the Fast Fluid Bed*
Yu Zhiqing, *Application Collocation*
Youchu Li, *Hydrodynamics*
Li Jinghai, *Modeling*
Yu Zhiqing and Jin Yong, *Heat and Mass Transfer*
Mooson Kwauk, *Powder Assessment*
Li Hongzhong, *Hardware Development*
Youchu Li and Xuyi Zhang, *Circulating Fluidized Bed Combustion*
Chen Junwu, Cao Hanchang, and Liu Taiji, *Catalyst Regeneration in Fluid Catalytic Cracking*

**Volume 21 (1995)**
Christopher J. Nagel, Chonghum Han, and George Stephanopoulos, *Modeling Languages: Declarative and Imperative Descriptions of Chemical Reactions and Processing Systems*
Chonghun Han, George Stephanopoulos, and James M. Douglas, *Automation in Design: The Conceptual Synthesis of Chemical Processing Schemes*
Michael L. Mavrovouniotis, *Symbolic and Quantitative Reasoning: Design of Reaction Pathways through Recursive Satisfaction of Constraints*
Christopher Nagel and George Stephanopoulos, *Inductive and Deductive Reasoning: The Case of Identifying Potential Hazards in Chemical Processes*
Keven G. Joback and George Stephanopoulos, *Searching Spaces of Discrete Soloutions: The Design of Molecules Processing Desired Physical Properties*

**Volume 22 (1995)**
Chonghun Han, Ramachandran Lakshmanan, Bhavik Bakshi, and George Stephanopoulos, *Nonmonotonic Reasoning: The Synthesis of Operating Procedures in Chemical Plants*
Pedro M. Saraiva, *Inductive and Analogical Learning: Data-Driven Improvement of Process Operations*

Alexandros Koulouris, Bhavik R. Bakshi and George Stephanopoulos, *Empirical Learning through Neural Networks: The Wave-Net Solution*
Bhavik R. Bakshi and George Stephanopoulos, *Reasoning in Time: Modeling, Analysis, and Pattern Recognition of Temporal Process Trends*
Matthew J. Realff, *Intelligence in Numerical Computing: Improving Batch Scheduling Algorithms through Explanation-Based Learning*

**Volume 23 (1996)**
Jeffrey J. Siirola, *Industrial Applications of Chemical Process Synthesis*
Arthur W. Westerberg and Oliver Wahnschafft, *The Synthesis of Distillation-Based Separation Systems*
Ignacio E. Grossmann, *Mixed-Integer Optimization Techniques for Algorithmic Process Synthesis*
Subash Balakrishna and Lorenz T. Biegler, *Chemical Reactor Network Targeting and Integration: An Optimization Approach*
Steve Walsh and John Perkins, *Operability and Control inn Process Synthesis and Design*

**Volume 24 (1998)**
Raffaella Ocone and Gianni Astarita, *Kinetics and Thermodynamics in Multicomponent Mixtures*
Arvind Varma, Alexander S. Rogachev, Alexandra S. Mukasyan, and Stephen Hwang, *Combustion Synthesis of Advanced Materials: Principles and Applications*
J. A. M. Kuipers and W. P. Mo, van Swaaij, *Computional Fluid Dynamics Applied to Chemical Reaction Engineering*
Ronald E. Schmitt, Howard Klee, Debora M. Sparks, and Mahesh K. Podar, *Using Relative Risk Analysis to Set Priorities for Pollution Prevention at a Petroleum Refinery*

**Volume 25 (1999)**
J. F. Davis, M. J. Piovoso, K. A. Hoo, and B. R. Bakshi, *Process Data Analysis and Interpretation*
J. M. Ottino, P. DeRoussel, S., Hansen, and D. V. Khakhar, *Mixing and Dispersion of Viscous Liquids and Powdered Solids*
Peter L. Silverston, Li Chengyue, Yuan Wei-Kang, *Application of Periodic Operation to Sulfur Dioxide Oxidation*

**Volume 26 (2001)**
J. B. Joshi, N. S. Deshpande, M. Dinkar, and D. V. Phanikumar, *Hydrodynamic Stability of Multiphase Reactors*
Michael Nikolaou, *Model Predictive Controllers: A Critical Synthesis of Theory and Industrial Needs*

**Volume 27 (2001)**
William R. Moser, Josef Find, Sean C. Emerson, and Ivo M, Krausz, *Engineered Synthesis of Nanostructure Materials and Catalysts*
Bruce C. Gates, *Supported Nanostructured Catalysts: Metal Complexes and Metal Clusters*
Ralph T. Yang, *Nanostructured Absorbents*
Thomas J. Webster, *Nanophase Ceramics: The Future Orthopedic and Dental Implant Material*
Yu-Ming Lin, Mildred S. Dresselhaus, and Jackie Y. Ying, *Fabrication, Structure, and Transport Properties of Nanowires*

**Volume 28 (2001)**
Qiliang Yan and Juan J. DePablo, *Hyper-Parallel Tempering Monte Carlo and Its Applications*
Pablo G. Debenedetti, Frank H. Stillinger, Thomas M. Truskett, and Catherine P. Lewis, *Theory of Supercooled Liquids and Glasses: Energy Landscape and Statistical Geometry Perspectives*
Michael W. Deem, *A Statistical Mechanical Approach to Combinatorial Chemistry*

Venkat Ganesan and Glenn H. Fredrickson, *Fluctuation Effects in Microemulsion Reaction Media*
David B. Graves and Cameron F. Abrams, *Molecular Dynamics Simulations of Ion–Surface Interactions with Applications to Plasma Processing*
Christian M. Lastoskie and Keith E, Gubbins, *Characterization of Porous Materials Using Molecular Theory and Simulation*
Dimitrios Maroudas, *Modeling of Radical-Surface Interactions in the Plasma-Enhanced Chemical Vapor Deposition of Silicon Thin Films*
Sanat Kumar, M. Antonio Floriano, and Athanassiors Z. Panagiotopoulos, *Nanostructured Formation and Phase Separation in Surfactant Solutions*
Stanley I. Sandler, Amadeu K. Sum, and Shiang-Tai Lin, *Some Chemical Engineering Applications of Quantum Chemical Calculations*
Bernhardt L. Trout, *Car-Parrinello Methods in Chemical Engineering: Their Scope and potential*
R. A. van Santen and X. Rozanska, *Theory of Zeolite Catalysis*
Zhen-Gang Wang, *Morphology, Fluctuation, Metastability and Kinetics in Ordered Block* Copolymers

**Volume 29 (2004)**
Michael V. Sefton, *The New Biomaterials*
Kristi S. Anseth and Kristyn S. Masters, *Cell–Material Interactions*
Surya K. Mallapragada and Jennifer B. Recknor, *Polymeric Biomaterias for Nerve Regeneration*
Anthony M. Lowman, Thomas D. Dziubla, Petr Bures, and Nicholas A. Peppas, *Structural and Dynamic Response of Neutral and Intelligent Networks in Biomedical Environments*
F. Kurtis Kasper and Antonios G. Mikos, *Biomaterials and Gene Therapy*
Balaji Narasimhan and Matt J. Kipper, *Surface-Erodible Biomaterials for Drug Delivery*

**Volume 30 (2005)**
Dionisio Vlachos, *A Review of Multiscale Analysis: Examples from System Biology, Materials Engineering, and Other Fluids-Surface Interacting Systems*
Lynn F. Gladden, M.D. Mantle and A.J. Sederman, *Quantifying Physics and Chemistry at Multiple Length-Scales using Magnetic Resonance Techniques*
Juraj Kosek, Frantisek Steěpánek, and Miloš Marek, Modelling of Transport and Transformation Processes in Porous and Multiphase Bodies
Vemuri Balakotaiah and Saikat Chakraborty, *Spatially Averaged Multiscale Models for Chemical Reactors*

**Volume 31 (2006)**
Yang Ge and Liang-Shih Fan, *3-D Direct Numerical Simulation of Gas–Liquid and Gas–Liquid–Solid Flow Systems Using the Level-Set and Immersed-Boundary Methods*
M.A. van der Hoef, M. Ye, M. van Sint Annaland, A.T. Andrews IV, S. Sundaresan, and J.A.M. Kuipers, *Multiscale Modeling of Gas-Fluidized Beds*
Harry E.A. Van den Akker, *The Details of Turbulent Mixing Process and their Simulation*
Rodney O. Fox, *CFD Models for Analysis and Design of Chemical Reactors*
Anthony G. Dixon, Michiel Nijemeisland, and E. Hugh Stitt, *Packed Tubular Reactor Modeling and Catalyst Design Using Computational Fluid Dynamics*

**Volume 32 (2007)**
William H. Green, Jr., *Predictive Kinetics: A New Approach for the 21st Century*
Mario Dente, Giulia Bozzano, Tiziano Faravelli, Alessandro Marongiu, Sauro Pierucci and Eliseo Ranzi, *Kinetic Modelling of Pyrolysis Processes in Gas and Condensed Phase*
Mikhail Sinev, Vladimir Arutyunov and Andrey Romanets, *Kinetic Models of C1–C4 Alkane Oxidation as Applied to Processing of Hydrocarbon Gases: Principles, Approaches and Developments*
Pierre Galtier, *Kinetic Methods in Petroleum Process Engineering*

**Volume 33 (2007)**

Shinichi Matsumoto and Hirofumi Shinjoh, *Dynamic Behavior and Characterization of Automobile Catalysts*

Mehrdad Ahmadinejad, Maya R. Desai, Timothy C. Watling and Andrew P.E. York, *Simulation of Automotive Emission Control Systems*

Anke Güthenke, Daniel Chatterjee, Michel Weibel, Bernd Krutzsch, Petr Kočí, Miloš Marek, Isabella Nova and Enrico Tronconi, *Current Status of Modeling Lean Exhaust Gas Aftertreatment Catalysts*

Athanasios G. Konstandopoulos, Margaritis Kostoglou, Nickolas Vlachos and Evdoxia Kladopoulou, *Advances in the Science and Technology of Diesel Particulate Filter Simulation*

**Volume 34 (2008)**

C.J. van Duijn, Andro Mikelić, I.S. Pop, and Carole Rosier, *Effective Dispersion Equations for Reactive Flows with Dominant Peclet and Damkohler Numbers*

Mark Z. Lazman and Gregory S. Yablonsky, *Overall Reaction Rate Equation of Single-Route Complex Catalytic Reaction in Terms of Hypergeometric Series*

A.N. Gorban and O. Radulescu, *Dynamic and Static Limitation in Multiscale Reaction Networks, Revisited*

Liqiu Wang, Mingtian Xu, and Xiaohao Wei, *Multiscale Theorems*

**Volume 35 (2009)**

Rudy J. Koopmans and Anton P.J. Middelberg, *Engineering Materials from the Bottom Up – Overview*

Robert P.W. Davies, Amalia Aggeli, Neville Boden, Tom C.B. McLeish, Irena A. Nyrkova, and Alexander N. Semenov, *Mechanisms and Principles of 1 D Self-Assembly of Peptides into β-*Sheet Tapes

Paul van der Schoot, *Nucleation and Co-Operativity in Supramolecular Polymers*

Michael J. McPherson, Kier James, Stuart Kyle, Stephen Parsons, and Jessica Riley, *Recombinant Production of Self-Assembling Peptides*

Boxun Leng, Lei Huang, and Zhengzhong Shao, *Inspiration from Natural Silks and Their Proteins*

Sally L. Gras, *Surface- and Solution-Based Assembly of Amyloid Fibrils for Biomedical and Nanotechnology Applications*

Conan J. Fee, *Hybrid Systems Engineering: Polymer-Peptide Conjugates*

**Volume 36 (2009)**

Vincenzo Augugliaro, Sedat Yurdakal, Vittorio Loddo, Giovanni Palmisano, and Leonardo Palmisano, *Determination of Photoadsorption Capacity of Polychrystalline TiO2 Catalyst in Irradiated Slurry*

Marta I. Litter, *Treatment of Chromium, Mercury, Lead, Uranium, and Arsenic in Water by Heterogeneous Photocatalysis*

Aaron Ortiz-Gomez, Benito Serrano-Rosales, Jesus Moreira-del-Rio, and Hugo de-Lasa, Mineralization of Phenol in an Improved Photocatalytic Process Assisted with Ferric Ions: *Reaction Network and Kinetic Modeling*

R.M. Navarro, F. del Valle, J.A. Villoria de la Mano, M.C. Alvarez-Galván, and J.L.G. Fierro, *Photocatalytic Water Splitting Under Visible Light: Concept and Catalysts Development*

Ajay K. Ray, *Photocatalytic Reactor Configurations for Water Purification: Experimentation and Modeling*

Camilo A. Arancibia-Bulnes, Antonio E. Jiménez, and Claudio A. Estrada, *Development and Modeling of Solar Photocatalytic Reactors*

Orlando M. Alfano and Alberto E. Cassano, *Scaling-Up of Photoreactors: Applications to Advanced Oxidation Processes*

Yaron Paz, *Photocatalytic Treatment of Air: From Basic Aspects to Reactors*

**Volume 37 (2009)**

S. Roberto Gonzalez A., Yuichi Murai, and Yasushi Takeda, *Ultrasound-Based Gas–Liquid Interface Detection in Gas–Liquid Two-Phase Flows*

Z. Zhang, J. D. Stenson, and C. R. Thomas, *Micromanipulation in Mechanical Characterisation of Single Particles*

Feng-Chen Li and Koichi Hishida, *Particle Image Velocimetry Techniques and Its Applications in Multiphase Systems*
J. P. K. Seville, A. Ingram, X. Fan, and D. J. Parker, *Positron Emission Imaging in Chemical Engineering*
Fei Wang, Qussai Marashdeh, Liang-Shih Fan, and Richard A. Williams, *Electrical Capacitance, Electrical Resistance, and Positron Emission Tomography Techniques and Their Applications in Multi-Phase Flow Systems*
Alfred Leipertz and Roland Sommer, *Time-Resolved Laser-Induced Incandescence*

**Volume 38 (2009)**
Arata Aota and Takehiko Kitamori, *Microunit Operations and Continuous Flow Chemical Processing*
Anıl Ağıral and Han J.G.E. Gardeniers, *Microreactors with Electrical Fields*
Charlotte Wiles and Paul Watts, *High-Throughput Organic Synthesis in Microreactors*
S. Krishnadasan, A. Yashina, A.J. deMello and J.C. deMello, *Microfluidic Reactors for Nanomaterial Synthesis*

**Volume 39 (2010)**
B.M. Kaganovich, A.V. Keiko and V.A. Shamansky, *Equilibrium Thermodynamic Modeling of Dissipative Macroscopic Systems*
Miroslav Grmela, *Multiscale Equilibrium and Nonequilibrium Thermodynamics in Chemical Engineering*
Prasanna K. Jog, Valeriy V. Ginzburg, Rakesh Srivastava, Jeffrey D. Weinhold, Shekhar Jain, and Walter G. Chapman, *Application of Mesoscale Field-Based Models to Predict Stability of Particle Dispersions in Polymer Melts*
Semion Kuchanov, *Principles of Statistical Chemistry as Applied to Kinetic Modeling of Polymer-Obtaining Processes*

**Volume 40 (2011)**
Wei Wang, Wei Ge, Ning Yang and Jinghai Li, *Meso-Scale Modeling—The Key to Multi-Scale CFD Simulation*
Pil Seung Chung, Myung S. Jhon and Lorenz T. Biegler, *The Holistic Strategy in Multi-Scale Modeling*
Milo D. Meixell Jr., Boyd Gochenour and Chau-Chyun Chen, *Industrial Applications of Plant-Wide Equation-Oriented Process Modeling—2010*
Honglai Liu, Ying Hu, Xueqian Chen, Xingqing Xiao and Yongmin Huang, *Molecular Thermodynamic Models for Fluids of Chain-Like Molecules, Applications in Phase Equilibria and Micro-Phase Separation in Bulk and at Interface*

**Volume 41 (2012)**
Torsten Kaltschmitt and Olaf Deutschmann, *Fuel Processing for Fuel Cells*
Adam Z.Weber, Sivagaminathan Balasubramanian, and Prodip K. Das, *Proton Exchange Membrane Fuel Cells*
Keith Scott and Lei Xing, *Direct Methanol Fuel Cells*
Su Zhou and Fengxiang Chen, *PEMFC System Modeling and Control*
François Lapicque, Caroline Bonnet, Bo Tao Huang, and Yohann Chatillon, *Analysis and Evaluation of Aging Phenomena in PEMFCs*
Robert J. Kee, Huayang Zhu, Robert J. Braun, and Tyrone L. Vincent, *Modeling the Steady-State and Dynamic Characteristics of Solid-Oxide Fuel Cells*
Robert J. Braun, Tyrone L. Vincent, Huayang Zhu, and Robert J. Kee, *Analysis, Optimization, and Control of Solid-Oxide Fuel Cell Systems*

**Volume 42 (2013)**
T. Riitonen, V. Eta, S. Hyvärinen, L.J. Jönsson, and J.P. Mikkola, *Engineering Aspects of Bioethanol Synthesis*
R.W. Nachenius, F. Ronsse, R.H. Venderbosch, and W. Prins, *Biomass Pyrolysis*
David Kubička and Vratislav Tukač, *Hydrotreating of Triglyceride-Based Feedstocks in Refineries*

Tapio Salmi, *Chemical Reaction Engineering of Biomass Conversion*
Jari Heinonen and Tuomo Sainio, *Chromatographic Fractionation of Lignocellulosic Hydrolysates*

**Volume 43 (2013)**
Grégory Francois and Dominique Bonvin, *Measurement-Based Real-Time Optimization of Chemical Processes*
Adel Mhamdi and Wolfgang Marquardt, *Incremental Identification of Distributed Parameter Systems*
Arun K. Tangirala, Siddhartha Mukhopadhyay, and Akhilananand P. Tiwari, *Wavelets Applications in Modeling and Control*
Santosh K. Gupta and Sanjeev Garg, *Multiobjective Optimization Using Genetic Algorithm*

**Volume 44 (2014)**
Xue-Qing Gong, Li-Li Yin, Jie Zhang, Hai-Feng Wang, Xiao-Ming Cao, Guanzhong Lu, and Peijun Hu, *Computational Simulation of Rare Earth Catalysis*
Zhi-Jun Sui, Yi-An Zhu, Ping Li, Xing-Gui Zhou, and De Chen, *Kinetics of Catalytic Dehydrogenation of Propane over Pt-Based Catalysts*
Zhen Liu, Xuelian He, Ruihua Cheng, Moris S. Eisen, Minoru Terano, Susannah L. Scott, and Boping Liu, *Chromium Catalysts for Ethylene Polymerization and Oligomerization*
Ayyaz Ahmad, Xiaochi Liu, Li Li, and Xuhong Guo, *Progress in Polymer Nanoreactors: Spherical Polyelectrolyte Brushes*

**Volume 45 (2014)**
M.P. Dudukovic and P.L. Mills, *Challenges in Reaction Engineering Practice of Heterogeneous Catalytic Systems*
Claudia Diehm, Hüsyein Karadeniz, Canan Karakaya, Matthias Hettel, and Olaf Deutschmann, *Spatial Resolution of Species and Temperature Profiles in Catalytic Reactors:* In Situ *Sampling Techniques and CFD Modeling*
John Mantzaras, *Catalytic Combustion of Hydrogen, Challenges, and Opportunities*
Ivo Roghair, Fausto Gallucci, and Martin van Sint Annaland, *Novel Developments in Fluidized Bed Membrane Reactor Technology*

**Volume 46 (2015)**
Wolfgang Peukert, Doris Segets, Lukas Pflug, and Günter Leugering, *Unified Design Strategies for Particulate Products*
Stefan Heinrich, Maksym Dosta, and Sergiy Antonyuk, *Multiscale Analysis of a Coating Process in a Wurster Fluidized Bed Apparatus*
Johan T. Padding, Niels G. Deen, E.A.J.F. (Frank) Peters, and J.A.M. (Hans) Kuipers, *Euler–Lagrange Modeling of the Hydrodynamics of Dense Multiphase Flows*
Qinfu Hou, Jieqing Gan, Zongyan Zhou, and Aibing Yu, *Particle Scale Study of Heat Transfer in Packed and Fluidized Beds*
Ning Yang, *Mesoscale Transport Phenomena and Mechanisms in Gas–Liquid Reaction Systems*
Harry E.A. Van den Akker, *Mesoscale Flow Structures and Fluid–Particle Interactions*

FSC
www.fsc.org
MIX
Papier | Fördert gute Waldnutzung
FSC® C083411

Kolibri 360 GmbH
Ferdinand-Jühlke-Straße 7
99095 Erfurt, Deutschland
produktsicherheit@kolibri360.de